Statistical Analysis in Forensic Science

Statistical Analysis in Forensic Science

Evidential Value of Multivariate Physicochemical Data

Grzegorz Zadora
Institute of Forensic Research, Kraków, Poland

Agnieszka Martyna
Faculty of Chemistry, Jagiellonian University, Kraków, Poland

Daniel Ramos
Escuela Politécnica Superior, Universidad Autonoma de Madrid, Spain

Colin Aitken
School of Mathematics, University of Edinburgh, UK

WILEY

This edition first published 2014
© 2014 John Wiley & Sons, Ltd

Registered office

John Wiley & Sons Ltd, The Atrium, Southern Gate, Chichester, West Sussex, PO19 8SQ, United Kingdom

For details of our global editorial offices, for customer services and for information about how to apply for permission to reuse the copyright material in this book please see our website at www.wiley.com.

Library of Congress Cataloging-in-Publication Data

Statistical analysis in forensic science : evidential value of multivariate physicochemical data / Grzegorz Zadora, Agnieszka Martyna, Daniel Ramos, Colin Aitken.
     p. cm.
  Includes bibliographical references and index.
  ISBN 978-0-470-97210-6 (cloth)
  1. Chemistry, Forensic.   2. Forensic statistics.   3. Chemometrics.   I. Zadora, Grzegorz.   II. Martyna, Agnieszka.   III. Ramos, Daniel.   IV. Aitken, Colin.
  RA1057.S73 2014
  614'.12–dc23

              2013031698

A catalogue record for this book is available from the British Library.

ISBN: 978-0-470-97210-6

Set in 10/12pt Times by Aptara Inc., New Delhi, India.

1   2014

To our families

Contents

Preface

An increase in the danger from new forms of crime and the need by those who administer justice for higher standards of scientific work require the development of new methods for measuring the evidential value of physicochemical data obtained during the analysis of various kinds of trace evidence.

The physicochemical analysis of various types of evidence by the application of various analytical methods (Chapter 1) returns numerous types of information including multivariate quantitative data (Chapter 3, Appendix B), for example, concentrations of elements or the refractive index of a glass fragment. The role of the forensic expert is to evaluate such physicochemical data (evidence, E) in the context of two competing propositions H_1 and H_2 (Chapter 2, Appendix A). The propositions H_1 and H_2 may be put forward by the police, prosecutors, defenders or the courts and they concern:

- *comparison problems* (Chapter 4), for example where H_1 states that the glass samples being compared originate from the same object, and H_2 that the glass samples being compared originate from different objects;

- *classification problems* (Chapter 5), for example where H_1 states that the glass sample which has been analysed originates from a car or building window, and H_2 states that the glass sample analysed originates from a container glass.

Bayesian models have been proposed for the evaluation of the evidence in such contexts. Statistical analysis is used to evaluate the evidence. The value of the evidence is determined by the likelihood ratio (LR). This is the ratio of the probability of the evidence if H_1 is true, $P(E \mid H_1)$, to the probability of the evidence if H_2 is true, $P(E \mid H_2)$. For evidence in the form of continuous data these probabilities are replaced with probability density functions, $f(E \mid H_1)$ and $f(E \mid H_2)$.

The LR approach (Chapter 2, Appendix A) has become increasingly popular for evidence evaluation in forensic sciences. For physicochemical data, the approach enables an objective evaluation of the physicochemical information about the analysed object(s) obtained from an analytical run, and about the rarity of the determined physicochemical features for recovered and/or control samples within a relevant population (Chapters 2, 4 and 5). The most common application of the LR approach in forensic science is in DNA profiling. The LR approach has also been applied to other evidence categories including earprints, fingerprints, firearms and toolmarks, hair, documents, envelopes and handwriting, and speaker recognition. In recent years, much has also been published on LR approaches for multivariate data. Some of these ideas (including examples from practice in forensic science) are discussed in this book.

The performance of each statistical approach should be subjected to critical analysis, not only in the form of error rates but also through the use of other formal frameworks which provide a measure of the quality of a method for the evaluation of evidence based on a likelihood ratio. There is a need not only for the measurement of the discriminating power of the LR models as represented by false positive and false negative rates, but for the information that the LR provides to the inference process in evidence evaluation, where the important concept of calibration plays a significant role. One of the objectives of this book is to consider the problem of the assessment of the performance of LR-based evidence evaluation methods. Several methods found in the literature are extensively described and compared, such as Tippett plots, detection error trade-off plots (DET) and empirical cross-entropy (ECE) plots (Chapter 6, Appendix C).

One reason for the slow implementation of these LR models is that there is a lack of commercial software to enable the calculation of the LR relatively easily by those without experience in programming (like most forensic experts). Therefore, in order to use these methods case-specific routines have to be written using an appropriate software package, such as the **R** software (www.r-project.org). Based on information gathered during workshops on statistics for forensic scientists (e.g. the "FORSTAT – Forensic Statistics" project under the auspices of the European Network of Forensic Sciences Institutes), the present authors believe that there is a need for a book that provides descriptions of the models in more detail than in published papers, as well as of the software routines, together with practical examples. Therefore, the aims of this book are to present and discuss recent LR approaches and to provide suitable software toolboxes with annotation and examples to illustrate the use of the approaches in practice. The routines included in the book are available from the website `www.wiley.com/go/physicochemical`. These include routines in **R** (Appendix D), pre-computed Bayesian networks for **Hugin Researcher**™ (Appendix E), and the **calcuLatoR** software (Appendix F) for the computation of likelihood ratios for univariate data. Manuals (Appendices D–F) including examples and recommendations of the use of all of these assessment methods in practice are included, as well as software and practical examples to enable forensic experts to begin to work with them immediately (Chapters 3–6).

Note also that the LR approaches presented in the book can be used whenever evidence is to be evaluated under the circumstances of two propositions. Therefore, the models described in the book can also be applied in other areas of analytical chemistry. Special emphasis is placed on the solution of problems where a decision made on the basis of results of statistical analyses of physicochemical data could have serious legal or economical consequences; thus, for example, one of these other areas of analytical chemistry could be that of food authenticity analysis.

Many people have helped in many ways in the preparation of this book, too many to enable us to acknowledge them all individually. However, we wish to acknowledge in particular Rafal Borusiewicz, Jakub M. Milczarek, David Lucy, Tereza Neocleous, Beata M. Trzcinska, and Janina Zieba-Palus for many helpful discussions and a great deal of collaborative work, from which we have been able to take much inspiration for the content of the book.

We also wish to thank Christopher J. Rogers. He checked the examples from the perspective of a beginner in the determination of the evidential value of physicochemical data. His suggestions helped to improve the quality of the practical examples contained herein.

Finally, we express our appreciation to the Institute of Forensic Research, Kraków, Poland, the Jagiellonian University, Kraków, Poland, the Escuela Politécnica Superior, Universidad Autónoma de Madrid, Spain, and the University of Edinburgh, UK, for their support of the research presented in this book.

1

Physicochemical data obtained in forensic science laboratories

1.1 Introduction

Various materials can be subjected to physicochemical examination by forensic experts. Such materials include illegal substances, blood and other body fluids, and transfer evidence (e.g. small fragments of glass, paint, fibres, plastics, organic and inorganic gunshot residues, fire debris). The size of samples subjected to analysis is very small, for example, fragments of glass with a linear dimension below 0.5 mm. Therefore, the analysis of morphological features, such as thickness and colour, is of no value for solving a comparison (Chapter 4) or classification problem (Chapter 5). Thus, it is necessary to employ the physicochemical features of the analysed fragments. When choosing an analytical method for analysis of microtraces for forensic purposes an expert should take into account not only the fact that the amount of material is very small but also that the method chosen should be non-destructive, leaving the material available for reuse. Examinations performed by the application of various analytical methods return several kinds of information including:

- qualitative data, for example, information on compounds detected in fire debris samples based on a chromatogram, information obtained from the spectrum of an unknown sample, and morphological information such as the number and thicknesses of layers in a cross-section of car paints;

- quantitative data, for example, the concentration of elements or value of the refractive index in a glass fragment, peak areas of a drug profile chromatogram or gasoline detected in fire debris, and the concentration of ethanol in blood samples.

In general, the fact finders (i.e. judges, prosecutors, policemen) are not interested in the physicochemical composition of a particular material (e.g. the elemental composition

Statistical Analysis in Forensic Science: Evidential Value of Multivariate Physicochemical Data, First Edition.
Grzegorz Zadora, Agnieszka Martyna, Daniel Ramos and Colin Aitken.
© 2014 John Wiley & Sons, Ltd. Published 2014 by John Wiley & Sons, Ltd.
Companion website: www.wiley.com/go/physicochemical

of glass) except in situations where such information could have a direct influence on the legal situation of a suspect (e.g. information on the level of ethyl alcohol in a blood sample). Questions raised by the police, prosecutors and the courts relate to the association between two or more items (which is known in the forensic sphere as a *comparison problem*; Chapter 4) and/or identification and classification of objects into certain categories (known in the forensic sphere as a *classification problem*; Chapter 5). These problems can be solved by the application of various statistical methods.

There are two main roles of statistics in forensic science. The first is during the investigation stage of a crime before a suspect has been identified, where statistics can be used to assist in the investigation (Aitken 2006a). The second is during the trial stage (Aitken 2006b), where statistics can be used to assist in the evaluation of the evidence. This last role of statistics is described in detail in this book.

When the evaluation of evidence is based on analytical data obtained from physicochemical analysis, careful attention to the following considerations is required:

- possible sources of uncertainty (sources of error), which should at least include variations in the measurements of characteristics within the recovered and/or control items, and variations in the measurements of characteristics between various objects in the relevant population (e.g. the population of glass objects);

- information about the rarity of the determined physicochemical characteristics (e.g. elemental and/or chemical composition of compared samples) for recovered and/or control samples in the relevant population;

- the level of association (correlation) between different characteristics when more than one characteristic has been measured;

- in the case of the comparison problem, the similarity of the recovered material and the control sample.

In this book it is advocated that the best way to include all these factors in the evidence evaluation process is by the application of likelihood ratio (LR) approach (Chapter 2).

It was mentioned that results of physicochemical analysis of various types of forensic evidence can be enhanced using statistical methods. Nevertheless such methods should always be treated as a supportive tool and any results should be subjected to critical analysis. In other words, statistical methods do not deliver the absolute truth as the possibility of obtaining false answers is an integral part of these methods. Therefore, sensitivity analysis (an equivalent of the validation process for analytical methods) should be performed in order to determine the performance of these methods and their influence on the next step, that of making a decision (Chapter 6).

With the aim of fully understanding the processes of the evaluation of the evidential value of physicochemical data it is necessary to first understand the origin of these data. Therefore, some details concerning the analysis of glass, flammable liquids, car paints, inks, and fibres for forensic purposes are presented in this chapter. The data obtained in the course of these analyses are used later in this book.

1.2   Glass

Glass is a material that is used in many areas of human activity. In domestic and commercial construction it appears most frequently as window glass, whereas in automotive transport

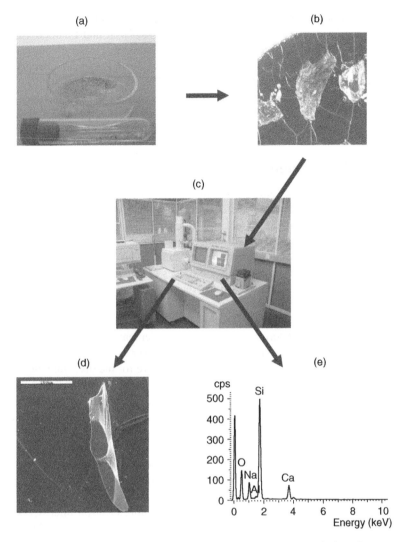

Figure 1.1 Principles of determination of elemental composition of glass fragments by the SEM-EDX technique: (a) debris collected from suspect clothes; (b) glass fragments located on an SEM stub; (c) view of SEM-EDX equipment; (d) SEM image of an analysed glass sample; (e) the spectrum of a glass sample obtained from an EDX detector.

it can form car windows and windscreens, car headlamps, car mirrors, and light bulbs. It is also used to make bottles, jars, tableware, and decorative items. Fragments of glass with a maximum linear dimension of 0.5 mm or less can be formed during events such as car accidents, burglaries and fights. These fragments may be recovered from the scene of the incident, as well as from the clothes and bodies of participants in any event of forensic interest (Figures 1.1(a),(b)). Such fragments may provide evidence of activity as well as the source of an object (Curran *et al.* 2000). The glass refractive index measurement (GRIM) method and scanning electron microscopy coupled with an energy dispersive X-ray spectrometer

(SEM-EDX) are routinely used in many forensic institutes for the investigation of glass and other trace evidence (Aitken *et al.* 2007; Evett 1977, 1978; Evett and Lambert 1982; Kirk 1951; Koons *et al.* 1988; Latkoczy *et al.* 2005; Lucy and Zadora 2011; Neocleous *et al.* 2011; Ramos and Zadora 2011; Zadora 2007a, 2009; Zadora *et al.* 2010; Zadora and Brozek-Mucha 2003; Zadora and Neocleous 2009a; Zadora and Neocleous 2009b; Zadora 2010).

Other methods used to determine the elemental composition of glass include: μ-X-ray fluorescence (μ-XRF) (Hicks *et al.* 2003) and laser ablation inductively coupled plasma mass spectrometry (LA-ICP-MS) (Latkoczy *et al.* 2005; Trejos and Almirall 2005a,b).

1.2.1 SEM-EDX technique

During the production of glass (Caddy 2001), many different elements are incorporated into the molten mixture. Certain elements are crucial for glass production and are always present. These major components are oxides of silica, sodium, calcium, magnesium, and potassium. Sodium is present, usually in the form of sodium carbonate, to reduce the softening point of silica, while calcium oxide and magnesium oxide make glass more chemically resistant. Minor components are also present, such as the oxides of aluminium and iron. Iron oxides are used to impart colour. Trace elements are also included, mostly depending on the required properties of the glass, particularly for any specialist uses.

The elements analysed using SEM-EDX are the major and minor elements found in glass. SEM-EDX does not allow for determination of the trace elements. The presence of the major and minor elements does not have great discriminating power as they are commonly present in glass. The trace elements are often regarded as imperative for the discrimination of glass, and suitable techniques are available for analysing the trace elements such as μ-XRF and LA-ICP-MS. However, in the field of forensic science the available equipment must often be used for as many purposes as possible, therefore if the concentrations of the major and minor elements alone can give correct and reliable data to solve a comparison and classification problem (Chapters 4 and 5), then the SEM-EDX method would be sufficient and useful for this purpose. The difference in the concentrations of these elements present in the sample is likely to be small. Therefore a statistical approach is imperative to detect any significant differences in the amounts of these elements.

The first stage of the SEM-EDX is the use of a scanning electron microscope (SEM; Figure 1.1(c)), which gives detailed three-dimensional images of a specimen (Figure 1.1(d)). The SEM works by using a beam of electrons as the source of illumination. A filament (e.g. made of tungsten) provides the source of electrons. As the filament is heated the electrons escape (thermionic emission) and a high voltage is applied to accelerate the negatively charged electrons away from the positively charged filament. The electrons interact with the specimen (e.g. glass sample) in various ways.

The X-rays produced by the SEM can provide information on elemental composition (Figure 1.1(e)), which is of interest here. The X-rays collected from the SEM are processed by an independent instrument (detector). Nowadays, an energy dispersive X-ray (EDX) detector is commonly used. This analyses the energy of the X-rays. The detector for the EDX system relies on a semiconductive crystal.

Quantitative analysis by the SEM-EDX method requires a surface of the sample be flat and smooth. An embedding procedure in resin could be used for sample preparation. This process is rather impractical for very small glass fragments (e.g. with linear dimensions less than 0.5 mm). Therefore, a question arises: is it possible to obtain useful information for forensic purposes when small fragments are prepared without the application of an embedding procedure for SEM-EDX analysis? The study of this problem was described in Falcone *et al.* (2006). The results presented by Falcone *et al.* showed that the accuracy and precision of the results (wt. % of SiO_2, Al_2O_3, Na_2O, K_2O, MgO, CaO, Fe_2O_3, Cr_2O_3) obtained for a non-embedded sample were not as good as those reported for an embedded sample. Nevertheless, this experiment does not reveal if data obtained for non-embedded glass fragments are reliable when solving comparison and classification problems, for example by applying an LR approach. It was shown that a simple procedure of glass preparation could be applied and this procedure allows the user to obtain reliable data for solving comparison and classification tasks for forensic purposes (Aitken *et al.* 2007; Neocleous *et al.* 2011; Ramos and Zadora 2011; Zadora 2009; Zadora 2010). In this procedure glass fragments were selected under an optical microscope in such a manner that their surfaces were as smooth and flat as possible (Figure 1.1(d)). A comparison of the elemental analysis results obtained for both embedded and non-embedded samples of glass standards (NIST 620, 621, 1830, 1831, USA) was carried out by one of the authors (unpublished results of validation process) with the aim of checking that such a procedure delivers useful information for solving forensic problems. The accuracy and precision of the results obtained for these prepared glass samples were the subject of analysis. Moreover, the results were used with the aim of solving a comparison problem by the application of the LR approach. No significant difference between results (accuracy, precision, and likelihood ratio values) was observed. It was concluded that the proposed method of small glass fragment preparation for SEM-EDX analysis (which excludes the embedding process) could be satisfactory for use in the forensic practice.

1.2.2   GRIM technique

Refraction is the phenomenon of light bending as it travels through a medium with a different optical density (Figure 1.2). The refractive index of a material is the degree at which a light wave bends upon passage through the transparent material. The light bends due to the change in its velocity as it passes from one material to a material of differing density (e.g. from air to glass). The speed of light is slowed as the light enters the denser material, and this causes the light wave to bend.

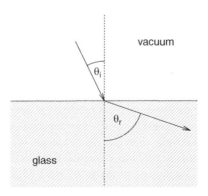

Figure 1.2   Refraction of light when passing from vacuum to glass.

A refractive index (RI) measurement quantifies the change in either the angle or the velocity of light and can be described by Snell's law:

$$RI = \frac{\sin \theta_i}{\sin \theta_r} = \frac{V_{vacuum}}{V_{glass}},$$

where θ_i is the angle of incidence, θ_r the angle of refraction, and V_{vacuum}, V_{glass} are the velocity of the light wave in the vacuum and glass material, respectively. The refractive index value effectively measures the ratio of the velocity of light in a vacuum to the velocity of light travelling within the transparent medium, or the ratio of the sine of the incident angle to the sine of the angle of refraction.

The RI of glass can be valuable as its value is affected by many factors. These include the substances present in the glass, the manufacturing process, any subsequent heating and cooling, and any stress present in the glass. This means that the RI value is highly discriminatory for glass samples and can be used for solving comparison problems (Chapter 4).

The RI of glass may be determined using a GRIM instrument (Figure 1.3(a)), which has a hot stage (Figure 1.3(b)) controlling the temperature of the immersion oil in which the sample has been mounted with a precision of at least 0.1°C. The glass fragment to be measured must be submerged in a suitable medium such as silicone oil. GRIM exploits the fact discovered

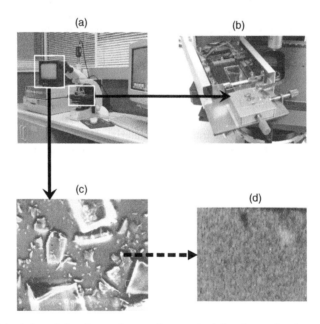

Figure 1.3 Principles of the thermoimmersion method for determination of the refractive index of glass microtraces: (a) GRIM2 set; (b) the appearance of a glass sample in an immersion oil on a microscopic slide located at the hot stage; (c) an image of glass edges in an immersion oil at the temperature when its refractive index is different from the refractive index of the measured glass sample; (d) an image of glass edges in an immersion oil at a matching temperature when its refractive index is equal to the refractive index of the measured glass sample.

by Emmons in the 1930s that the RI of a liquid varies with changing temperature, but the RI of glass changes very little with changing temperature. The technique also relies on the fact that when the RI of the liquid and the glass are the same the glass fragment submerged in the liquid is no longer visible (Figures 1.3(c),(d)). Therefore, when a piece of glass is submerged in a suitable medium (such as immersion oil) changing the temperature changes the RI of the immersion oil, but not the RI of the glass. The temperature can therefore be changed until the RI values are the same and the glass is no longer visible. In the past, the operator measuring the RI had to monitor for the disappearance of the glass fragment by eye. However, the development of the GRIM by Foster and Freeman meant that an instrument could replace the eyes of the operator. To do this the GRIM software identifies the point at which the contrast between the edge of the glass fragment and the oil is lowest and obtains a match temperature for this point. The match point temperature is then converted into the RI value by referring to a calibration equation which is compiled on the GRIM instrument using the same oil and reference glass fragments of known RI values.

If the amount of material available for analysis is large enough (in general, objects larger than 0.5 mm), then the fragment can also be annealed (Figure 1.4(a)). During annealing, tensions present in the glass object (which have an influence on the value of the RI) are removed. Stresses occur in glass because of the limited thermoconduction of glass, which means that the outer glass layers can cool significantly faster than the inner layers. This effect leads to the establishment of internal stresses during the manufacturing process. The stresses take the form of compression in the outer layers, whereas the inner layers are subject to tearing forces. Annealing eliminates or reduces internal stresses in glass. The annealing process works in such a way that the stresses upon the inner layers are removed slowly during controlled heating at high temperature. Afterwards, slow cooling is carried out to allow the glass layers to relocate to those positions where the internal/external stresses are minimised. Commonly produced glass types such as building windows and container glass are most often annealed as part of the manufacturing process. However, some glass types are not annealed at all, and in others stresses are deliberately introduced as part of a toughening process (e.g. toughened glass used for car windows). During the analysis of glass objects, annealing is often undertaken in muffle furnaces (Figure 1.4(b)). The heating/cooling programme can, however, vary in a manner that is dependent upon the glass and the preferences of the individual scientist. A typical temperature programme (Caddy 2001; Casista and Sandercock 1994; Locke and

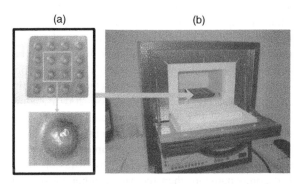

(a)   (b)

Figure 1.4 Annealing of glass fragments: (a) a metal holder containing glass fragments prepared for annealing; (b) a Nabertherm L3/11 muffle furnace with P320 programmer.

Underhill 1985; Newton *et al.* 2005) contains a step of fast heating up to 550°C, called the maximum temperature, at which most glass objects begin to melt. The glass fragment is kept at this temperature for some pre-determined period in order to eliminate the stresses present in the glass. In some laboratories, short temperature programs for annealing are conducted in tube furnaces. The glass fragment is heated up to 590°C, kept at this temperature for 12 minutes and then cooled at a rate of 4.5°C min^{-1} down to 425°C, at which it is held for 1 minute before being cooled down to room temperature (Caddy 2001). Long temperature programs, which employ fast heating of the specimen up to a high temperature, can be used, but this temperature is retained for a longer time (typically 10–15 hours) before the glass is slowly cooled down.

In general, for toughened glass, differences between RI values measured after annealing and those observed before the annealing process should be larger than differences for non-toughened glass. This is because of structural stresses introduced into the glass object during the toughening process, and removed during the annealing process. This information is used in a classification problem presented by Zadora and Wilk (2009) and in Section 5.4.2.

1.3   Flammable liquids: ATD-GC/MS technique

Arson is a frequently observed criminal offence (Almirall and Furton 2004; Mark and Sandercock 2007; Nic Daéid 2004; Zadora and Borusiewicz 2010). In cases where circumstances suggest that a fire might have been started deliberately, the scene of the putative offence is subject to the most comprehensive examination to recover materials and trace evidence possibly associated with the offence. The identification of flammable liquids, which can be used to start a fire, is one of the aims of forensic fire examination (Figure 1.5).

These compounds are flammable because they are volatile and they very quickly evaporate from the scene of the fire (Figure 1.5(a)). Success, therefore, depends on how quickly fire debris is collected, and how it is stored (Borusiewicz 2002). The correct method of fire debris packing and storage is in tightly closed metal cans, glass jars, or volatile compound free plastic bags. One of the difficulties in the interpretation of fire debris by gas chromatography (GC) is that the chemical composition of the debris often significantly differs from the composition of the primary liquid. This is because of partial evaporation, burning and contamination by pyrolysis products of other burned materials such as wood and carpets. Another problem is that volatile compounds are only present in trace amounts. Therefore, a desirable first stage of the analysis of flammable liquids in fire debris is the isolation and concentration of the traces of any flammable liquids which may be present in the sample.

Traditional methods of isolating volatile compounds in fire debris are based on procedures such as distillation and solvent extraction with substances such as carbon disulfate. These methods have been replaced by headspace isolation methods (Almirall and Furton 2004; Borusiewicz and Zieba-Palus 2007; Nic Daéid 2004; Zadora 2007b; Zadora and Borusiewicz 2010), in which a short tube packed with an adsorbent (e.g. Tenax TA^{TM}, Carbotrap 300^{TM}; Figure 1.5(b)) is inserted into a container with the fire debris. A fixed volume of headspace vapour is pulled through the adsorbent tube using a pump (dynamic headspace method), or compounds from the sample migrate to the adsorbent by diffusion only (static headspace method). The adsorbed compounds are then recovered either by solvent extraction or by heating (Figure 1.5(c)). A direct headspace method, based on the collection and direct dosage of the fixed volume of the headspace phase, is often additionally employed. Automated

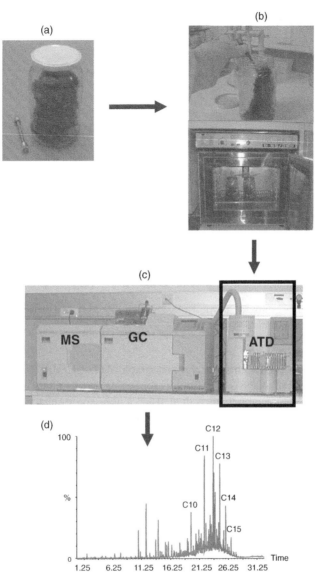

Figure 1.5   Principles of analysis of fire debris by the ATD-GC/MS technique: (a) a sample collected at the scene of a fire; (b) a metal tube filled with adsorbent Tenax TA™ is placed in a jar with fire debris (top) and the jar put in an oven at a pre-concentration stage of analysis (bottom); (c) the metal tube is placed in the automatic thermal desorber (ATD; black box) and the sample analysed by the GC/MS technique; (d) a chromatogram, showing a result that allows the investigator to determine the category (here: kerosene) from which the flammable liquid originates.

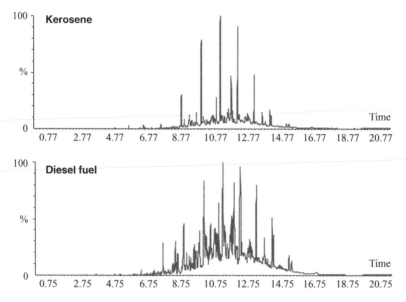

Figure 1.6 Chromatograms of partially evaporated samples of kerosene and diesel fuel revealing similarities in their composition.

thermal desorption coupled with a gas chromatograph (ATD-GC/MS, Figure 1.5(c)) is used as the secondary stage of analysis of fire debris separation on a capillary column. Nowadays, the detection of separated compounds is done using a mass spectrometer.

Flammable liquids detected in any fire debris sample require classification. Usually the American Society for Testing and Materials (ASTM E 1618) classification system is employed (ASTM 2006; Stauffer and Lentini 2003). The ASTM standard divides all ignitable liquids into nine classes: gasoline, petroleum distillate, isoparaffinic products, aromatic products, naphthenic paraffinic products, n-alkanes, de-aromatised distillates, oxygenated solvents, and a miscellaneous class of compounds. Each class, except gasoline, is in turn divided into three subclasses based on n-hydrocarbon range: light, medium, and heavy. Light indicates that the main components of the detected mixture fall in a carbon range from butane (C_4H_{10}) to nonane (C_9H_{20}), medium from octane (C_8H_{18}) to tridecane ($C_{13}H_{28}$), and heavy from octane (C_8H_{18}) to icosane ($C_{20}H_{42}$).

One of the analytical problems is the differentiation between evaporated kerosene and diesel fuel samples (Section 5.4.1) because there are only small quantitative differences in the composition of these chemical mixtures (Figure 1.6). The differences in their physicochemical properties result only from different amounts of individual components in each of these liquids. This is the reason why chromatograms obtained for weathered samples cannot be distinguished visually (Borusiewicz et al. 2006; Zadora and Zuba 2010).

1.4   Car paints: Py-GC/MS technique

The examination of car paints is an important aspect of road accident investigations (Caddy 2001; Thornton and Crim 2002). The paint fragments (Figure 1.7(a)) are collected from the

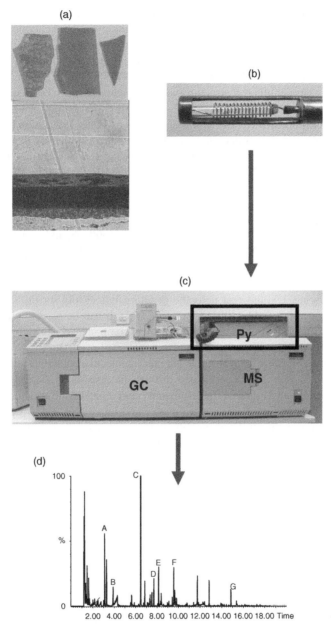

Figure 1.7 Principles of the analysis of car paints by the Py-GC/MS technique: (a) an image of a sample collected at the scene of a car accident (top) and a cross-section of car paint (bottom); (b) a car paint sample (50–100 μg) put into a quartz tube located in the platinum coil of a CDS Pyroprobe 2000 pyroliser (Py); (c) an analysis of pyrolysis products by GC/MS; (d) an example of a chromatogram obtained for styrene-acrylic-urethane car paint.

road or from the victim's clothes and compared with paint originating from the suspected car. It is necessary to obtain information on both the morphology and the chemical composition of the analysed samples in order to solve the comparison problem. First of all, different techniques of optical microscopy are used to define colour, surface texture and layer sequence (i.e. morphological features). Unfortunately, most paint specimens subjected to forensic examination do not have a full layer structure, so other methods have to be used to characterise the samples.

Fourier transform infrared (FTIR) spectroscopy (Zieba-Palus 1999) is an analytical technique routinely applied to determine the type of paint, that is, to identify the binder and the main inorganic pigments and fillers. Inorganic pigments can also be identified on the basis of their elemental content using X-ray methods, such as SEM-EDX (Caddy 2001; Zadora and Brozek-Mucha 2003). However, if the paint samples belong to the same class, that is, they contain a similar polymer binder and pigments, then further individualisation is required. This could involve the application of a more sensitive and discriminating analytical methods, such as Raman spectroscopy (Massonnet and Stocklein 1999; Suzuki and Carrabba 2001; Zieba-Palus and Borusiewicz 2006; Zieba-Palus et al. 2011) and pyrolytic gas chromatography mass spectrometry (Py-GC/MS) (Challinor 2007; Zieba-Palus 1999; Zieba-Palus et al. 2008a,b).

Py-GC/MS (Figure 1.7) has real potential as an analytical tool for the forensic examination of polymer traces (Caddy 2001; Challinor 2007). It can be used to characterise and compare polymer materials by determination of their components. During pyrolysis of the polymer sample (Figures 1.7(b),(c)), molecular fragments are produced which are usually characteristic of the composition of the original macromolecular material. The identification of pyrolysis products may be accomplished through the use of GC/MS. In paint examination, the chromatogram (Figure 1.7(d)) obtained may contain information about binder compounds as well as additives, organic pigments, and impurities, which could be helpful in the classification of paints or discrimination between paints with a similar polymer composition (i.e. similar FTIR spectra). Moreover, Py-GC/MS is able to find subtle structural or trace compositional variations within a matrix that is overwhelmingly similar among closely similar samples of the same category. The disadvantage of using Py-GC/MS for forensic purposes is that the sample is destroyed during analysis. It should be remembered that the analysed paint sample is the evidence and it should be available for various analyses at any time.

Numerous forensic applications have proved that Py-GC/MS can be used to characterise and compare polymer materials like paints by determination of their components (Milczarek et al. 2005). Several pyrolysis systems and techniques were discussed in overviews by Challinor (2007). Some authors tried to use this method for the discrimination between car paints (Burns and Doolan 2005a,b). Results presented in these papers showed that Py-GC/MS analysis confirmed results of FTIR spectroscopy analysis and allowed some subcategories within the particular category to be created. However, the distinction between samples of very similar chemical compositions remains a problem. An attempt was made to apply a likelihood ratio approach (Zadora 2010; Zieba-Palus et al. 2008b) to elaborate on the results obtained by Py-GC/MS analysis of styrene-acrylic-urethane paint samples that were indistinguishable on the basis of their infrared spectra and elemental composition (see also Section 4.4.3). The differences observed in the obtained pyrograms of the compared samples were small. Therefore, the aim of the analysis was to check if these samples could be distinguished on the basis of selected variables, which represented the main compounds presented on their pyrograms.

1.5 Fibres and inks: MSP-DAD technique

Textile fibres (Figures 1.8(a),(b)) are useful pieces of evidence that can help to reconstruct criminal events (Robertson and Grieve 1999). Forensic fibre analysis is usually of a comparative kind, whereby a recovered sample is examined in order to determine if it could share a common origin with control fibres.

Inks are a type of evidence widely examined in the forensic field (Olson 1986). They are mainly collected in criminal cases such as suicides, document and will forgeries, and blackmail. In most cases such ink evidence is the subject of comparative analysis, the aim of which is to determine whether or not it shares a common origin with the ink sample from, for example, a document constituting a control material.

Microscopic observation (Palenik 1999) of the morphology is the first step in the analysis of fibres. It allows a screening of numerous items at the scientist's disposal and the selection of those that should be analysed further by instrumental methods. These include the characterisation of the polymer composition by FTIR spectroscopy (Adolf and Dunlop 1999; Kirkbride and Tungol 1999) and, for fibres which are dyed, a microspectrophotometric comparison of colour (Adolf and Dunlop 1999; Zadora 2010; Zieba-Palus 2005). Microspectrophotometry (MSP) in the ultraviolet and visible range (UV-VIS) or only in visible range (VIS, 380–800 nm) has been used as part of the analysis of coloured textile fibres in forensic fibre examination (Figure 1.8(c)), with the results being generated in the form of transmission and absorbance spectra.

One feature that differs among all objects is their colour (Shevell 2003). Colour recognition is only possible due to the eye's ability to receive light stimuli, which are further interpreted in the brain. Even though the human eye is considered to be a highly precise organ, each of us perceives the colours in a slightly different manner. Therefore, colour impression by the human eye has proved to be rather subjective and thus difficult for objective interpretation. Since the seventeenth century, when Isaac Newton explained the dualistic nature of light, the mathematical description of electromagnetic waves and colours has been a concern of physics. To eliminate such subjectivity, attempts at the standardisation of colour using a two-degree (2°) standard observer were introduced in 1931. The notation 2°, along with the 10° observer introduced in 1964, refers to the viewing angle under which the standard observer sees the colour.

Comparison of spectra (Figure 1.8(d)) routinely involves overlaying these to determine whether they match with respect to peak positions and general spectral shape. Nevertheless, these approaches do not provide information about the value of evidence (Chapter 2).

An alternative approach which has been employed in forensic science (Olson 1986) is the parametrisation of MSP spectra in terms of coordinates in a three-colour system. The International Commission on Illumination (CIE) defines standards for colour transmittance, or reflection, for any illumination/measurement system, where a measured spectrum is weighted by some standard colour, under standard conditions. This is undertaken for the standard primary colours red, green, and blue, which are known as the tristimulus values of the *CIE XYZ* system. For microspectrophotometry with diode array detector (MSP-DAD) spectra, the

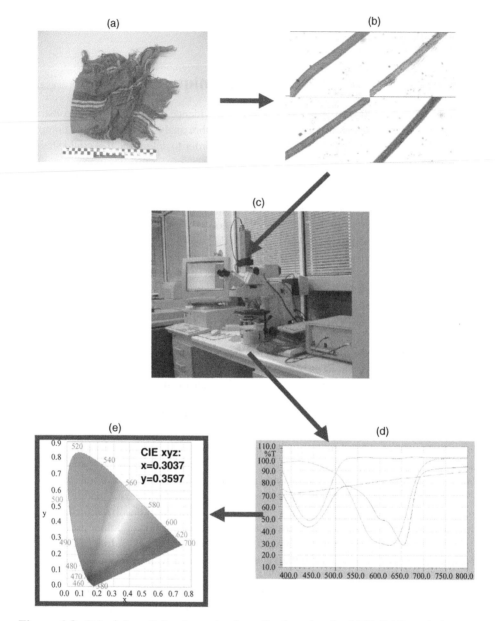

Figure 1.8 Principles of the determination of colour by the MSP-DAD technique – an example of fibres analysis: (a) textile sample; (b) fibre samples; (c) MSP-DAD set; (d) examples of spectra; (e) illustration of concepts of *CIE xyz* systems and an example of calculated chromaticity coordinates obtained for a fibre sample $(x + y + z = 1)$.

quantity of blue in any reflected or transmitted spectra is given by the component of the tristimulus values (Shevell 2003).

The tristimulus values X, Y, and Z can be used to provide coordinates in a three-dimensional colour space. However, standardised values can also be used to form the *CIE xyz* system (so-called chromaticity coordinates, Figure 1.8(e)):

$$x = \frac{X}{X+Y+Z}, \quad y = \frac{Y}{X+Y+Z}, \quad z = \frac{Z}{X+Y+Z}.$$

As an alternative, a set of weightings of the tristimulus values has been used to produce the *CIE Lab* system, where:

- L stands for lightness, ranging from black ($L = 0\%$) to white ($L = 100\%$);

- a is a chromatic value changing from green to red;

- b is a chromatic value changing from blue to yellow.

References

Adolf FP, Dunlop J 1999 Microspectrophotometry/colour measurements. In *Forensic Examination of Fibres*, 2nd edn, Robertson J, Grieve M (eds). Taylor & Francis, London.

Aitken CGG 2006a Statistics in forensic science. Part I. An aid to investigation. *Problems of Forensic Sciences* **65**, 53–67.

Aitken CGG 2006b Statistics in forensic science. Part II. An aid to evaluation of evidence. *Problems of Forensic Sciences* **65**, 68–81.

Aitken CGG, Zadora G, Lucy D 2007 A two-level model for evidence evaluation. *Journal of Forensic Sciences* **52**, 412–419.

Almirall JR, Furton KG 2004 *Analysis and interpretation of fire scene evidence*. CRC Press, Boca Raton, FL.

ASTM Standard Test Method for Ignitable Liquid Residues in Extracts from Fire Debris Samples by Gas Chromatography-Mass Spectrometry 2006. *Annual Book of ASTM Standards*.

Borusiewicz R 2002 Fire debris analysis – a survey of techniques used for accelerants isolation and concentration. *Problems of Forensic Sciences* **50**, 44–63.

Borusiewicz R, Zieba-Palus J 2007 Comparison of the effectiveness of Tenax TA and Carbotrap 300 in concentration of flammable liquids compounds. *Journal of Forensic Sciences* **52**, 70–74.

Borusiewicz R, Zieba-Palus J, Zadora G 2006 The influence of the type of accelerant, type of burned material, time of burning and availability of air on the possibility of detection of accelerants traces. *Forensic Science International* **160**(2–3), 115–126.

Burns DT, Doolan KP 2005a A comparison of pyrolysis–gas chromatography–mass spectrometry and Fourier transform infrared spectroscopy for the characterisation of automative paint samples. *Analytica Chimica Acta* **539**, 145–155.

Burns DT, Doolan KP 2005b The discrimination of automotive clear coat paints indistinguishable by Fourier transform infrared spectroscopy via pyrolysis–gas chromatography–mass spectrometry. *Analytica Chimica Acta* **539**, 157–164.

Caddy B 2001 *Forensic examination of glass and paints*. Taylor & Francis, London and New York.

Casista AR, Sandercock PML 1994 Effects of annealing on toughened and non-toughened glass. *Journal of Canadian Society of Forensic Science* **27**, 171–177.

Challinor JM 2007 Examination of forensic evidence. In *Applied Pyrolysis Handbook*, Wampler TP (ed). CRC Press, Boca Raton, FL, 207–239.

Curran JM, Hicks TN, Buckleton JS 2000 *Forensic interpretation of glass evidence.* CRC Press, Boca Raton, FL.

Evett IW 1977 The interpretation of refractive index measurement. *Forensic Sciences International* **9**, 209–217.

Evett IW 1978 The interpretation of refractive index measurement II. *Forensic Sciences International* **12**, 37–47.

Evett IW, Lambert JA 1982 The interpretation of refractive index measurement III. *Forensic Sciences International* **20**, 237–245.

Falcone R, Sommariva G, Merita M 2006 WDXRF, EPMA and SEM/EDX quantitative chemical analyses of small glass samples. *Microchimica Acta* **155**, 137–140.

Hicks TN, Monard-Sermier F, Goldmann T, Brunelle A, Champod C, Margot P 2003 The classification and discrimination of glass fragments using non destructive energy dispersive X-ray fluorescence. *Forensic Science International* **137**, 107–118.

Kirk PL 1951 *Density and Refractive Index: Their Application in Chemical Identification.* Charles C. Thomas, Springfield, IL.

Kirkbride KP, Tungol MW 1999 Infrared microspectroscopy of fibres. In *Forensic Examination of Fibres*, 2nd edn, Robertson J, Grieve M (eds.), Taylor & Francis, London.

Koons RD, Fiedler C, Rawalt R 1988 Classification and discrimination of sheet and container glasses by inductively coupled plasma – atomic emission spectrometry and pattern recognition. *Journal of Forensic Sciences, JFSCA* **33**, 49–67.

Latkoczy C, Becker S, Dücking M, Günther D, Hoogewerff J, Almirall JR, Buscaglia J, Dobney A, Koons RD, Montero S, van der Peijl G, Stoecklein W, Trejos T, Watling J, Zdanowicz V 2005 Development and evaluation of a standard method for the quantitative determination of elements in float glass samples by LA-ICP-MS. *Journal of Forensic Sciences* **50**(6), 1327–1341.

Locke J, Underhill M 1985 Automatic refractive index measurement of glass particles. *Forensic Science International* **27**, 247–260.

Lucy D, Zadora G 2011 Mixed effects modelling for glass category estimation from glass refractive indices. *Forensic Science International* **212**, 189–197.

Mark P, Sandercock L 2007 Fire investigation and ignitable liquid residue analysis. A review: 2001–2007. *Forensic Science International* **176**, 93–110.

Massonnet G, Stocklein W 1999 Identification of organic pigments in coatings: applications to red automotive topcoats: Part III: Raman spectroscopy (NIR FT-Raman). *Science and Justice* **39**, 181–187.

Milczarek JM, Zieba-Palus J, Koscielniak P 2005 Application of pyrolysis-gas chromatography to car paint analysis for forensic purposes. *Problems of Forensic Sciences* **61**, 7–18.

Neocleous T, Aitken CGG, Zadora G 2011 Transformations for compositional data with zeros with an application to forensic evidence evaluation. *Chemometrics and Intelligent Laboratory Systems* **109**, 77–85.

Newton AWN, Kitto L, Buckleton JS 2005 A study of the performance and utility of annealing in forensic glass analysis. *Forensic Science International* **115**, 119–125.

Nic Daéid N 2004 *Fire Investigation.* CRC Press, Boca Raton, FL.

Olson LA 1986 Color comparison in questioned document examination using MSP. *Journal of Forensic Sciences* **4**, 1330–1340.

Palenik SJ 1999 Microscopical examination of fibres. In *Forensic Examination of Fibres*, 2nd edn, Robertson J, Grieve M (eds). Taylor & Francis, London.

Ramos D, Zadora G 2011 Information-theoretical feature selection using data obtained by scanning electron microscopy coupled with an energy dispersive X-ray spectrometer for the classification of glass traces. *Analytica Chimica Acta* **705**, 207–217.

Robertson J and Grieve M (eds) 1999 Forensic examination of fibres. In *Forensic Examination of Fibres*, 2nd edn. Taylor & Francis, London.

Shevell SK 2003 *The science of color*. Elsevier, Oxford.

Stauffer E, Lentini JJ 2003 ASTM standards for fire debris analysis: a review. *Forensic Science International* **132**, 63–67.

Suzuki EM, Carrabba M 2001 In situ identification and analysis of automotive paint pigments using line segment excitation Raman spectroscopy: I. Inorganic topcoat pigments. *Journal of Forensic Sciences* **46**, 1053–1069.

Thornton JI, Crim D 2002 Forensic paint examination. In *Forensic Science Handbook* Saferstein R (ed.). Prentice Hall, Upper Saddle River, NJ.

Trejos T, Almirall JR 2005a Sampling strategies for the analysis of glass fragments by LA-ICP-MS Part I: Micro-homogeneity study of glass and its application to the interpretation of forensic evidence. *Talanta* **67**, 388–395.

Trejos T, Almirall JR 2005b Sampling strategies for the analysis of glass fragments by LA-ICP-MS Part II: Sample size and sample shape considerations. *Talanta* **67**, 396–401.

Zadora G 2007a Glass analysis for forensic purposes – a comparison of classification methods. *Journal of Chemometrics* **21**, 174–186.

Zadora G 2007b Laundering of 'illegal' fuels – a forensic chemistry perspective. *Acta Chimica Slovenica* **54**, 110–113.

Zadora G 2009 Classification of glass fragments based on elemental composition and refractive index. *Journal of Forensic Sciences* **54**, 49–59.

Zadora G 2010 Evaluation of evidential value of physicochemical data by a Bayesian network approach. *Journal of Chemometrics* **24**, 346–366.

Zadora G, Borusiewicz R 2010 Fire scene and fire debris analysis. In *INTERPOL's Forensic Science Review*, Nic Daéid N, Hauck MM (eds). CRC Press, Boca Raton, FL, 562–589.

Zadora G, Brozek-Mucha Z 2003 SEM-EDX – a useful tool for forensic examinations. *Material Chemistry and Physics* **81**, 345–348.

Zadora G, Neocleous T 2009a Likelihood ratio model for classification of forensic evidences. *Analytica Chimica Acta* **64**, 266–278.

Zadora G, Neocleous T 2009b Evidential value of physicochemical data – comparison of methods of glass database creation. *Journal of Chemometrics* **24**, 367–378.

Zadora G, Wilk D, 2009 Evaluation of evidence value of refractive index measured before and after annealing of container and float glass fragments. *Problems of Forensic Sciences* **80**, 365–377.

Zadora G, Zuba D 2010 Gas chromatography in forensic science, *Encyclopedia of Analytical Chemistry*. Wiley, Chichester.

Zadora G, Neocleous T, Aitken CGG 2010 A two-level model for evidence evaluation in the presence of zeros. *Journal of Forensic Sciences* **55**, 371–384.

Zieba-Palus J 1999 Application of micro-FTIR to the examination of paint samples. *Journal of Molecular Structure* **511–512**, 327–335.

Zieba-Palus J 2005 Microspectrophotometry in forensic science. In *Encyclopedia of Analytical Chemistry*. Wiley, Chichester.

Zieba-Palus J, Borusiewicz R 2006 Examination of multilayer paint coats by the use of infrared, Raman and XRF spectroscopy for forensic purposes. *Journal of Molecular Structure* **792–793**, 286–292.

Zieba-Palus J, Zadora G, Milczarek JM, Koscielniak P 2008a Pyrolysis-gas chromatography/mass spectrometry analysis as a useful tool in forensic examination of automotive paint traces. *Journal of Chromatography A* **1179**, 41–46.

Zieba-Palus J, Zadora G, Milczarek JM 2008b Differentiation and evaluation of evidence value of styrene acrylic urethane topcoat car paints analysed by pyrolysis-gas chromatography. *Journal of Chromatography A* **1179**, 47–58.

Zieba-Palus J, Michalska A, Weselucha-Birczynska A, 2011 Characterisation of paint samples by infrared and Raman spectroscopy for criminalistic purposes. *Journal of Molecular Structure* **993**, 134–141.

2

Evaluation of evidence in the form of physicochemical data

2.1 Introduction

The application of numerous analytical methods to the analysis of evidence samples returns various kinds of data (Chapter 1), which should be reliable. Confirmation that the data are reliable can be obtained when the particular analytical method is validated, that is, it is confirmed by the examination and provisions of objective evidence that the particular requirements for a specific intended use are fulfilled. Parameters such as repeatability, intermediate precision, reproducibility, and accuracy should be determined during the validation process for a particular quantitative technique. Therefore, statistical quantities such as standard deviations and relative standard deviations are usually calculated, and a regression/correlation analysis as well as an analysis of variance are usually carried out during the determination of the above-mentioned parameters (Chapter 3). This means that the use of statistical tools in the physicochemical analysis of evidence should not be viewed as a passing fad, but as a contemporary necessity for the validation process of analytical methods as well as for measuring the value of evidence in the form of physicochemical data.

It should also be pointed out that in general, representatives of the administration of justice, who are not specialists in chemistry, are not interested in details such as the composition of the analysed objects, except in a situation such as the concentration of alcohol in a driver's blood sample or the content of illegal substances in a consignment of tablets or body fluids. Therefore, results of analyses should be presented in a form that can be understood by non-specialists, but at the same time the applied method of data evaluation should express the role of a forensic expert in the administration of justice. This role is to evaluate physicochemical data (evidence, E) in the context of the prosecution proposition H_1 and defence proposition H_2, that is, to estimate the conditional probabilities $P(E|H_1)$ and $P(E|H_2)$ (some basic

Statistical Analysis in Forensic Science: Evidential Value of Multivariate Physicochemical Data, First Edition.
Grzegorz Zadora, Agnieszka Martyna, Daniel Ramos and Colin Aitken.
© 2014 John Wiley & Sons, Ltd. Published 2014 by John Wiley & Sons, Ltd.
Companion website: www.wiley.com/go/physicochemical

information on probability can be found in Appendix A). H_1 and H_2 are raised by the police, prosecutors, and the courts and they concern:

(a) the association between two or more items (referred to as a *comparison problem*; Chapter 4), for which H_1 could be that the compared samples originate from the same object, and H_2 that the compared samples originate from different objects;

(b) the classification of objects into certain categories (referred to as a *classification problem*; Chapter 5), for which H_1 could be that the analysed sample originates from category A, and H_2 could be that the analysed sample originates from category B.

The definition of the hypotheses H_1 and H_2 is an important part of the process in case assessment and interpretation methodologies considering likelihood ratios (Cook *et al.* 1998a; Evett 2011), as it conditions the entire subsequent evidence evaluation process. In particular, propositions can be defined at different *levels* in a so-called *hierarchy of propositions* (Aitken *et al.* 2012; Cook *et al.* 1998b). The first level in the hierarchy is the *source* level, where the inferences of identity are made considering the possible sources of the evidence. An example of propositions at the source level for a comparison problem is as follows:

- H_1: the source of the glass fragments found in the jacket of the suspect is the window at the crime scene;

- H_2: the source of the glass fragments found in the jacket of the suspect is some other window in the population of possible sources (windows).

Notice that at the source level the hypotheses make no reference to whether the suspect smashed the window or not, which should be addressed at the *activity* level. The hypotheses considered at this level are as follows:

- H_1: the person (from whose clothes the glass fragments were recovered) broke a window at the crime scene;

- H_2: the person (from whose clothes the glass fragments were recovered) did not break a window at the crime scene.

These hypotheses concern the problem of whether the person did something or not. A reliable answer to this question should also include the probabilities of incidental occurrence of glass fragments on the garments of the person due to, for example, contact with the person who actually did commit the crime or simple contamination.

The activity level makes no reference to whether the suspect committed the crime inside the house, which should be addressed at the *offence* level. The hypotheses considered at this level are as follows:

- H_1: the person committed the crime,

- H_2: the person did not commit the crime.

These are in fact the questions for the court involving some legal consequences. Thus, a forensic examiner typically starts to work with propositions at a *source* level. As more information is available in the case, it is possible to go up in the hierarchy of propositions and state propositions to the *activity* level. Nevertheless the *offence* level is reserved for the representatives of the court.

In this book, all the likelihood ratio computation methods consider the propositions defined at a *source* level, as these are the basic propositions in a case if no information other than the evidence is available.

2.2   Comparison problem

The evaluation of the evidential value of physicochemical data within the comparison problem requires taking into account the following aspects:

- possible sources of uncertainty (sources of error), which should at least include variations in the measurements of characteristics within the recovered and/or control items, and variations in the measurements of characteristics between various objects in the relevant population (e.g. the population of glass objects);

- information about the rarity of the determined physicochemical characteristics (e.g. elemental and/or chemical composition of compared samples) for recovered and/or control samples in the relevant population;

- the level of association (correlation) between different characteristics when more than one characteristic has been measured;

- in the case of the comparison problem, the similarity of the recovered material and the control sample.

Adopting a likelihood ratio (LR) approach allows the user to include all these factors in one calculation run. The LR is a well-documented measure of evidence value in the forensic field frequently applied in order to obtain the evidential value of various data (Chapter 1) being evaluated by a forensic expert (Aitken and Taroni 2004). Nevertheless, another so-called *two-stage* approach has been proposed. Both approaches are discussed in this section.

2.2.1   Two-stage approach

Within a *two-stage* approach (Aitken 2006a,b; Aitken and Taroni 2004; Lucy 2005) it is possible to compare two sets of evidence by using the data from these sets along with the application of suitable tests from the frequentist theory of hypothesis testing (Section 3.4). These are readily available in commercial statistical software. Nevertheless, the application of such an approach ignores information regarding the variability of the determined physicochemical values in a general population (rarity) and between-object variability. Therefore, this approach is a two-stage one, as it contains a comparison stage and then a rarity stage.

In the first stage, the recovered sample is compared with the control sample (e.g. physicochemical data obtained for a glass fragment recovered from a suspect's clothing and a glass fragment collected from the scene of a crime) by application of Student's t-test for univariate data or Hotelling's T^2-test for multivariate data (Section 3.4.7). The null hypothesis (H_0) states that mean values of particular physicochemical features are equal for compared objects (e.g. means of univariate data such as glass refractive index values; Section 1.2.2) or vectors of mean values of multivariate data are equal (e.g. the elemental composition of an analysed object; Section 1.2.1) determined for control and recovered samples. H_0 fails to be rejected when the calculated value of the significance probability (p) is larger than the assumed significance level (α; Section 3.4). Therefore, there is a binary outcome of this comparison as it could be decided that the control and recovered samples are similar or dissimilar.

If the objects are deemed dissimilar, then the analysis is stopped and it is decided to act as if the two pieces of evidence came from different sources. If the samples are deemed similar, the second stage is the assessment of the rarity of the evidence. This is an important part of the evaluation of evidence in the form of physicochemical data, as the above-mentioned significance tests take into account only information about within-object variation and the similarity of the compared items. Thus, the tests provide an answer to the question: *are the determined physicochemical properties similar for the compared samples?* As already mentioned, representatives of the administration of justice require the answer to the question: *did two compared samples originate from the same source or not?* Thus, an answer to this question also requires some knowledge about the rarity of the measured physicochemical properties in a population representative of the analysed casework, called the relevant population (e.g. the population of car windows in the case of a hit-and-run accident) and the between-object variability. For instance, one would expect refractive index values (Section 1.2.2) from different locations of the same glass object to be very similar. However, equally similar RI values could also be observed for different glass items. Therefore, information about the rarity of a determined RI value has to be taken into account. The value of the evidence in support of the proposition that the recovered glass fragments and the control sample have a common origin is greater when the determined values are similar and rare in the relevant population, than when the physicochemical values are equally similar but common in the same population. The information about the rarity of the physicochemical data could be obtained from relevant databases.

Discrimination based on the Welch test (Section 3.4.5), a modification of Student's t-test, as well as Hotelling's T^2 test, the equivalent test for multivariate data, was applied by Hicks *et al.* (2003). The research was performed on glass fragments from windows collected from a crime scene. It was possible to distinguish 6892 pairs based on univariate characteristics (i.e. RI, Section 1.2.2). A further 129 pairs were distinguished by results obtained from energy dispersive μ-X-ray fluorescence (μ-XRF; Kellner *et al.* 2004). Curran *et al.* (2000) also suggested the use of Hotelling's T^2-test for gathering evidence against the null hypothesis that glass fragments recovered from a suspect and a control glass sample from the crime scene share a common source. The control and recovered samples were described by concentrations of several different elements analysed by inductively coupled plasma atomic emission spectrometry (Kellner *et al.* 2004). Nevertheless, the authors concluded that the proposed method was designed for laboratory-based pre-screening or case pre-assessment of multivariate data as it does not consider any additional information relating to the samples, such as the rarity of the data in a relevant population. This is in contrast to methods based on the likelihood ratio where rarity of the particular physicochemical data in a relevant population is considered.

A further problem with the application of the classical approach for forensic purposes is an effect which has been likened to *falling off a cliff* (Robertson and Vignaux 1995). For $\alpha = 0.05$, the difference between a result of comparison which is significant at the $p = 0.051$ level and a difference which is significant at the $p = 0.049$ level has a very large effect on the suspect. In the first case, the evidence is taken to the second stage and is still considered as evidence against the suspect. In the second case, the evidence is discounted and is not considered as evidence against the suspect. At one point, one is safely positioned at the edge of a cliff, yet a very small step will take one over the edge. Therefore, a very small step has a very large effect. Such a small difference in significance probability values could be caused by a very slight difference in the obtained physicochemical results which sometimes could be lower than the assumed analytical error of an applied method (Section 2.2.3). A related problem is that there is a subjective choice in the *cut-off point* (or cliff edge) for the significance tests, represented by the arbitrary choice of the significance level α.

Yet another problem is that the burden of proof is placed on the defence rather than the prosecution as the significance test is based on an assumption that the evidence from the crime scene and from the environment of the suspect are from the same source (H_0 : mean measurement for crime scene equals mean measurement for environment of the suspect). This is unless there is a difference between characteristics of the evidence which is significant at some pre-specified level, say $\alpha = 0.05$. It is the responsibility of the defence to show that the difference is significant at this level. Thus, there is an assumption that the suspect is associated with the crime unless demonstrated otherwise.

These problems do not occur when the likelihood ratio approach is used. The LR approach treats the position of the prosecution and defence equally.

2.2.2   Likelihood ratio approach

A *random variable*, in a circular definition, is a variable that varies at random. For example, a DNA profile is a variable that varies at random among members of a population. A physicochemical measurement is a variable that varies at random among measurements performed on some chemical material, such as glass. A DNA profile is an example of a discrete random variable, one that takes only discrete values such as base-pair counts for an allele (Evett and Weir 1998). A concentration of a particular drug is an example of a continuous measurement, a measurement that can be represented on a continuum such as a straight line. An actual base-pair count is known as a *realisation* of a random variable. An actual concentration is a realisation of a random variable. Mathematical notation is introduced as short-hand to enable discussion of the analysis and interpretation of results. A random variable, representing for example a base-pair count or a chemical composition, is conventionally denoted by an upper-case Latin letter, such as X or Y. A realisation of a random variable, which is a number or group of numbers, such as (16, 17) for a base-pair count, or 3.2 wt. % for a chemical composition, is conventionally denoted with a lower-case Latin letter, such as x or y. Thus, for discrete random variables, it makes sense to write $P(X = x)$ as short-hand for the expression 'the probability that the DNA profile takes the value x' (Appendix A). For continuous random variables, the distribution of the probabilities across the appropriate continuum is represented by a continuous function, known as a *probability density function*, often denoted by $f(x)$ (Section 3.3.5). This is a non-negative function that integrates to one and for which the probability that X lies in a certain interval, (x_1, x_2) say, is the integral of $f(x)$ between x_1 and x_2. The function is non-negative since probabilities cannot be negative and integrates to one since it is certain that X takes some value and events that are certain have probability 1.

Consider a hypothesis H_1. For discrete measurements, the probability that X takes the value x if H_1 is true is denoted by $P(X = x \mid H_1)$. Similarly, $P(X = x \mid H_2)$ denotes the probability that X takes the value x when H_2 is true. The likelihood ratio compares the probability that $X = x$ when H_1 is true with the probability that $X = x$ when H_2 is true. It is the statistic

$$LR = \frac{P(X = x \mid H_1)}{P(X = x \mid H_2)}.$$

This measures the strength of the evidence in favour of H_1 compared with H_2 when $X = x$. The likelihood ratio is not a probability, but a ratio of probabilities, and it takes values between 0 and infinity (∞).

For continuous measurements, a similar argument holds. The probabilities are replaced by probability density functions $f(x \mid H_1)$ and $f(x \mid H_2)$ which are known in this context as likelihoods, hence the name *likelihood ratio* (Appendix A). In this case

$$LR = \frac{f(x \mid H_1)}{f(x \mid H_2)}.$$

Thus it can be seen that values of LR above 1 support H_1, while values of LR below 1 support H_2. A value of LR equal to 1 does not provide support for either proposition. The higher (lower) the value for LR, the stronger the support of X for $H_1(H_2)$. Evett *et al.* (2000) gives verbal equivalents of the calculated LR values: for $1 < LR \leq 10$ there is limited support for H_1, for $10 < LR \leq 100$ moderate support, for $100 < LR \leq 1000$ moderately strong support, for $1000 < LR \leq 10\,000$ strong support, and for $LR > 10\,000$ very strong support. Further discussion of these ideas is given in Appendix A.

The most successful application of the LR approach in the forensic sphere is found in the evaluation of the results of DNA profiling (Aitken and Taroni 2004; Evett and Weir 1998). This approach has been also used in the analysis of earprints, fingerprints, firearms and toolmarks, hair, documents, and handwriting, as well as speaker recognition (Ramos 2007). There are also an increasing number of applications of this approach in the evaluation of physicochemical data for univariate and multivariate data. Suitable LR models for solving comparison problems are presented in Chapter 4 along with examples of how to use them in practice. They are mostly based on the work of Aitken and Lucy (2004), where five methods of assessment of the evidential value of multivariate data were proposed. Two are based on significance tests and three on the evaluation of likelihood ratios. One of the likelihood ratio approaches transforms the data to a univariate projection based on the first principal component. The other two versions of the likelihood ratio for multivariate data account for the correlation among the variables and for two levels of variation: that between objects and that within objects (Section 3.3.2). One version assumes that between-object variability is modelled by a multivariate normal distribution; the other version models the variability with a multivariate kernel density estimate. Results are compared with the analysis of measurements on the elemental composition (ratios Ca/K, Ca/Si, and Ca/Fe) of 62 glass samples by inductively coupled plasma mass spectrometry (ICP-MS; Koons and Buscaglia 1999, 2002). The LR model proposed in Aitken and Lucy (2004), which assumed that between-object variability could be estimated by a multivariate kernel density estimate, was tested using a database comprising measurements of eight major elements (O, Na, Mg, Al, Si, K, Ca, and Fe) for 200 glass objects by the SEM-EDX technique (Aitken *et al.* 2007).

Another example of the use of LR models given by Aitken and Lucy (2004) can be found in Pierrini *et al.* (2007). The authors applied various multivariate LR models for the evaluation of the evidential value of the results of analysis of 26 samples of Semtex H by carbon and nitrogen isotope-ratio measurements. The lowest false positive rates (5.5%) were obtained for a model which applied a multivariate Hotelling's T^2 distance and kernel density estimation in the context of the following propositions: that of the prosecutor (H_1), that the control and recovered Semtex samples came from the same bulk source; and that of the defence (H_2), that the control and recovered Semtex samples came from different bulk sources.

In Zieba-Palus *et al.* (2008), 36 car paint samples were taken from new and repainted cars belonging to the IR category of styrene-acrylic-urethane topcoat paint samples and were examined by pyrolytic gas chromatography mass spectrometry (Py-GC/MS, Section 1.4). All the car paint samples contained styrene-acrylic-urethane binders and their chemical composition was very similar. All the chromatograms obtained had peaks from styrene and other products of polystyrene degradation as well as 1,6-diisocyanatohexane suggesting urethane modification of car paint. Some differences between the paints concerned only variation in acrylic compounds (i.e. occurrence of some small peaks or differences in their intensities). Nevertheless, it was difficult to conclude how important the relatively small differences were, especially when they were related to small peaks. Therefore, statistical analysis was performed. An LR model, in which a kernel density estimation for between-object variability was applied, based on seven proposed variables expressed as logarithms to base 10 of the ratio of the peak areas of seven organic compounds (methylmethacrylate, toluene, butylmethacrylate, methylstyrene, 2-hydroxyethylmethacrylate, 2-hydroxypropylmethacrylate, 1,6-diisocyanatohexane) to the peak area of styrene. It also took into account information delivered after the application of a graphical model (Section 3.7.2). Relatively low numbers of false positive (3%) and false negative (11%) answers were observed when the LR model, which applied kernel density estimation for estimating the between-object variability, was used (Section 4.3). It was also concluded that the LR model used seemed to be efficient for evaluating the evidence value of paint samples on the basis of Py-GC/MS analysis.

LR models applied in research published in Aitken *et al.* (2006, 2007) considered that, for instance, the concentration of iron oxides in non-coloured glass objects (which are present in each glass object because of their natural occurrence in sand) is below the detection limit of SEM-EDX detection (Section 1.2.1). Thus, it is reasonable to treat zero concentrations of iron as structural zeros, as they reflect trace quantities that are undetectable by this analytical method. A similar argument can be applied to traces of other chemical elements, allowing researchers to treat the values below the detection limit of SEM-EDX as structural zeros for the purposes of statistical analysis. Based on this assumption, a two-level multivariate LR model for comparison and classification of physicochemical data with zeros was proposed (Zadora *et al.* 2010). In this model, the presence of zeros is modelled by binomial (Bernoulli) distributions (Appendix A), whilst the non-zero sub-compositions are considered independently, and the LR is calculated by the model presented, for example, in Aitken and Lucy (2004). Different methods of data pre-treatment (obtained during the analysis of 320 samples by SEM-EDX) before the application of the LR approach were described in Neocleous *et al.* (2011). These techniques were the log-ratio transformation, a complementary log-log type transformation, and a hyperspherical transformation (Section 3.2). The application of the last method ensured that the compositions of glass objects lay on the unit hypersphere. This transformation essentially maps Cartesian coordinates to polar coordinates.

Models which allow for consideration of three levels of variability were also proposed by Aitken *et al.* (2006). These three levels of variability considered variability of measurements made on the analysed glass piece originating from a larger glass object (called measurement error), variability of measurements made within the glass object (within-object variability), and variability of measurements made on various glass objects from a relevant population (between-object variability).

2.2.3 Difference between an application of two-stage approach and likelihood ratio approach

The difference between the likelihood ratio approach and the two-stage approach is illustrated by an example. The following refractive index values (Section 1.2.2) were determined for control (A) and recovered (B) samples:

- A: 1.51907, 1.51909, 1.51903, 1.51898, 1.51896 ($\overline{RI}_A = 1.519026$);

- B: 1.51907, 1.51906, 1.51908, 1.51911, 1.51913 ($\overline{RI}_B = 1.519090$),

where $\overline{RI}_A$ and $\overline{RI}_B$ are the corresponding sample means. It should be mentioned that samples A and B originated from the same object. Student's t-test (Section 3.4) was used to test the null hypothesis $H_0 : \overline{RI}_A = \overline{RI}_B$. A significance probability p of 0.053 was obtained (the relevant **R** code is available in Section 3.4). This means that H_0 failed to be rejected at the 0.05 level of significance (Section 3.4). This is a positive answer to the question: *do the samples compared have the same RI values?* But, the question of interest is: *could they have come from the same object?* Thus, a rarity stage should be applied to evaluate the evidential value of an observed match between samples A and B. In this aim a suitable database should be analysed (Figure 2.1).

Taking into account information from a database, one might conclude that the frequency of the $\overline{RI} = 1.51906$ (the mean value of all RI values determined for samples A and B) is rare. But a further question is: *how rare is this value?* The answer to this question could be very subjective.

At the same time, the LR value obtained is 25 (equations (4.7) and (4.8), Appendix F). According to Evett *et al.* (2000) the verbal equivalent explaining the calculated LR value is

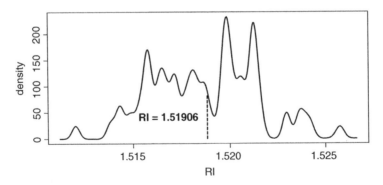

Figure 2.1 Distribution of RI values in the glass database presented in a file: RI_ database.txt[1].

---

[1] All the files are available from www.wiley.com/go/physicochemical

that the observed RI values provide limited support to the H_1 hypothesis, that the samples originate from the same object, as opposed to the H_2 hypothesis, that samples originate from different objects.

When a very small change in analytical results is made (e.g. from 1.51911 to 1.51914 in the fourth measurement of sample B), the significance probability changes from 0.053 to 0.047 and then the null hypothesis is rejected and a rarity stage is not included. Such a small change is lower than the precision of the applied analytical method (e.g. 5×10^{-5} units of refractive index). This allows a forensic expert working with a two-stage approach to conclude that the compared samples originate from different objects, which is a substantially different conclusion than the original one obtained for data with a small change. For this example LR = 20. It could be stated that the observed RI values provide limited support to the H_1 hypothesis that samples originate from the same object.

2.3   Classification problem

The evaluation of the evidential value of physicochemical data within the classification problem requires taking into account the following aspects:

- possible sources of uncertainty (sources of error) which should at least include variation of the measurements of characteristics within the analysed item, and variation of the measurements of characteristics between various objects in the relevant population (e.g. the population of glass objects);

- information about the rarity of the determined physicochemical characteristics (e.g. elemental and/or chemical composition) in the relevant population;

- the level of association (correlation) between different characteristics when more than one characteristic has been measured.

It should be explained that in the forensic sphere when the LR approach is used the word *classification* refers to solving a problem which in statistics and chemometrics is called *discrimination*. This is because discrimination in the forensic sphere is related to the problem of *comparison* (Section 2.2). In statistics and chemometrics, *discrimination* is the method by which a sample is ascribed to one of a number of previously determined classes of objects. In contrast, *classification* is the method by which the number and description of classes of objects are determined on the basis of results of the application of a particular classifier on the data, that is, the number of possible classes to which an object could be ascribed is not determined *a priori*.

2.3.1   Chemometric approach

Until now, a limited number of chemometric methods have been applied to solve the classification task in the forensic sphere. The most commonly applied methods have been the following:

(a) Discriminant analysis (DA), which is a classic method of discrimination into one of two or more groups. In general, an object is assigned to one of a number of pre-determined groups based on variables determined for the object. In discriminant analysis, the dependent variable is the group and the independent variables are the object features (e.g. physicochemical data) that might describe the group. If it can be assumed that the groups are linearly separable, a linear discriminant model can be used (LDA), that is, the groups can be separated by a linear combination of features that describe the objects. If there are only two features, the separators between groups become lines; if there are three features, the separator is a plane; and if the number of features is more than three, the separators become hyperplanes.

(b) Partial least-squares discriminant analysis (PLS-DA) is performed in order to sharpen the separation between groups of observations. This is done by rotating principal components (Section 3.7.1) so that maximal separation among classes is obtained, in order to understand which variables carry the class separating information. Classification in PLS is performed by the soft independent modelling of class analogy (SIMCA) approach.

(c) Support vector machines (SVM). This technique has been considered by many researchers to be the best-performing discriminative classifier, especially in the case of non-linearly separable cases. The SVM model selects an optimal boundary line that maximises the distance between the classes. This distance is known as the margin, while the objects at the border are called support vectors, because the optimisation task involving the maximisation of the margin depends only on their displacement in the data space.

(d) The naïve Bayesian classifier (NBC) is a generative type model, meaning that it represents a joint probability distribution over the data and the considered classes. Such classifiers are called naïve because they make the assumption that variables are generated independently of the others, which is almost always false. This assumption reduces a p-dimensional task to p one-dimensional tasks. However, a naïve Bayesian classifier is robust and generally performs well.

(e) The artificial neural network (ANN, commonly shortened to neural network or neural net (NN)) is a processing device, an algorithm, whose design was inspired by the design and functioning of animal brains and components thereof. The network is a weighted directed graph built with nodes (neurons). Each node sends an impulse along its outgoing arcs to its neighbour nodes. All the nodes sum their incoming impulses, and when this sum exceeds a threshold, they instantly turn on. In contrast, if the sum falls below the threshold, the nodes remain switched off. Computation proceeds by setting some input nodes, waiting for the network to reach a steady state, and then reading the output nodes, providing information as to which class a particular object belongs.

Principal component analysis (PCA, Section 3.7.1) and cluster analysis (CA, Section 3.6) are often applied as part of the classification task in the forensic sphere. These methods are so-called unsupervised techniques and they should only be used as exploratory techniques. They allow users to analyse the structure of multivariate data, present them in a graphical form, find relations between objects or variables in the multivariate sphere, or reduce the dimensionality of a problem.

More details on the theory of these methods can be found in Otto (2007) and Bishop (2006). Examples illustrating these techniques are presented below.

Results obtained for glass samples have been the subject of chemometric analysis. Attempts were made to apply CA to classify the glass sample on the basis of the results of inductively coupled plasma atomic emission spectrometry (Hickman 1987; Koons *et al.* 1988) or SEM-EDX (Zadora *et al.* 2003). The performance of this approach was satisfactory, but the number of incorrectly classified objects was sometimes high (i.e. the clusters obtained did not contain only glass objects originating from the same type (category of glass)). It was previously mentioned that CA is better used to analyse the structure of multivariate data (Section 3.6). Therefore, the task of classifying glass fragments for forensic purposes (e.g. those analysed by SEM-EDX) may be done by the application of methods such as SVM or NBC (Zadora 2007). The application of SVM has been especially successful in the case of classifying glass objects into *car window* or *building window* glass, categories whose compositions are similar as they are produced in a very similar way.

The use of μ-XRF for the analysis of small glass fragments is described by Hicks *et al.* (2003). Classification of fragments has been achieved using Fisher's linear discriminant analysis and neural networks. The results show that neural networks and LDA using qualitative and semi-quantitative data establish a classification of glass samples (headlamps, containers, windows, vehicle windows, light bulbs, eyeglasses) with a high degree of reliability.

PCA was used to discriminate between premium and regular gasoline based on Py-GC/MS data obtained from gasoline sold in Canada over one calendar year (Doble *et al.* 2003). Around 80–93% of the samples were correctly classified as either premium or regular gasoline using Mahalanobis distances (Section 3.6.1) calculated from the principal components scores. However, further classification into winter and summer subgroups gave unsatisfactory results (only 48–62% of the samples were correctly classified). An application of ANN to the second classification task resulted in approximately 97% of the premium and regular samples being correctly classified according to their winter or summer sub-group.

Differentiation of inks is one of the problems which a forensic chemist is asked to solve for determination of the document authenticity (Section 1.5). Kher *et al.* (2006) analysed the results from the analysis of blue inks by high-performance liquid chromatography with diode array detector (HPLC-DAD; Kellner *et al.* 2004) and infrared spectroscopy with PCA followed by the SIMCA model. It was concluded that the best classification rates were obtained when PCA/SIMCA was performed on data from HPLC-DAD analysis rather than data from IR spectra. Similar results were obtained by application of LDA. Nevertheless, the authors concluded that the results obtained based on IR spectra were also satisfactory and, taking into account the non-destructive character of this method, further research on its application for ink discrimination should be done. A further example relating to inks is provided by Thanasoulias *et al.* (2003).

IR spectra of different types of paper were collected by Kher *et al.* (2001) with the use of attenuated total reflectance (ATR) and diffuse reflectance (DRIFTS) techniques (Kellner *et al.* 2004). The spectral data were classified by the application of SIMCA. PCA was used for the selection of discrete spectral features with the highest discriminating abilities and DA was used to obtain a probabilistic classification. The use of PCA scores as DA variables provided the best resolution (100% of correct classifications) for the DRIFTS spectra. PCA on the ATR spectra resulted in the best discrimination, separating 68% of paper pairs completely. The results of the presented studies showed that IR spectroscopy coupled with multivariate statistical methods of analysis could provide a powerful discriminating tool for the document examiner.

The chemometric methods presented above are also used in drug profiling in forensic laboratories. According to the Drugs Working Group of the European Network of Forensic Science Institutes (ENFSI, www.enfsi.eu), drug profiling is the use of methods to define the chemical and/or physical properties of a drug seizure to compare seizures for intelligence (strategic, tactical) and evidential purposes. The standard procedure of drug profiling is based on GC/MS or the use of a gas chromatography-flame ionisation detector (GC/FID). These techniques analyse the presence of important impurities which are residues of substrates, semi-substrates or substances which are products of extraneous reactions etc. The detected compounds are present in the analysed samples (and therefore on their chromatograms) in various concentrations. Sometimes these concentrations vary up to some level of magnitude. Therefore, it is very rare for raw data to be used in chemometric analysis, and instead they are subjected to data pre-treatment (Section 3.2). This allows for the elimination of interference of applied analytical techniques and equipment. The methods of pre-treatment that are most commonly used are normalisation (e.g. in comparison to the peak area of the internal standard or the main sample component), standardisation (e.g. by dividing the peak area of a particular impurity by the standard deviation estimated from the results of the analysed group of samples), weighting (e.g. by calculation of the ratio of the peak area of a particular impurity to the sum of peak areas of all considered impurities), calculation of the logarithm to base 10, calculation of the fourth root, or a combination of two or more of these methods. In the next step a similarity measure between the compared samples is determined. Commonly used distance (similarity) measures (Section 3.6.1) include Euclidean distance and its square, Canberra distance, Manhattan distance, Pearson correlation coefficient, square cosine function, and similarity index. Visualisation of results is an integral part of the profiling procedure. This is most often done by the application of CA or PCA.

Amphetamine was the first illegal drug to be subjected to profiling (Andersson *et al.* 2007). The first attempt (based on results from GC/FID analysis), was made in the 1970s. The application of various chemometric methods for amphetamine profiling (on the basis of detected impurities) was also a subject of research. Twelve different samples of amphetamine were synthesised by the application of various methods (i.e. Leuckart's method, reductive amination of benzylmethylketone, and the nitrostyrene method). Lactose or caffeine was added to some samples to synthesize amphetamine similar to that sold on the market. This created 44 samples in total. Samples from both the same batch and various batches (obtained from a repeatable synthesis procedure) were treated as linked samples (i.e. from the same seizure). Samples from different synthesis batches and obtained in non-repeatable conditions were treated as non-linked (from different seizures). The comparison of distances obtained between linked and non-linked samples led to the conclusion that the most effective procedures of data pre-treatment (Section 3.2) were normalisation and normalisation with fourth root degree for Euclidean distance, Manhattan distance, Pearson correlation coefficient and cosine, normalisation plus logarithm of Canberra distance, and similarity index.

A method of heroin profiling was discussed by Klemenc (2001). Heroin samples from seizures were analysed using gas chromatography to separate the major alkaloids present in illicit heroin. Statistical analysis was then performed on 3371 samples. Initially, PCA was performed as a preliminary screening to identify samples of a similar chemical profile. About 20 chemical classes with similar chemical profiles were identified. The normalised peak areas of six target compounds were then used to train an ANN to classify each sample into its appropriate class. The training data set consisted of 468 samples. Sixty samples were treated as blind and 370 as non-linked samples. The results show that in 96% of cases the ANN attributed the seizure to the correct chemical class.

2.3.2   Likelihood ratio approach

The likelihood ratio approach can also be used to solve the classification problem. For two classes, two hypotheses are formulated: H_1, that the object came from class 1; and H_2, that the object came from class 2. Suitable LR models for continuous type physicochemical data are described in Chapter 5.

The LR model was first satisfactorily applied to the classification of glass fragments analysed by SEM-EDX and the GRIM technique (Section 1.2) for 111 objects (23 building windows, 32 car windows, 56 containers; Zadora 2009). The analysis showed that the best results (i.e. the lowest false classification rates) were obtained when variables derived from elemental compositions measured by SEM-EDX and refractive index values determined before (RI_b) and after (RI_a) the annealing process (Section 1.2.2), expressed as $dRI = \log_{10} |RI_a - RI_b|$, were used. Additionally, this analysis allowed for samples to be categorised into car or building windows, whose elemental composition and manufacture are similar. Moreover, research from these data (Zadora 2007; Zadora and Neocleous 2009) shows some evidence that the performance of the LR model is comparable to some existing classification techniques (such as SVM and NBC). The application of SVM and NBC gave slightly better results than application of the LR model. However, the observed differences in misclassification rates were not great, and no single classification method was clearly more effective than any other. Therefore, LR models can be recommended as more suitable for evaluation of evidential value of physicochemical data.

Another example is the application of a Bayesian networks model to solve a problem relating to the presence of kerosene or diesel fuel in a fire debris sample. This is described in detail in Section 5.4.1 as well as in Zadora (2010).

Finally, the problem of verifying food products authenticity is a classification problem that can be solved using physicochemical data obtained from various analytical methods. The results of the research presented in Martyna *et al.* (2013) show that LR models can be successfully applied to this problem.

2.4   Likelihood ratio and Bayes' theorem

Equation (A.1), obtained from Bayes' theorem (Appendix A), can be written as

$$\frac{P(H_1|E)}{P(H_2|E)} = \frac{P(H_1)}{P(H_2)} \cdot \frac{P(E|H_1)}{P(E|H_2)}. \tag{2.1}$$

$P(H_1)$ and $P(H_2)$ are called *prior probabilities* and their quotient is called the *prior odds*.[2] Their estimation lies within the competence of the fact finders (judge, prosecutor, or police) expressing their opinions on the hypotheses of interest before the evidence is analysed, thus

---

[2] Strictly speaking, the term *odds* should only be used when the propositions H_1 and H_2 are mutually exclusive and exhaustive. In practice, H_1 and H_2 need only be mutually exclusive for equation (2.1) to hold. In such a situation, the term *odds* is used somewhat loosely to refer to the ratio of the probabilities of H_1 and H_2.

without the availability of further information. It is the duty of a fact finder or the court to determine whether the objects are deemed to stem from the same or different sources, and this decision is based on the results expressed in the form of conditional probabilities, $P(H_1|E)$ and $P(H_2|E)$, namely *posterior probabilities* and their quotient, the *posterior odds*. These could be assigned by taking into account the prior odds and the information delivered by the forensic expert in the form of LR. Therefore, it is important that the method used for the evaluation of evidence provides strong support for the correct hypothesis (i.e. $LR \gg 1$ when H_1 is correct and $LR \ll 1$ when H_2 is correct). Additionally, it is desired that if an incorrect hypothesis is supported by LR (i.e. $LR > 1$ when H_2 is correct and $LR < 1$ when H_1 is correct) then LR should be close to 1, providing weak misleading evidence. Roughly speaking, according to equation (2.1), it is of great importance to obtain LR values that do not provide misleading information for the court or police. This implies the need to evaluate the performance of the data evaluation method used. This can be done by carrying out suitable experiments which allow experts to obtain information on the level of:

(a) false positive and false negative answers within a comparison problem. A false positive occurs when fragments originating from different objects are thought to originate from the same object. A false negative occurs when fragments originating from the same object are thought to originate from different objects.

(b) incorrect classifications within a classification problem.

Nevertheless, as will be discussed later, these rates are not sufficient for an adequate performance measurement of likelihood ratios. Suitable measures of the performance of LR methods are described in detail in Chapter 6.

References

Aitken CGG 2006a Statistics in forensic science. Part I. An aid to investigation. *Problems of Forensic Sciences* **65**, 53–67.

Aitken CGG 2006b Statistics in forensic science. Part II. An aid to evaluation of evidence. *Problems of Forensic Sciences* **65**, 68–81.

Aitken CGG, Lucy D 2004 Evaluation of trace evidence in the form of multivariate data. *Applied Statistics* **53**, 109–122.

Aitken CGG, Taroni F 2004 *Statistics and the evaluation of evidence for forensic scientists*. In 2nd edn. Wiley, Chichester.

Aitken CGG, Lucy D, Zadora G, Curran JM 2006 Evaluation of trace evidence for three-level multivariate data with the use of graphical models. *Computational Statistics and Data Analysis* **50**, 2571–2588.

Aitken CGG, Zadora G, Lucy D 2007 A two-level model for evidence evaluation. *Journal of Forensic Sciences* **52**, 412–419.

Aitken CGG, Roberts P, Jackson G 2012 *Fundamentals of Probability and Statistical Evidence in Criminal Proceedings*. Royal Statistical Society, London.

Andersson K, Lock E, Jalava K, Huizer H, Jonson S, Kaa E, Lopes A, Poortman-van der Meer A, Sippola E, Dujourdy L, Dahlen J 2007 Development of a harmonized method for the profiling of amphetamines VI. Evaluation of methods for comparison of amphetamine. *Forensic Science International* **169**, 86–99.

Bishop CM 2006 *Pattern Recognition and Machine Learning*. Springer, New York.

Cook R, Evett IW, Jackson G, Jones PJ and Lambert JA 1998a A model for case assessment and interpretation. *Science and Justice* **38**, 151–156.

Cook R, Evett IW, Jackson G, Jones PJ and Lambert JA 1998b A hierarchy of propositions: Deciding which level to address in casework. *Science and Justice* **38**, 231–239.

Curran JM, Hicks TN, Buckleton JS 2000 *Forensic Interpretation of Glass evidence*, CRC Press, Boca Raton, FL.

Doble P, Sandercock M, Du Pasquier E, Petocz P, Roux C, Dawson M 2003 Classification of premium and regular gasoline by gas chromatography/mass spectrometry, principal component analysis and artificial neural networks. *Forensic Science International* **132**, 26–39.

Evett IW, *et al.* 2011 Expressing Evaluative Opinions: A Position Statement. *Science and Justice* **51**, 1–2.

Evett IW, Weir BS 1998 *Interpreting DNA Evidence*. Sinauer Associates, Sunderland, MA.

Evett IW, Jackson G, Lambert JA, McCrossan S 2000 The impact of the principles of evidence interpretation and the structure and content of statements. *Science and Justice* **40**, 233–239.

Hickman DA 1987 Glass types identified by chemical analysis. *Forensic Science International* **33**, 23–46.

Hicks TN, Monard-Sermier F, Goldmann T, Brunelle A, Champod C, Margot P 2003 The classification and discrimination of glass fragments using non destructive energy dispersive x-ray fluorescence. *Forensic Science International* **137**, 107–118.

Kellner R, Mermet JM, Otto M, Valcarcel M, Widmer HM (eds.) 2004 *Analytical Chemistry: A Modern Approach to Analytical Science*, 2nd edn. Wiley VCH, Weinheim.

Kher A, Mulholland M, Reedy B, Maynard P 2001 Classification of document papers by infrared spectroscopy and multivariate statistical techniques. *Applied Spectroscopy* **55**, 1192–1198.

Kher A, Mulholland M, Green E, Reedy B 2006 Forensic classification of ballpoint pen inks using high performance liquid chromatography and infrared spectroscopy with principal components analysis and linear discriminant analysis. *Vibration Spectroscopy* **40**, 270–277.

Klemenc S 2001 In common batch searching of illicit heroin samples – evaluation of data by chemometric methods. *Forensic Science International* **115**, 43–52.

Koons RD, Buscaglia J 1999 The forensic significance of glass composition and refractive index measurements. *Journal of Forensic Sciences* **44**, 496–503.

Koons RD, Buscaglia J 2002 Interpretation of glass composition measurements: The effects of match criteria on discrimination capability. *Journal of Forensic Sciences* **47**, 505–512.

Koons RD, Fiedler C, Rawalt R 1988 Classification and discrimination of sheet and container glasses by inductively coupled plasma-atomic emission spectrometry and pattern recognition. *Journal of Forensic Sciences* **33**, 49–67.

Lucy D 2005 *Introduction to Statistics for Forensic Scientists*. John Wiley & Sons, Ltd, Chichester.

Martyna A, Zadora G, Stanimirova I, Ramos D 2013 Wine authenticity verification as a forensic problem. An application of likelihood ratio approach to label verification. *Food Chemistry*.

Neocleous T, Aitken CGG, Zadora G 2011 Transformations for compositional data with zeros with an application to forensic evidence evaluation. *Chemometrics and Intelligent Laboratory Systems* **109**, 77–85.

Otto M 2007 *Statistics and Computer Application in Analytical Chemistry*. Wiley VCH, Weinheim.

Pierrini G, Doyle S, Champod C, Taroni F, Wakelin D, Lock C 2007 Evaluation of preliminary isotopic analysis (^{13}C and ^{15}N) of explosives. A likelihood ratio approach to assess the links between Semtex samples. *Forensic Science International* **167**, 43–48.

Ramos D 2007 Forensic evaluation of the evidence using automatic speaker recognition systems. PhD thesis, Depto. de Ingeniería Informática, Escuela Politécnica Superior, Universidad Autónoma de Madrid. Available at http://atvs.ii.uam.es.

Robertson B, Vignaux GA 1995 *Interpreting Evidence: Evaluating Forensic Science in the Courtroom*. John Wiley & Sons, Ltd, Chichester.

Thanasoulias NC, Parisis NA, Evmiridis NP 2003 Multivariate chemometrics for the forensic discrimination of blue ball-point pen inks based on their VIS spectra. *Forensic Science International* **138**, 75–84.

Zadora G 2007 Glass analysis for forensic purposes – a comparison of classification methods. *Journal of Chemometrics* **21**, 174–186.

Zadora G 2009 Classification of glass fragments based on elemental composition and refractive index. *Journal of Forensic Sciences* **54**, 49–59.

Zadora G, Neocleous T 2009 Likelihood ratio model for classification of forensic evidences. *Analytica Chimica Acta* **64**, 266–278.

Zadora G 2010 Evaluation of the evidence value of physicochemical data by a Bayesian network approach. *Journal of Chemometrics* **24**, 346–366.

Zadora G, Piekoszewski W, Parczewski A 2003 An application of chosen similarity measurements of objects for forensic purposes. In *Bulletin of the International Statistical Institute 54th Session* **60**, 364–367.

Zadora G, Neocleous T, Aitken CGG 2010 A two-level model for evidence evaluation in the presence of zeros. *Journal of Forensic Sciences* **55**, 371–384.

Zieba-Palus J, Zadora G, Milczarek JM 2008 Differentiation and evaluation of evidence value of styrene acrylic urethane topcoat car paints analysed by pyrolysis-gas chromatography. *Journal of Chromatography A* **1179**, 47–58.

3

Continuous data

3.1 Introduction

This chapter presents some general background information on continuous data that is necessary to fully understand the topics on which the book focuses in other chapters (see also Aitken and Taroni 2004; Curran 2011; Lucy 2005). Examinations performed by the application of various analytical methods (Chapter 1) return various kinds of information:

- qualitative data, for example, information on compounds detected in fire debris samples based on a chromatogram, information obtained from the spectrum of an unknown sample, and morphological information such as the number and thicknesses of layers in a cross-section of car paints;

- quantitative data, for example, the concentration of elements or value of the refractive index in a glass fragment, peak areas of a drug profile chromatogram or gasoline detected in fire debris, and the concentration of ethanol in blood samples.

Another way to classify data is connected with the possible values they may generate. If the analysis produces data that can take any value within a specified range, they are considered to be *continuous*, as opposed to *discrete* data, which may take certain values. A Typical examples of continuous data are human height and the elemental composition of glass fragments. A typical example of discrete data is the number of children in a family, as it may take only integer values, and cannot take values such as 3.3. Most of the data that chemists have to deal with are of continuous type and so this book focuses on them.

Physicochemical data analysis requires some data organisation. A recommended way to do so is to present the data generated during the analysis in the form of a rectangular table as

Statistical Analysis in Forensic Science: Evidential Value of Multivariate Physicochemical Data, First Edition.
Grzegorz Zadora, Agnieszka Martyna, Daniel Ramos and Colin Aitken.
© 2014 John Wiley & Sons, Ltd. Published 2014 by John Wiley & Sons, Ltd.
Companion website: www.wiley.com/go/physicochemical

No. object, m	No. measurement, n	Variables, p $X_1 \quad X_2 \quad X_3 \quad \cdots \quad X_p$
1	1	
1	2	
1	3	
2	1	
2	2	
2	3	
.	1	
.	2	
.	3	
m	1	
m	2	
m	3	

Figure 3.1 Data organisation in the form of a matrix. Here the data refer to p variables (X_p) measured three times for each of m objects.

$$\text{(a) } \mathbf{X} = \begin{bmatrix} x_{11} & \cdots & x_{1j} \\ \vdots & \ddots & \vdots \\ x_{i1} & \cdots & x_{ij} \end{bmatrix}, \qquad \text{(b) } \mathbf{x} = [x_1, \ldots, x_j], \qquad \text{(c) } \mathbf{x} = \begin{bmatrix} x_1 \\ \vdots \\ x_i \end{bmatrix}$$

Figure 3.2 An example of (a) a matrix, (b) a row vector, and (c) a column vector.

in Figure 3.1 (see also the example of a database in the file glass_data.txt).[1] Such tables containing data may take the form of what is generally called a matrix, as in Figure 3.2(a).

A matrix is filled with so-called elements, usually of numeric type. They are arranged in i rows and j columns, which determine the size of the matrix. Each element is described by two subscripts: the first denotes the number of the row and the second the number of the column the element belongs to. Thus the element denoted by x_{13} is the element in the first row and third column.

If either $i = 1$ or $j = 1$ we speak of vectors rather than matrices. A matrix of size $1 \times j$ is called a row vector (Figure 3.2(b)), while a matrix of size $i \times 1$ is called a column vector (Figure 3.2(c)).

Matrices are usually denoted by bold capital letters (e.g. $\mathbf{X}$), with the matrix size indicated by means of subscripts (e.g. $\mathbf{X}_{[i,j]}$ or $\mathbf{X}_{ij}$). Vectors are usually represented by bold lower-case letters (e.g. $\mathbf{x}$). The elements of matrices and vectors are denoted by non-bold lower-case letters (e.g. x).

Figure 3.3 presents a special type of matrix, the symmetric matrix. A symmetric matrix is a matrix that is equal to its transpose: $\mathbf{X} = \mathbf{X}^T$ (Appendix B). Note that this means that $x_{mn} = x_{nm}$, for $m = 1, \ldots, i$ and $n = 1, \ldots, j$. A matrix with equal numbers of rows and columns, $i = j$, is called a square matrix. All symmetric matrices are square as well (Figure 3.3). Two matrices are equal when their corresponding elements are equal.

---

[1] All the files are available from www.wiley.com/go/physicochemical

(a)   (b)   (c)

$$
X = \begin{bmatrix} 2 & 0 & 1 & 3 \\ 0 & 1 & 2 & -2 \\ 1 & 2 & 1 & 1 \\ 3 & -2 & 1 & -3 \end{bmatrix} \quad X = \begin{bmatrix} 2 & 0 & 0 & 0 \\ 0 & 1 & 0 & 0 \\ 0 & 0 & 1 & 0 \\ 0 & 0 & 0 & -3 \end{bmatrix} \quad X = \begin{bmatrix} 1 & 0 & 0 & 0 \\ 0 & 1 & 0 & 0 \\ 0 & 0 & 1 & 0 \\ 0 & 0 & 0 & 1 \end{bmatrix}
$$

Figure 3.3 Examples of symmetric matrices: (a) a simple symmetric matrix; (b) a diagonal matrix; (c) an identity matrix.

A diagonal matrix (Figures 3.3(b),(c)) is a symmetric matrix, in which elements not lying on the diagonal are equal to 0. An example of such a matrix is the identity matrix (Figure 3.3(c)), commonly denoted as I, whose diagonal elements (i.e. elements lying on the main diagonal of the matrix) are all equal to unity and whose off-diagonal elements are all equal to zero. When the matrix X is multiplied by the unity matrix I, X matrix is returned: $XI = IX = X$. More information on the operations performed on matrices is presented in Appendix B.

3.2   Data transformations

Many chemometric and statistical methods require specific dataset features such as a particular data distribution or a linear relationship between variables. If the original data, obtained in the course of the analysis, do not fulfil the requirements of the method, data transformation methods can be employed.

Some chemometric techniques put some restraints on the mean or data dispersion. For example, PCA (Section 3.7.1) requires that the data have mean 0 and each variable has a variance equal to 1. This can be achieved by *autoscaling*, the most widely used form of data transformation, which addresses the inconvenience caused by data measured in different data scales or units. A significant feature of autoscaling is due to the equalisation (to unity) of the variance estimates for each of the variables considered. In such cases each of the variables has the same contribution to the general variance (Section 3.3.2). Autoscaling can be performed by using the equation

$$ z_{ij} = \frac{x_{ij} - \bar{x}_i}{s_i}, \tag{3.1} $$

where x_{ij} is the jth observation of the ith variable, $\bar{x}_i$ the mean of the ith variable, and s_i the standard deviation of the ith variable.

If the data distribution is far from normal, a *logarithmic transformation* may be helpful. The new variable x' is given as $x' = \log_{10}(x)$. Figure 3.4 presents Q-Q plots (Section 3.3.5) illustrating the distribution of the data before and after a logarithmic transformation.

Some modifications of logarithmic transformation are also permissible. Data describing glass elemental content (Chapter 1) are often subjected to the logarithmic transformation after taking the ratios of a particular element content to the oxygen content, for example $\log_{10} NaO = \log_{10}\left(\frac{Na}{O}\right)$. This approach not only brings the data closer to normality but also reduces the instrumental stochastic measurement fluctuations.

In the case of glass elemental composition data, it is quite common to obtain zero values within the results of the analysis. Such an outcome means that the sample lacks this

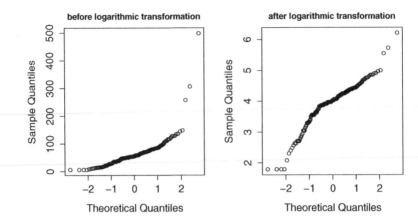

Figure 3.4   Q-Q plots for a dataset (left) before and (right) after logarithmic transformation.

particular element or its content is below the measuring instrument's limits of detection. When zeros occur it is impossible to take log ratios, so other transformation techniques must be employed. One of these is the *hyperspherical transformation*, which converts Cartesian coordinates into polar coordinates (Stephenson 1973). For data expressed as a vector of p variables, $\mathbf{x} = (x_1, \ldots, x_p)$, the square root of the original data is taken and given in the form

$$\mathbf{s} = \left(\sqrt{x_1}, \ldots, \sqrt{x_p}\right) = \left(s_1, \ldots, s_p\right).$$

Such new variables are further transformed following the pattern:

$$\omega_p = \arccos\left(s_p\right),$$

$$\omega_{p-1} = \arccos\left(\frac{s_{p-1}}{\sin \omega_p}\right),$$

$$\omega_{p-2} = \arccos\left(\frac{s_{p-2}}{\sin \omega_p \sin \omega_{p-1}}\right),$$

$$\vdots$$

$$\omega_2 = \arccos\left(\frac{s_2}{\sin \omega_p \sin \omega_{p-1} \ldots \sin \omega_3}\right).$$

It is crucial that owing to the procedure of the hyperspherical transformation, zeros are replaced by a value different from zero and equal to $\omega = \arccos(0) = \pi/2$. The hyperspherical transformation reduces the dimensionality by 1.

Details of other methods of data transformation frequently encountered in data analysis can be found in Otto (2007).

3.3   Descriptive statistics

Each population (which is the whole group of objects under consideration) is characterised by values referred to as parameters, which describe for example measures of location (e.g. central tendency measures) and dispersion. However, due to the fact that sometimes the whole population may be unknown, their true values can only be estimated from the observations within a dataset (sample), being a subset of the population, and expressed by *estimators*. The parameters are denoted by Greek letters and their estimators by Latin letters. For example, the variance of the whole population is written as σ^2, but its estimator as s^2.

3.3.1   Measures of location

Among the measures of location, the central tendency measures are the most frequently used. They provide information about the value μ around which the population is centred in some sense. Often μ refers to the population mean. There are three common estimators of central tendency.

The *arithmetic mean* of n observations of the variable x is given by

$$\bar{x} = \frac{1}{n}(x_1 + \cdots + x_n) = \frac{1}{n}\sum_{i=1}^{n} x_i.$$

Various properties of the mean should be noted:

- The mean need not be a member of the sample dataset from which it was calculated. For example, the mean of a set of integers need not be an integer.

- The sum of differences between each observation and the mean of the observations is identically equal to 0: $\sum_{i=1}^{n}(x_i - \bar{x}) \equiv 0$.

- The mean is affected by the extreme values of a variable, referred to as *outliers* (Figure 3.5(a)).

- The mean can be a misleading summary of the data when the data are multimodal (Figure 3.5(b)) or asymmetric (Figures 3.5(c),(d)).

The *mode* is the most frequently occurring value among all the observations.

The *median* of a sample is the value which divides the sample into two equal-sized halves. If n is odd, then the sample median is the middle value of the order statistics $x_{\left(\frac{n+1}{2}\right)}$. If n is even, then the sample median is mid-way between the two values in the middle of the list of order statistics $\frac{1}{2}\left(x_{\left(\frac{n}{2}\right)} + x_{\left(\frac{n}{2}+1\right)}\right)$. The parentheses ( ) around the subscripts denote that these are *order statistics*. The *order statistics* of a sample are the data points ranked in order of increasing magnitude. Given a sample $\{x_1, \ldots, x_n\}$, the order statistics are $\{x_{(1)}, \ldots, x_{(n)}\}$ where $x_{(1)}$ is the minimum value of $\{x_1, \ldots, x_n\}$ and $x_{(n)}$ is the maximum value of $\{x_1, \ldots, x_n\}$.

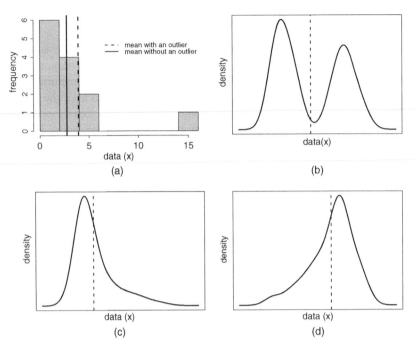

Figure 3.5 Illustration of cases where the mean (*dashed line*) is a misleading summary of the data: (a) an outlier; (b) multimodal distribution; (c) positively skewed distribution; (d) negatively skewed distribution.

This concept of dividing the data can be generalised to give *quantiles*. A sample quantile $Q(p)$, for $0 \leq p \leq 1$, divides the ordered observations in the ratio $p : 1 - p$.

$$Q(p) = x_{(np+\frac{1}{2})}, \quad 1 \leq np + \frac{1}{2} \leq n;$$

$$= x_{(1)}, \quad np < \frac{1}{2};$$

$$= x_{(n)}, \quad np > n - \frac{1}{2}.$$

If the subscripts referring to the order statistics are fractional, then the value of the relevant quantile is given from the weighted expression involving the two nearest order statistics (e.g. for $Q = x_{(2\frac{3}{4})}$ we take $\frac{1}{4}x_{(2)} + \frac{3}{4}x_{(3)}$).

The *median*, Q_2, has special notation with

$$Q_2 = Q\left(\frac{1}{2}\right) = x_{(\frac{1}{2}n+\frac{1}{2})} = x_{(\frac{2n+2}{4})}$$

$$= x_{(\frac{n+1}{2})}, \quad \text{for } n \text{ odd},$$

$$= \frac{1}{2}x_{(\frac{n}{2})} + \frac{1}{2}x_{(\frac{n}{2}+1)}, \quad \text{for } n \text{ even}.$$

The points which divide the data into quarters are known as *quartiles*, the median being the second quartile. The first, or lower, quartile and third, or upper, quartile, are defined as

$$Q_1 = Q\left(\frac{1}{4}\right) = x_{(\frac{1}{4}n+\frac{1}{2})} = x_{(\frac{n+2}{4})};$$

$$Q_3 = Q\left(\frac{3}{4}\right) = x_{(\frac{3}{4}n+\frac{1}{2})} = x_{(\frac{3n+2}{4})}.$$

The so-called *five-point* summary of the data is given by $x_{(1)}$ (minimum), Q_1, Q_2, Q_3, $x_{(n)}$ (maximum).

Example

Consider a set of 11 observations: $\mathbf{x} = (4, 2, 1, 6, 2, 3, 4, 3, 1, 2, 2)$.

The mean $\bar{x} = \frac{1}{11} \cdot (4 + 2 + 1 + 6 + 2 + 3 + 4 + 3 + 1 + 2 + 2) = 2.73$. Note that this is not an integer.

The mode is equal to 2, a value which occurs 4 times within the dataset.

The number of observations is odd, so the median can be calculated from $Q_2 = x_{((n+1)/2)} = x_{(6)}$. The median is 2, as can be viewed from order statistics ($\mathbf{x} = (1, 1, 2, 2, 2, 2, 3, 3, 4, 4, 6)$). The mean is higher.

Inspection of Figure 3.6 shows that the distribution of the data is positively skewed. Data that are positively skewed are data in which the distribution has a long tail to the right (Figure 3.5(c)). Data that are negatively skewed (Figure 3.5(d)) are data in which the distribution has a long tail to the left. For positively skewed data, the mean is greater than the median. For negatively skewed data, the mean is less than the median.

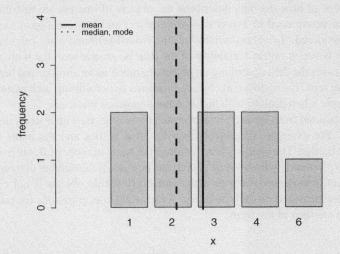

Figure 3.6 The illustration of central tendency measures for the dataset (note that the median is equal to the mode for this dataset).

Suitable **R** functions for calculating the central tendency measures are available in Appendix D.

3.3.2   Dispersion: Variance estimation

The measure of dispersion which is easiest to work with mathematically and the only one considered here is the sum of squares of deviations of the observations $\{x_i, \ i = 1, \ldots, n\}$ from their overall mean $\bar{x} = \sum_{i=1}^{n} x_i/n$, divided by 1 less than the sample size $(n - 1)$. This is known as the *variance*. More formally, the variance σ^2 of a population is a measure of spread (dispersion). The variance of a population from which a sample $\{x_i, \ i = 1, \ldots, n\}$ is taken is estimated by

$$s^2 = \frac{1}{n-1} \sum_{i=1}^{n} (x_i - \bar{x})^2. \tag{3.2}$$

The term $(n - 1)$, known as the degrees of freedom (usually denoted as ν or *df* in **R** software; Appendix D), is the number of independent pieces of information that are used for the estimation of parameters. For example, when estimating the variance for a dataset of n observations, there are assumed to be n independent pieces of information. Estimation of the sample mean uses up one piece of information. This leaves $(n - 1)$ pieces of information for estimation of the variance, hence the divisor $(n - 1)$ in its estimation. Alternatively, given $(n - 1)$ out of n observations and the mean of all n observations, the value of the nth observation is automatically determined.

In order to have a measure of dispersion in the original units of measurement, the square root of the variance is used. This is known as the *standard deviation* and denoted by σ for a population standard deviation and by s for a sample standard deviation.

For solving comparison (Chapter 4) or classification (Chapter 5) problems it is crucial to have some knowledge of how the data describing the objects of interest are distributed. In particular, there is a strong need to assess the *within-object* and *between-object* variability of the variables considered. To obtain reliable results from likelihood ratio calculations it is desirable that the between-object variability of the data be greater than the within-object variability. In such cases the data describing an object are much more similar, and hence less dispersed, than those referring to different objects. Problems with fulfilling such assumptions can occur, for example when dealing with highly inhomogeneous materials.

Consider data obtained from an analysis performed on m objects, with n items measured within each object. For example, the objects may be glass bottles and the items may be fragments from the bottles. The analysis here assumes the same number of items from each object. Adjustments proposed by Bozza *et al.* (2008) can be made if the number of items varies from object to object (unbalanced data) or if the variability within objects is not constant. There are p variables measured on each item (e.g. p elemental concentrations for each glass fragment). The data are then of the form

$$\mathbf{x}_{ij} = (x_{ij1}, \ldots, x_{ijp})^T,$$

where $i = 1, \ldots, m$ and $j = 1, \ldots, n$.

The within-object variance estimate, recording the p variances and $p(p - 1)/2$ covariances (Section 3.3.4) for the variation within objects, is denoted by $\mathbf{U}$ and is assumed constant from object to object. Its estimate, $\hat{\mathbf{U}}$, takes into account the sum of squares ($\mathbf{S}_w$) of deviations of each measurement ($\mathbf{x}_{ij}$) from the m object means ($\bar{\mathbf{x}}_i$) and the relevant number of degrees of freedom. Thus, the value of $\hat{\mathbf{U}}$ is obtained as

$$\hat{\mathbf{U}} = \frac{\mathbf{S}_w}{m(n-1)},\tag{3.3}$$

where

$$\mathbf{S}_w = \sum_{i=1}^{m}\sum_{j=1}^{n}(\mathbf{x}_{ij} - \bar{\mathbf{x}}_i)(\mathbf{x}_{ij} - \bar{\mathbf{x}}_i)^T,$$

in which $\mathbf{x}_{ij}$ is a vector of the values of the p variables obtained in the jth measurement for the ith object, and $\bar{\mathbf{x}}_i$ is a vector of the means of the p variables calculated using n measurements for the ith object in the database, $\bar{\mathbf{x}}_i = \frac{1}{n}\sum_{j=1}^{n}\mathbf{x}_{ij}$.

The between-object variance matrix, recording the p variances and $p(p-1)/2$ covariances for the variation between objects, is denoted by $\mathbf{C}$. Its estimate, $\hat{\mathbf{C}}$, involves the sum of squares ($\mathbf{S}^*$) of the mean deviations of the object means ($\bar{\mathbf{x}}_i$), $i = 1, \ldots, m$, from the overall mean ($\bar{\mathbf{x}}$), reduced by the within-object variance estimate ($\mathbf{S}_w$) and normalised by the relevant number of degrees of freedom:

$$\hat{\mathbf{C}} = \frac{\mathbf{S}^*}{m-1} - \frac{\mathbf{S}_w}{mn(n-1)},\tag{3.4}$$

where

$$\mathbf{S}^* = \sum_{i=1}^{m}(\bar{\mathbf{x}}_i - \bar{\mathbf{x}})(\bar{\mathbf{x}}_i - \bar{\mathbf{x}})^T$$

and

$$\bar{\mathbf{x}} = \frac{1}{mn}\sum_{i=1}^{m}\sum_{j=1}^{n}\mathbf{x}_{ij}.$$

In the case of univariate data (when $p = 1$) all data vectors become scalars and we have

$$u^2 = \frac{s_w^2}{m(n-1)},$$

where

$$s_w^2 = \sum_{i=1}^{m}\sum_{j=1}^{n}(x_{ij} - \bar{x}_i)^2$$

and

$$c^2 = \frac{s^{*2}}{m-1} - \frac{s_w^2}{mn(n-1)},$$

where

$$s^{*2} = \sum_{i=1}^{m}(\bar{x}_i - \bar{x})^2.$$

R code (Appendix D) suitable for within- and between-object variance estimation is available in the file UC_comparison_calculations.R on the website. Estimates of the within- and between-object variance are also widely used in analysis of variance (ANOVA) methods (Section 3.5) for testing the equality of means of several datasets.

3.3.3   Data distribution

3.3.3.1   Box-plots

Box-plots are a very effective visual device for comparing several sets of data. They are especially helpful when analysing the dispersion, symmetry, or skewness of the data. Box-plots (Figure 3.7) highlight the *five-point* summary of the data:

- the smallest observation ($x_{(1)}$, minimum),
- the largest observation ($x_{(n)}$, maximum),
- lower quartile (Q_1),
- median (Q_2),
- upper quartile (Q_3).

They enable a researcher to quickly compare and gain an insight into the data structure.

It should be pointed out that there are several different types of box-plots differing in the interpretation of particular elements. Most commonly the top and bottom of a box-plot show the location of upper (Q_3) and lower (Q_1) quartiles, respectively. The line drawn in the middle of the box depicts the median (Q_2). The whiskers may present the maximum

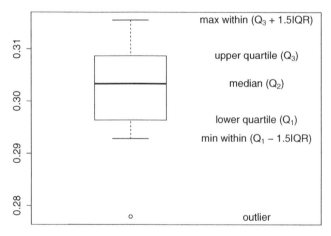

Figure 3.7   Box-plot (the description of Q_1, Q_2, Q_3, and *IQR* is provided in the text).

and minimum values of the dataset, or the highest value within the region $Q_3 + 1.5IQR$ and the lowest value within the region $Q_1 - 1.5IQR$ (where $IQR = Q_3 - Q_1$). Sometimes the location of one standard deviation above and below the mean of the data may be shown. If the data include some outside data points, they are marked by a circle and may in some cases be treated as outliers.

The relevant **R** code for drawing box-plots is available in Appendix D (Section D.7.1).

3.3.3.2 Histograms

Histograms (Figure 3.8) are a useful tool for examining the distribution of continuous data. They present the data distribution, which is constructed from adjacent intervals (not necessarily of equal size) shown as narrow bars called *bins*. Each bin has a specified location and width and its area refers to the frequency (count) of occurrence of values lying within the range determined by the base of the bin. The height of the rectangle of a histogram is therefore a measure of the frequency per unit interval which may be thought of as a frequency density. Sometimes the frequency is replaced by the probability of obtaining an observation within the interval. In this case the area under the histogram is equal to 1.

The probability density function of a data distribution cannot be estimated from a histogram because histograms are step-like in appearance and mostly affected by the location and width of the bins, which are subjectively chosen by a researcher. That said, histograms are helpful in noting multimodal distributions.

The relevant **R** code for drawing histograms is available in Appendix D (Section D.7.4).

3.3.4 Correlation

The vast majority of analyses produce more than one variable in a single measurement. For example, when the elemental content of glass fragments is determined (Section 1.2.1), the data yielded during the analysis consist of p values, each assigned to the relevant elemental content.

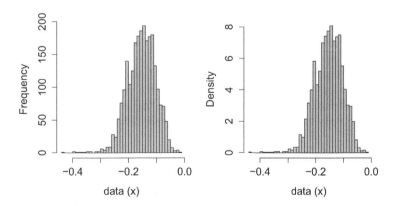

Figure 3.8 Histograms illustrating (left) the frequency and (right) probability density of occurrence of particular values of silicone content in glass (glass_data_package.txt). The heights of the rectangles in the diagram on the left may be represented as a frequency (instead of a frequency density) because the bin widths are all equal.

When more than one variable is considered the question of their mutual association arises. Two parameters that describe the relationship between multidimensional variables are *covariance* and *Pearson's correlation coefficient*.

From the mathematical point of view the covariance is defined as the pairwise product of each variable's deviation from its general mean divided by $n - 1$ (where n is the number of observations of each of the variables). The only problem with expressing the relationship between variables as a covariance is connected with its dependence on the scale in which the values are expressed. This disadvantage is removed by introducing a correlation coefficient, which is a dimensionless, symmetric measure of the linear association between the variables. It is computed by standardisation of the covariance: dividing the covariance by the standard deviations of the variables concerned. Given a set of n pairs of variables $\{(x_j, y_j), j = 1, \ldots, n\}$, with $\bar{x} = \sum_{j=1}^{n} x_j/n$, $\bar{y} = \sum_{j=1}^{n} y_j/n$, $s_x^2 = \sum_{j=1}^{n}(x_j - \bar{x})^2/(n - 1)$, and $s_y^2 = \sum_{j=1}^{n}(y_j - \bar{y})^2/(n - 1)$, the sample statistics for covariance (cov) and correlation coefficient (r) are defined by

$$cov(x, y) = \frac{1}{n - 1} \sum_{j=1}^{n} (x_j - \bar{x})(y_j - \bar{y})$$

and

$$r(x, y) = r(y, x) = \frac{cov(x, y)}{s_x s_y}.$$

The correlation coefficient can be also expressed by *autoscaled* variables z_1 and z_2 (Section 3.2):

$$r(x, y) = \frac{cov(x, y)}{s_x s_y} = \frac{1}{n - 1} \sum_{j=1}^{n} \left(\frac{x_j - \bar{x}}{s_x}\right)\left(\frac{y_j - \bar{y}}{s_y}\right) = \frac{1}{n - 1} \sum_{j=1}^{n} z_{1j} z_{2j},$$

where $z_{1j} = (x_j - \bar{x})/s_x$ and $z_{2j} = (y_j - \bar{y})/s_y$, $j = 1, \ldots, n$.

Covariance may theoretically take any value in the real numbers $(-\infty, \infty)$. However, in practice, for a given dataset it is within $[-s_x s_y, s_x s_y]$, since the range of the correlation coefficient falls in the interval $[-1, 1]$.

For totally uncorrelated variables the covariance and correlation coefficient are equal to 0, as shown in the *scatter plot* in Figure 3.9. However, the converse (that if $r = 0$, then the correlation does not exist) does not hold, since the correlation coefficient describes only the linear dependence between variables; it does not specify whether a non-linear correlation exists or not.

When the covariance (or correlation coefficient) takes positive values, the value of y increases as x increases (Figure 3.9(b)). Analogously, for negative correlation, the value of y decreases as x increases (Figure 3.9(c)). The extreme values of $|r| = 1$ exhibit perfect correlation, which means that there exists some mathematically defined linear function relating the considered variables as in Figures 3.9(d),(e). Variables with $r = 0$ may have a strong non-linear association as with a sinusoid curve, see Figure 3.9(f). Therefore, an analysis of correlation between two variables (x, y) should include both a calculation of r and a visual

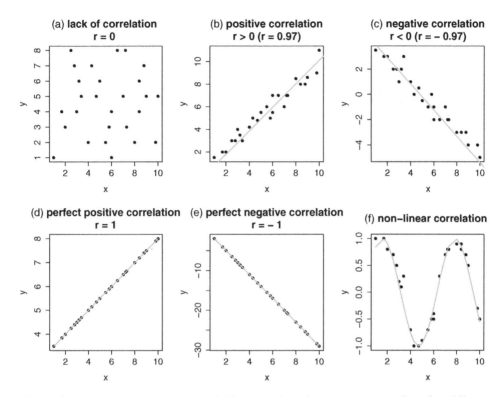

Figure 3.9 Correlation between variables x and y shown as scatter plots for different samples. The correlation coefficient r is shown in the figure titles for (a) to (e). It is not shown for (f) as there is no linear correlation. (Reproduced by permission of Elsevier.)

analysis of object in this two-dimensional space. In the case of higher-dimensional data, plots like the one presented in Figure 3.10 can be used. A test has been devised, based on the t distribution, of a hypothesis that a population correlation coefficient for normally distributed data is zero; see Section 3.4.8 and Appendix D.7.6.

Apart from such interpretation, the *correlation coefficient* has a geometric interpretation. Its value determines the cosine of the angle between the vectors created by the variables (Figure 3.27).

The square of the correlation coefficient, $d = r^2$, is called the *coefficient of determination*. It represents the percentage of the variation in one variable y that is explained by the variation in another x in a regression line $y = \alpha + \beta x$. The coefficient of determination is commonly used in regression analysis and expresses how well the regression line fits the data.

Correlation coefficients for a set of autoscaled data (see equation (3.1)), for example in the form of a matrix $\mathbf{Z}$ (Figure 3.1), when the number of variables is more than 1 ($p > 1$), may be presented in the form of a symmetric correlation matrix defined as

$$\mathbf{r}(x, y) = \frac{1}{n-1}\mathbf{Z}^T\mathbf{Z}.$$

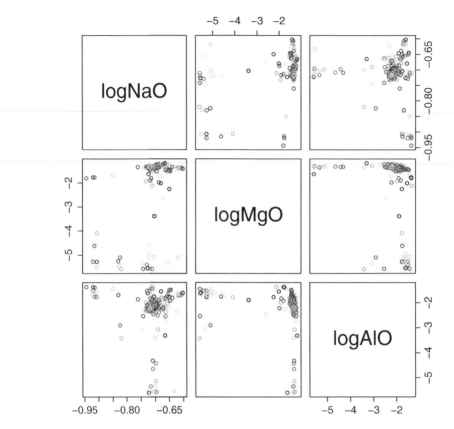

Figure 3.10 Graphical presentation of the correlation between pairs of variables *logNaO*, *logAlO*, and *logMgO* for glass data available in `glass_data.txt`. (Reproduced by permission of Elsevier.)

The diagonal elements equal 1 since the correlation between two identical variables is perfect. The correlation coefficients between two variables are given on the intersection of particular rows and columns corresponding to the relevant variables. For example, a correlation coefficient matrix for a dataset describing the elemental content of glass is given in Table 3.1.

The variance contained in the measured variables (Section 3.3.2) as well as their mutual association expressed by covariance may be presented in the form of a symmetric *variance–covariance* matrix, also called a *covariance* matrix:

$$
\begin{bmatrix}
s_{11}^2 & cov_{12} & cov_{13} & cov_{14} \\
cov_{21} & s_{22}^2 & cov_{23} & cov_{24} \\
cov_{31} & cov_{32} & s_{33}^2 & cov_{34} \\
cov_{41} & cov_{42} & cov_{43} & s_{44}^2
\end{bmatrix}.
\tag{3.5}
$$

Each element corresponds to the covariance between the relevant variables. The covariance of a variable i with itself (cov_{ii}) is known as variance. Therefore, the diagonal elements of

Table 3.1 Correlation coefficient matrix for glass data available in `glass_data.txt` (the lower half of the matrix is given by symmetry).

	logNaO	logMgO	logAlO	logSiO	logKO	logCaO	logFeO
logNaO	1.00	0.48	−0.08	0.10	−0.37	0.70	0.13
logMgO		1.00	−0.19	−0.12	−0.40	0.55	0.22
logAlO			1.00	−0.07	0.48	−0.17	−0.26
logSiO				1.00	0.05	−0.05	0.14
logKO					1.00	−0.36	−0.14
logCaO						1.00	0.13
logFeO							1.00

the matrix correspond to the variances of the variables (s_i^2) and the remaining entries give the covariances for each pair of variables (e.g. cov_{jk}). Since the covariance matrix is symmetric (e.g. $cov_{13} = cov_{31}$), usually only half of it is presented.

In order to quickly find any existing correlation between the variables, especially in the case of a large dataset, visualisation of the data is highly recommended. This is done by plotting one of the variables against the other in the form of a scatter plot. The mutual arrangement of the data points outlines the existing correlation, its direction, strength, and shape. It is sometimes useful to visualise the correlation between all pairs of variables in a single plot as in Figure 3.10 or D.11 (Appendix D).

Partial correlation coefficients are widely applied in methods dedicated to the reduction of data dimensionality (e.g. Section 3.7.1).

3.3.5 Continuous probability distributions

There are numerous data distributions referring to different datasets. However, the most widely used are the *normal* distribution, *t* distribution, and *F* distribution.

3.3.5.1 Normal distribution

The normal distribution (also called the *Gaussian distribution* after the German mathematician Carl Friedrich Gauss 1777–1855) is one of the most commonly used distributions giving the probability density function of a continuous random variable. The probability density function of the normal distribution for a single variable (univariate normal distribution) is specified by two parameters: the mean (μ) as a measure of central location (Section 3.3.1) and the variance (σ^2) as a measure of dispersion (Section 3.3.2). Given a normally distributed variable X, with mean μ and variance σ^2, its distribution is denoted by $N(\mu, \sigma^2)$ and its probability density function $f(x)$ is expressed as

$$f(x) = \frac{1}{\sigma\sqrt{2\pi}} \exp\left(-\frac{(x-\mu)^2}{2\sigma^2}\right).$$

The bell-curve shape of the normal distribution curve is shown in Figure 3.11.

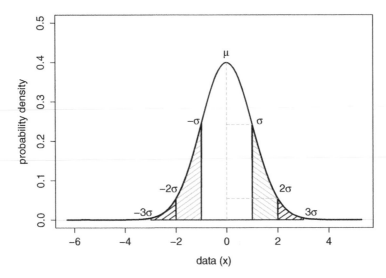

Figure 3.11   A probability density function for a normal distribution with mean $\mu = 0$ and $\sigma = 1$ (standard normal distribution). The probabilities of observing a value x in a specified range are: (a) unshaded for $P(-\sigma < x < \sigma)$, (b) shaded light grey for $P(-2\sigma < x < -\sigma)$ and $P(\sigma < x < 2\sigma)$, (c) shaded dark grey for $P(-3\sigma < x < -2\sigma)$ and $P(2\sigma < x < 3\sigma)$.

For some applications it may be advantageous to transform the data obtained in the course of the analysis. One of the most popular transformation methods is standardisation (Section 3.2). Standardisation is commonly chosen such that the mean is set equal to 0 and the standard deviation is set equal to 1 (and hence the variance is also equal to 1). The resultant distribution, denoted by $N(0, 1)$, is known as the *standard normal distribution* with associated probability density function

$$f(z) = \frac{1}{\sqrt{2\pi}} \exp\left(-\frac{z^2}{2}\right),$$

where z is the variable x transformed according to the equation $z = (x - \mu)/\sigma$. The standard normal distribution curve is presented in Figure 3.11.

For n observations from a $N(\mu, \sigma^2)$ distribution, it can be shown that the random variable $\bar{X}$ corresponding to the mean $\bar{x}$ of a particular sample has a normal distribution:

$$\bar{X} \sim N\left(\mu, \frac{\sigma^2}{n}\right).$$

Thus, by standardisation,

$$Z = \frac{\bar{X} - \mu}{\sigma/\sqrt{n}} \sim N(0, 1).$$

It is also the case, from a result known as the *central limit theorem*, that for any distribution with a mean μ and variance σ^2, as $n \to \infty$, the distribution of the sample mean $\bar{X}$ of n observations tends to $N(\mu, \sigma^2/n)$. This result contributes to the importance of the normal distribution in statistics.

The normal distribution is unimodal and symmetric about the mean and its probability density function takes its maximum value at the mean, μ. The distance between the mean and the point of inflexion of the curve is specified by σ. The smaller the standard deviation, the less variation there is in the population and the more the distribution is concentrated around the mean. A change in μ modifies the location of the distribution (and the bell curve as a graphical representation of this) by a translation. A change in σ^2 modifies the dispersion of the distribution (and the bell curve as a graphical representation of this) with a small value of the standard deviation leading to a less dispersed distribution than occurs with a large value of the standard deviation. Figure 3.12 presents normal distribution curves for four sets of data, each described by different pairs of means and variances. For the distributions with equal means, the locations of the maxima of the probability density functions are the same (e.g. $N(0, 1)$ and $N(0, 9)$ or $N(8, 1)$ and $N(8, 0.25)$). The lower the standard deviation, the less dispersed the function.

For a normal distribution nearly all the area under the probability density function (approximately 99.7%) lies within a range of six standard deviations ($\mu - 3\sigma, \mu + 3\sigma$). This is referred to as the *three-sigma rule* (Figure 3.11). Furthermore, about 68% of the area under the probability density function falls within the range ($\mu - \sigma, \mu + \sigma$) and 95% falls within the range ($\mu - 1.96\sigma, \mu + 1.96\sigma$).

For continuous probability distributions in general with probability density function $f(x)$ for a random variable X, say, the probability of obtaining a value x of X lying within a

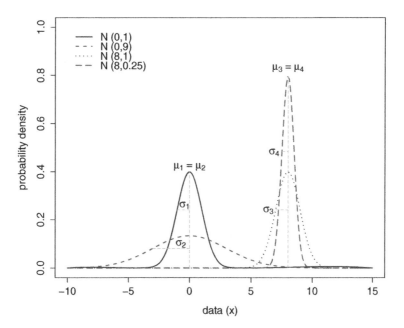

Figure 3.12 Normal distribution curves with different parameters μ and σ^2.

specified range (e.g. between a and b) is given by the area under the relevant part of the curve:

$$P(a \leq x \leq b) = \int_a^b f(x)dx$$

for $a < b$. The area under the whole curve represents the probability of observing any value. The probability of this certain event equals 1:

$$P(-\infty < x < +\infty) = \int_{-\infty}^{+\infty} f(x)dx = 1.$$

For $a = b$, $\int_a^a f(x)dx = 0$ as the probability of observing an exact value is 0. This is a perfectly reasonable theoretical result. In practice, the accuracy of measuring instruments ensures that measurements cannot be made beyond a certain specification and there will be a non-zero probability that a measurement will lie within that specification.

When dealing with two-dimensional data (bivariate data), it can be useful to visualise their distribution in one graph as in Figure 3.13. The probability density function for a so-called *bivariate normal distribution* takes into account the possible existence of the correlation ρ between the two variables, where ρ is the population correlation coefficient, the population analogue of the sample correlation coefficient defined in Section 3.3.4.

R code for illustrating the normal distribution for univariate data is available in Appendix D (Section D.7.3).

The generalisation of the normal distribution for multivariate data expressed in the form of a p-vector $\mathbf{x} = (x_1, \ldots, x_p)^T$ is given as $N \sim (\boldsymbol{\mu}, \boldsymbol{\Sigma})$, where $\boldsymbol{\Sigma}$ is the covariance matrix. Then the probability density function is expressed as:

$$f(\mathbf{x}) = (2\pi)^{-p/2} |\boldsymbol{\Sigma}|^{-\frac{1}{2}} \exp\left\{ -\frac{1}{2}(\mathbf{x} - \boldsymbol{\mu})^T (\boldsymbol{\Sigma})^{-1} (\mathbf{x} - \boldsymbol{\mu}) \right\}.$$

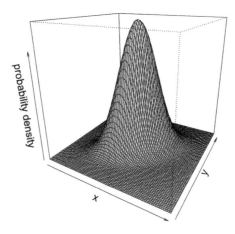

Figure 3.13 Multivariate (here: bivariate) normal distribution.

Many statistical methods are based on an assumption that the underlying probability density function is normal. In order to check quickly if the data under study are normally distributed or not, quantile–quantile (Q-Q) plots can be helpful. In Q-Q plots, the quantiles of the dataset (Section 3.3.1) are plotted against theoretical normal distribution quantiles. If the points lie close to or along the leading diagonal of the plot, the probability distribution of the data may be assumed normal. If the points deviate strongly from the diagonal, some other probability density function is appropriate (as an example of what may be done, see the discussion of *kernel density estimation* below. Figure 3.4 shows Q-Q plots the aim of which was to visualise and verify if the distribution for the transformed data was somehow closer to normal than that for the original untransformed data.

R code for drawing Q-Q plots is given in Appendix D (Section D.7.2).

3.3.5.2   t distribution

Assume the data still follow a $N(\mu, \sigma^2)$ distribution, with a sample size n and sample mean $\bar{x}$. If the variance σ were known, then as noted before, the corresponding random variable

$$Z = \frac{\bar{X} - \mu}{\sigma/\sqrt{n}}$$

would have a standard normal distribution. Suppose the variance σ of the distribution is unknown and the number of observations in the dataset is relatively small (say, $n < 30$). Replacement of σ by the estimate $s = \sqrt{\sum_{i=1}^{n} \frac{(x_i - \bar{x})^2}{n-1}}$ in the expression for Z above to give

$$\frac{\bar{X} - \mu}{s/\sqrt{n}}$$

provides a statistic which does not have a normal distribution. The distribution that this statistic follows is known as *Student's t distribution*.

Student's t distribution (or simply the *t distribution*) is symmetric about 0. Its curve is bell-shaped and similar to that of the normal distribution (Figure 3.14). Its variance is dependent on a parameter known as the degrees of freedom (Section 3.3.2), denoted in general by v (or *df*), with the variance equal to $v/(v - 2)$ as long as $v > 2$. The variance is greater than 1 but the larger v is, the closer the variance becomes to 1 and the closer the density function resembles that of the normal distribution (Figure 3.14). The probability density function of the t distribution with v degrees of freedom is

$$f(t) = \frac{\Gamma\left(\frac{v+1}{2}\right)}{\sqrt{v\pi}\,\Gamma\left(\frac{v}{2}\right)} \left(1 + \frac{t^2}{v}\right)^{-\frac{v+1}{2}},$$

where $\Gamma(\cdot)$ is the so-called gamma function

$$\Gamma(v) = \int_0^\infty t^{v-1} e^{-t} dt,$$

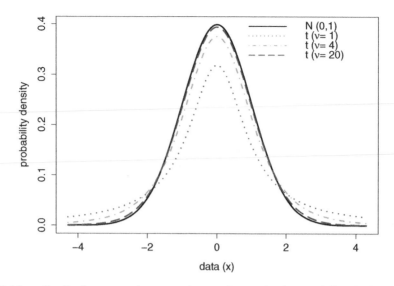

Figure 3.14 t distribution curves in comparison to the standard normal distribution ($N(0, 1)$) curve depending on the number of degrees of freedom (ν).

with $\Gamma(\nu) = (\nu - 1)!$ for integer ν (Evans *et al.* 2000), or

$$f(t) = \frac{1}{\sqrt{\nu}B\left(\frac{1}{2}, \frac{\nu}{2}\right)} \left(1 + \frac{t^2}{\nu}\right)^{-\frac{\nu+1}{2}},$$

where $B(x, y)$ is the so-called beta function (Evans *et al.* 2000)

$$B(x, y) = \int_0^1 t^{x-1}(1 - t)^{y-1}dt$$

or, for integer x and y,

$$B(x, y) = \frac{(x - 1)!(y - 1)!}{(x + y - 1)!}.$$

3.3.5.3 χ^2 distribution

The sum of squares of n independent random variables with a standard normal distribution has a continuous and asymmetric distribution known as the χ^2 distribution, read as a *chi-squared distribution* (Figure 3.15). The probability density function for ν degrees of freedom (Section 3.3.2) is given by

$$f(x) = \frac{x^{(\nu/2)-1}e^{-x/2}}{2^{\nu/2}\Gamma\left(\frac{\nu}{2}\right)}$$

for $x \geq 0$, where $\Gamma(\nu/2)$ is the gamma function as defined above.

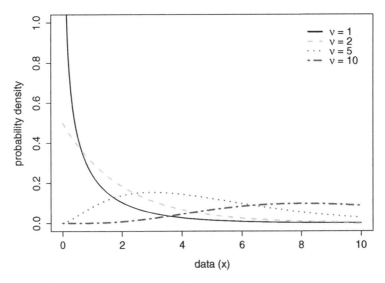

Figure 3.15 χ^2 distribution curves depending on the number of degrees of freedom ν.

3.3.5.4 *F* distribution

The *F* statistic is given as the quotient of two independent random variables characterised by two χ^2 distributions, χ_1^2 and χ_2^2 (Evans *et al.* 2000), divided by the respective numbers of degrees of freedom ν_1 and ν_2 (Section 3.3.2).

It is an asymmetric distribution (Figure 3.16) on the non-negative real line, skewed for low numbers of degrees of freedom. If there are more degrees of freedom, the distribution is

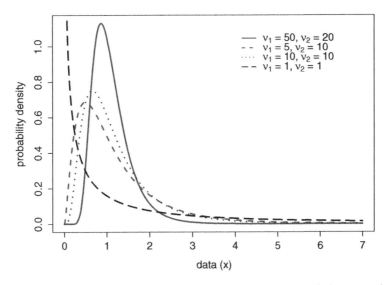

Figure 3.16 *F* distribution curves depending on the numbers of degrees of freedom (ν_1 and ν_2).

less skewed. The form of the F distribution is dependent on both the degrees of freedom for the numerator (ν_1) and for the denominator (ν_2) of the expression:

$$F_{(\nu_1,\nu_2)} = \frac{\chi_1^2/\nu_1}{\chi_2^2/\nu_2}.$$

The F statistic is frequently used in analysis of variance.

3.3.5.5 Kernel density estimation procedure

In many cases it is observed that the data distribution cannot be simply modelled by a parametric distribution. This situation is demonstrated by the histogram in Figure 3.17.

The naïve approach to establishing a probability density function is based on histograms (Section 3.3.3). They illustrate the frequency of observing values falling into strictly defined intervals (not necessarily of equal size) called *bins*. Their width and location are chosen subjectively and the shape of the final probability density function has a step-like character. Therefore, for data which are far from normal, kernel density estimation (KDE) is a useful non-parametric method for probability density function estimation (Silverman 1986).

KDE is a method that resembles the procedure of calculating probability density functions typical for histograms. However, it takes the procedure one step further. Instead of the bins characteristic of histograms, KDE creates a function with a shape similar to, for example, the normal distribution centred at each data point, x_i, considered. Other functions (called Parzen potentials) may also be applied such as rectangular, triangular, or exponential functions, as presented in Figure 3.18. When KDE is used, the estimated probability density function depends upon all the data points considered, hence its shape is adjusted to the data distribution and is not arbitrarily fixed. The width of bins is replaced by the bandwidth parameter referred

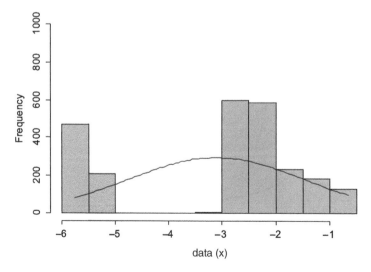

Figure 3.17 Multimodal data distribution represented by a histogram which cannot be modelled by a normal distribution illustrated by a solid line.

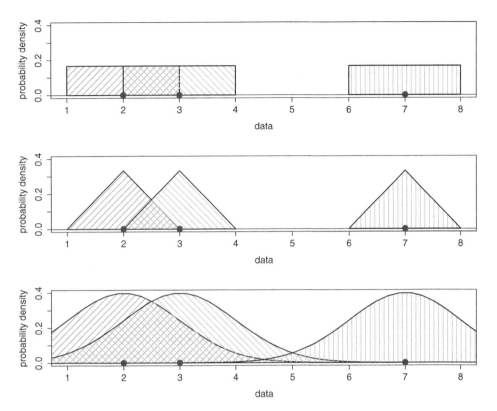

Figure 3.18 Parzen potentials of (top) rectangular, (middle) triangular, and (bottom) Gaussian shapes centred at the points (marked by dots) commonly used in KDE.

to as a smoothing parameter (h). In the case of Gaussian functions (Gaussian kernels) placed at each data point, Gaussian kernel functions (K) are defined by:

$$K\left(\frac{x - x_i}{h}\right) = \frac{1}{\sqrt{2\pi}\, h} \exp\left(-\frac{1}{2}\left(\frac{x - x_i}{h}\right)^2\right).$$

As already mentioned, these represent the normal distribution functions $N(\mu, \sigma^2)$, which can also be expressed as $N(x_i, h^2)$, located on each data point x_i with smoothing parameters h adjusting their widths.

The final KDE curve is obtained by placing the Gaussian curves on each of the data points considered, summing them over the whole range of the data values and dividing them by the number of observations, so that it integrates to 1 (Figure 3.19). The probability density function is then expressed as

$$f(x) = \frac{1}{n} \sum_{i=1}^{n} K\left(\frac{x - x_i}{h}\right).$$

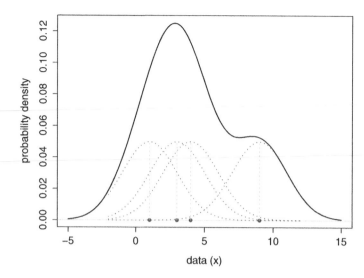

Figure 3.19 Illustration of a probability density function following the KDE procedure using Gaussian kernels.

The appropriate determination of the smoothing parameter (h) is crucial for further use of the probability density function obtained following the KDE procedure, for example in likelihood ratio calculations for forensic purposes (Chapters 4 and 5). If it is set too high, the information about the real distribution and some subtle features of the data under analysis will be lost (dotted curve for $h = 1$ in Figure 3.20). On the other hand, when h is too low, the probability density function has many spiky extremes making the distribution difficult to

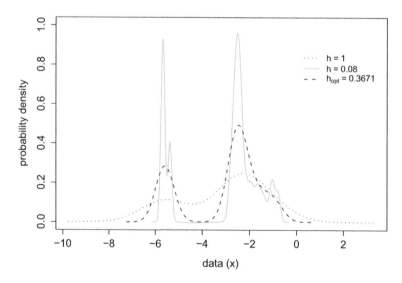

Figure 3.20 Illustration of the influence of the value of a smoothing parameter (h) on the shape of the probability density function computed by the KDE procedure.

interpret (solid light curve for $h = 0.08$ in Figure 3.20). Mathematical methods for calculating the optimal value of the smoothing parameters have been proposed, suitable for further LR calculations (Chapters 4 and 5). A way of computing h values (Silverman 1986) for LR calculations is given by the expression $h = h_{opt} = \left(\frac{4}{m(2p+1)}\right)^{\frac{1}{p+4}}$, where p stands for the number of variables taken for calculations using m objects in the database. The shape of the probability density function computed with the use of such an h parameter is illustrated by the dashed curve in Figure 3.20 (for $h_{opt} = 0.3671$).

R code suitable for drawing KDE functions is provided in Appendix D (Section D.7.5).

3.4   Hypothesis testing

3.4.1   Introduction

The aim of a statistical hypothesis test is to determine if a claim or conjecture about some feature of the population (a parameter) is strongly supported by the information obtained from the sample data.

When it is desired to establish an assertion with substantial support obtained from the sample, the negation of the assertion is taken to be the null hypothesis, conventionally denoted by H_0, and the assertion itself is taken as the alternative hypothesis, H_1. The null hypothesis may be stated as, for example, $H_0 : \mu = \mu_0$, where μ might be the level of blood alcohol concentration (BAC, see example below) for the individual tested and μ_0 might be the legal limit for BAC. An alternative hypothesis might be that $H_1 : \mu = \mu_1 > \mu_0$.

In this case a false rejection of H_0 is a decision to act as if H_1 is true when in fact H_0 is true and H_1 is false. In a trial context this would be a decision to convict a truly innocent person. A failure to reject H_0 when H_1 is true is a decision to act as if H_0 is true when in fact H_1 is true and H_0 is false. In a trial context this would be a decision to acquit a truly guilty person. There are thus two types of error in this context:

- A *type I error* is the error of rejecting H_0 when it is true.

- A *type II error* is the error of not rejecting H_0 when it is false.

Notice that a *failure to reject H_0* does not mean that H_0 is accepted as true. Failure to reject H_0 is a decision to act as if H_0 were true. A verdict of *not guilty* is not proof of innocence.

A test of the null hypothesis specifies a course of action by stating what sample information is to be used and how it is to be used in making a decision. The decision to be made is one of the following two:

- reject H_0 and conclude that H_1 is substantiated, that is, act as if H_1 is true;

- retain H_0 and conclude that H_1 fails to be substantiated, that is, act as if H_0 is true.

Note, again, that these are not definitive statements about the truth or otherwise of either hypothesis. Rejection of H_0 does not mean *acceptance* of H_1, and failure to reject H_0 does not mean *acceptance* of H_0 – it only means there is insufficient evidence to reject H_0.

Assume that $\mathbf{x} = (x_1, \ldots, x_n)$ is a set of n independent observations of a normally distributed random variable $X \sim N(\mu, \sigma^2)$. It can be shown that the random variable $\bar{X}$, corresponding to $\bar{x} = \sum_{i=1}^{n} x_i/n$, has a normal distribution $N(\mu, \sigma^2/n)$. The sample mean $\bar{X}$ is the statistic, known as the *test statistic*, that is the basis for deciding whether to reject

H_0 or not. The alternative hypothesis H_1 might be $\mu > 80$ mg of ethyl alcohol in 100 ml of blood, where 80 mg per 100 ml of blood is the legal limit of BAC in the UK for driving. Therefore, high values of $\bar{X}$ are of interest. Thus, a decision rule is of the form:

- reject H_0 if $\bar{X} > c$,

- do not reject H_0 if $\bar{X} < c$.

The set of outcomes $\{\mathbf{X} = (X_1, \ldots, X_n) : \bar{X} > c\}$ for which H_0 is rejected is known as the *rejection region* or *critical region*, conventionally denoted C. The cut-off point c is known as the *critical value*.

In order to determine the value of c, a low probability of rejecting H_0 when it is true (type I error) is chosen; conventionally 0.05 or 0.01. The consequences of a false rejection of H_0 are considered when making this choice; the more serious they are, the lower this probability should be chosen.

3.4.2   Hypothesis test for a population mean for samples with known variance σ^2 from a normal distribution

For the moment, assume the variance, σ^2, is known. The null hypothesis is stated as $H_0 : \mu = \mu_0$, where, for example, μ is the BAC (see example below) for the individual tested and μ_0 is the legal limit for BAC. The probability of a type I error is denoted by α. There are three possible hypotheses for H_1:

- $H_1 : \mu < \mu_0$. It is decided to reject H_0 in favour of H_1 with a significance level of α. Hence, choose a critical value, c, such that $P(\bar{X} \leq c \mid \mu = \mu_0) = \alpha$, with critical region $\{\mathbf{X} : \bar{X} \leq c\}$.

- $H_1 : \mu > \mu_0$. It is decided to reject H_0 in favour of H_1 with a significance level of α. Hence, choose a critical value, c, such that $P(\bar{X} \geq c \mid \mu = \mu_0) = \alpha$, with critical region $\{\mathbf{X} : \bar{X} \geq c\}$.

- $H_1 : \mu \neq \mu_0$. It is decided to reject H_0 in favour of H_1 with a significance level of α. Hence, choose critical values, c_1 and c_2, such that $P(\bar{X} \leq c_1 \mid \mu = \mu_0) = P(\bar{X} \geq c_2 \mid \mu = \mu_0) = \alpha/2$, with critical region $\{\mathbf{X} : \bar{X} \leq c_1 \text{ or } \bar{X} \geq c_2\}$.

The first two tests are known as *one-sided tests*. The third test is known as a *two-sided test*. The second test is the one to use for BAC measurements, as in the example below.

Example: Blood alcohol concentration

The current permitted level of BAC in the UK for driving is 80 mg of alcohol per 100 ml of blood. Five BAC readings are taken using a breathalyser from an individual within a few minutes: 79.5, 80.1, 80.6, 80.8, 82.4 (after ordering). How should these data be used towards a determination that the true BAC is greater than 80 mg per 100 ml?

Let μ be the true level of BAC for the individual tested. This is unknown. There are five readings from the breathalyser. In the current example, $\mathbf{x} = (79.5, 80.1, 80.6, 80.8, 82.4)$. These are all estimates of μ. Let $\mu_0 = 80$ be the legal limit for BAC. The alternative hypothesis in this example is $H_1 : \mu > 80$. It is of no interest to the legal authorities if

$\mu < 80$. The null hypothesis H_0 is written as the value for the legal limit. The testing problem can then be written as

$$H_0 : \mu = \mu_0 = 80, \quad H_1 : \mu > \mu_0.$$

This is a one-sided test of $\mu = \mu_0 = 80$ against $\mu > 80$. Take $\alpha = 0.05$ for the present. Then c is determined as that value for which

$$P(\bar{X} > c \mid \mu = 80) = 0.05.$$

When H_0 is true, $\mu = 80$. For the moment, assume that the standard deviation σ of the instrument for measuring BAC is assumed known from previous experiments and is equal to 1.1 mg per 100 ml. Thus, when $\mu = 80$,

$$Z = \frac{\bar{X} - 80}{1.1/\sqrt{5}} \sim N(0, 1).$$

The upper 5% point of $N(0, 1)$ is 1.645, obtainable from the **R** command qnorm(0.95,0,1) (Appendix D.7.3). Substitution of the desired critical value c for the random variable $\bar{X}$ and 1.645 for Z gives the equation

$$1.645 = \frac{c - 80}{1.1/\sqrt{5}},$$

which implies c is given by

$$c = 80 + 1.645 \cdot \left(\frac{1.1}{\sqrt{5}}\right) = 80 + 0.8 = 80.8.$$

To answer our initial question, we calculate the mean BAC for the five measurements, and denote it by $\bar{x}_{obs}$, with obs meaning *observed*. If $\bar{x}_{obs}$ is less than 80.8 there is insufficient evidence to suggest that the true mean reading μ from the instrument is greater than 80. If $\bar{x}_{obs}$ is greater than or equal to 80.8, reject H_0 and act as if the true mean of BAC is greater than 80. In this example, the mean $\bar{x}_{obs} = 80.68$. This is less than 80.8. Thus, there is insufficient evidence to suggest the mean level of BAC is greater than 80.

The design of the test is such that the probability of making a type I error is fixed in advance of the collection of the data. This probability is known as the *significance level* of the test and denoted conventionally as α; it is also sometimes known as the *size* of the test.

The probability of making a type II error is denoted conventionally as β. The complement of this, $1 - \beta$, is the probability of rejecting H_0 when it is false (a correct decision). This probability is known as the *power* of the test: it is the probability of detecting a departure from the null hypothesis when it exists. Consider $\bar{X} \sim N(\mu, \sigma^2/n)$, a null hypothesis $H_0 : \mu = \mu_0$, an alternative hypothesis $H_1 : \mu = \mu_1 > \mu_0$, and a significance level α. The critical value c of the test is chosen such that $P(\bar{X} > c \mid \mu = \mu_0) = \alpha$. Then

- the probability β of a type II error is $P(\bar{X} < c \mid \mu = \mu_1)$, and
- the power $1 - \beta = P(\bar{X} > c \mid \mu = \mu_1)$.

Example: Blood alcohol concentration (*continued*)

In order to determine power, it is necessary to know the exact value of μ specified by the alternative hypothesis. If the true BAC is 81, *what is the probability that a test based on five measurements and a critical value c of* 80.8 *would detect a value of* 81, *in the sense that the sample mean* $\bar{x}_{obs}$ *would be greater than* 80.8? This probability is the probability that a random variable $\bar{X}$ with a normal distribution, mean 81 and standard deviation $\frac{1.1}{\sqrt{5}}$ would take a value greater than 80.8. The standardised deviate z is given by $z = (80.8 - 81) \cdot \frac{\sqrt{5}}{1.1} = -0.407$. The probability that a standardised deviate is greater than -0.407 is 0.66. The power of the test to detect a BAC of 81 mg per 100 ml is 0.66. In other words if the true BAC were 81 mg per 100 ml a test based on a critical region of $\{\bar{X} > 80.8\}$ would register a significant result about 66% of the time. For the other 34% of the time, there would be a type II error and the hypothesis that the true BAC was 80 would not be rejected.

The concept of power is of use in clinical trials, for example, where the effectiveness of a new drug is compared with that of an old drug. If the new drug is more effective than the old drug (H_1 is true) it is good to have a test which has a high probability of detecting this.

Unfortunately, it is not possible to design a test that minimises the significance level and maximises the power simultaneously. If the critical region is adjusted to reduce the significance level then the power is also reduced. If the critical region is adjusted to reduce β then α is increased.

Example: Blood alcohol concentration (*continued*)

Given the observed value of the standardised deviate, how small a value of α can be used for which the conclusion would still be one of rejection of H_0?

Suppose that $\bar{x}_{obs} = 80.68$ and $H_0 : \bar{X} \sim N(80, 1.1^2/5)$. Then

$$P\left(\frac{\bar{X}_{obs} - 80}{1.1/\sqrt{5}} > \frac{80.68 - 80}{1.1/\sqrt{5}}\right) = P(Z > 1.382) = 0.083,$$

where $Z = \frac{\bar{X}_{obs} - 80}{1.1/\sqrt{5}}$. This probability can be determined using the **R** command `1 - pnorm(1.382,0,1)` (Appendix D).

The smallest value of α for $\bar{X} = 80.68$ that permits rejection of H_0 is therefore 0.083. This value is called the *significance probability* or *p-value*. This probability is the probability of observing a value as extreme as or more extreme than 80.68 if H_0 is true. It is a measure of compatibility of the data with the null hypothesis. The probability in this example is 0.083 (about 1 in 12) and the result is thus not significant at the 5% level (since $0.083 > 0.05$). The observed mean $\bar{x}_{obs}$ has to be greater than the critical value 80.8 to be significant at the 5% level.

The *p*-value serves as a measure of the strength of the evidence against H_0. Notice that it is a *small* value that leads one to believe the null hypothesis is false. A large value provides little evidence that the null hypothesis is false and the null hypothesis is not rejected.

Note also that the *p*-value is not a measure of the probability the null hypothesis is true. Nothing in the above material contains inferences about the truth or otherwise of either of the hypotheses. The probabilities that are considered are those of the data given the truth of H_0 or H_1.

If σ is not known but n is large ($n > 30$), the test can be conducted with the estimate s obtained from the observed values $x_1, \ldots, x_n$, replacing σ and inference made with reference to the standard normal distribution. As before, the level of significance is specified in advance, and the data are collected. The statistic

$$z = \frac{\bar{x}_{obs} - \mu_0}{s/\sqrt{n}}$$

is calculated and referred to the standard normal distribution. This statistic is known as a normalised *deviate*. It is the deviation of $\bar{x}$ from the mean (μ_0), as specified by H_0, standardised by division by the estimated standard error. If the normalised deviate is greater, in absolute terms, than the value specified as the critical value for the given level of significance then H_0 is rejected.

3.4.3   Hypothesis test for a population mean for small samples with unknown variance σ^2 from a normal distribution

Consider independent, identically distributed random variables $X_1, \ldots X_n$ with distribution $N(\mu, \sigma^2)$ with small n ($n < 30$). There is a null hypothesis H_0 that $\mu = \mu_0$. The variance σ^2 is unknown. Analogous to the situation where σ is known, the test statistic is

$$T = \frac{\bar{X} - \mu_0}{s/\sqrt{n}},$$

where the sample standard deviation s replaces σ and the random variable T replaces Z. If H_0 is true, T has a distribution known as the t distribution (Section 3.3.5), denoted t_{n-1}, where the subscript $n - 1$ indicates the dependence on the sample size. The quantity $n - 1$ is the divisor in the estimate s^2 of the variance σ^2 and is known as the *degrees of freedom* (usually denoted as ν or df, Section 3.3.2).

The same principles hold for hypothesis testing as when σ is known. Let $(t_{n-1,\alpha})$ denote the upper $100(1 - \alpha)\%$ point of the t_{n-1} distribution; this is the value of a random variable T_{n-1} which has a t distribution with $n - 1$ degrees of freedom such that

$$P(T_{n-1} > t_{(n-1,\alpha)}) = \alpha.$$

As when the standard deviation σ is known, there are three possible alternative hypotheses H_1 to be considered with the corresponding critical regions C for the test statistic (random variable) T_{n-1}:

- when $H_1 : \mu > \mu_0$, $C : T_{n-1} \geq t_{(n-1,\alpha)}$, which is a one-sided test,
- when $H_1 : \mu < \mu_0$, $C : T_{n-1} \leq -t_{(n-1,\alpha)}$, which is a one-sided test,
- when $H_1 : \mu \neq \mu_0$, $C :| T_{n-1} | \geq t_{(n-1,\alpha/2)}$, which is a two-sided test,

where $| T_{n-1} |$ denotes the absolute value of the test statistic T_{n-1}. The form of the last test is chosen to ensure the size of the test is α with $\alpha/2$ assigned to each of the possible directions for an extreme value. The sample mean can be an extreme distance from μ_0 by the value of $\bar{x}$ being very much smaller or very much larger than μ_0 and so far as to consider rejection of $H_0 : \mu = \mu_0$.

Example: Blood alcohol concentration (*continued*)

The BAC readings taken from the individual were $79.5, 80.1, 80.6, 80.8, 82.4$. Is this sufficient evidence, at the 5% level of significance, that blood alcohol level is greater than 80 mg per 100 ml?

The standard deviation σ of readings from the breathalyser is now assumed unknown. It is estimated by s, the standard deviation of the five readings given.

Let μ_0 be 80 mg per 100 ml. Let μ be the true level.

The hypotheses are as follows:

$$H_0 : \mu = \mu_0, \quad H_1 : \mu > \mu_0.$$

The sample size is $n = 5$ so there are $\nu = n - 1 = 4$ degrees of freedom for a t-test. Denote the five readings as x_1, x_2, x_3, x_4, x_5. Then $\bar{x} = 80.68$ and s, the standard deviation of x, is 1.085. The test statistic is

$$t = \frac{\bar{x} - \mu_0}{s}\sqrt{n} = \frac{(80.68 - 80)\sqrt{5}}{1.085} = 1.401.$$

The 95% point of a t distribution with 4 degrees of freedom is 2.132 (obtainable from the R command qt(0.95,4); Appendix D). Thus, there is insufficient evidence (at the 5% level) that the true mean is greater than 80 mg per 100 ml. The significance probability of the result is 0.12; the probability of obtaining a value as large as 80.68 if the true level is 80 mg per 100 ml is 0.12 (1-pt(1.401,4)). Thus approximately one in eight of mean readings for people with a true blood alcohol reading of 80 mg per 100 ml will give a sample mean reading of 80.68 or more. This result can be compared with the p-value of 0.083 when it was assumed that σ was known and equal to 1.1. When σ^2 is unknown, the p-value is greater than when σ^2 is known, a result explained by the extra dispersion of the t distribution so that large deviances from the mean under the null hypothesis are more likely than with a normal distribution.

The t-test applied in this example could also be performed by using **R** code:

```
> BAC = c(79.5, 80.1, 80.6, 80.8, 82.4)


> t.test(BAC, mu=80, alternative = "greater")


        One Sample t-test

data: BAC
t = 1.4015, df = 4, p-value = 0.1168
alternative hypothesis: true mean is greater than 80
95 percent confidence interval:
 79.64567 Inf
sample estimates:
mean of x
  80.68
```

Note that the confidence interval is $(79.65, \infty)$ due to the one-sided test used (the alternative hypothesis was $H_1 : \mu > \mu_0$; Section 3.4.4).

In analytical chemistry, there is a strong need to control whether the analytical procedure used works properly. The process of assessing whether the analytical method meets the requirements set is called *validation* (see ISO 17025). It is concerned, among other things, with verifying the precision, accuracy, reproducibility, and intermediate precision of the results of measurements. In the example presented below, the accuracy is measured, which is defined as a degree of the agreement of the value measured and the true value of the feature. This is frequently done by the analysis of suitable standard, such as a sample for which the value of a particular feature is well known.

Example

In order to check whether an instrument was working properly, an analysis was performed which involved measuring the refractive index of a standard glass fragment for which the true RI value is known.

A total of $n = 40$ measurements of the RI of a standard glass fragment using the GRIM method (Section 1.2.2) were analysed. The RI declared for the standard sample was $\mu_0 = 1.52214$. All 40 observations come from a normally distributed population $N(\mu, \sigma^2)$, where μ and σ are estimated as $\bar{x} = 1.52219$ and $s = 5.036 \times 10^{-5}$.

The hypotheses are

$$H_0 : \mu = \mu_0, \quad H_1 : \mu \neq \mu_0.$$

If both RI values (the RI determined in the course of the analysis and the declared value for the standard glass fragment) are equal based on the t-test results, this suggests that there is insufficient evidence to conclude that the instrument works improperly.

For the data under consideration the t-value is equal to

$$t = \frac{\bar{x} - \mu_0}{s}\sqrt{n} = \frac{1.52219 - 1.52214}{5.036 \cdot 10^{-5}} \cdot \sqrt{40} = 6.3.$$

The critical t-value limiting the regions of rejecting or failing to reject the null hypothesis (Figure 3.21), when the significance level is set at $\alpha = 0.05$ and $\nu = 40 - 1 = 39$, is equal to 2.02 (obtainable from **R** command qt(0.975,39); Appendix D). Due to the fact that the calculated t-value (6.3) exceeds the critical value, the null hypothesis stating the equality of both means is rejected, which means that there is sufficient evidence to indicate that the instrument does not work properly.

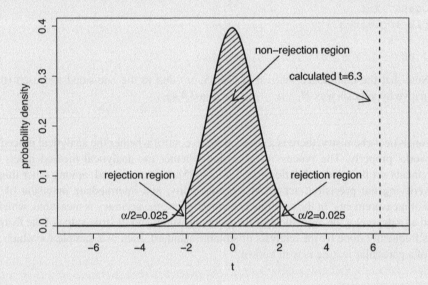

Figure 3.21   Regions in which the null hypothesis is not rejected (shaded) and in which it is (unshaded) when the significance level $\alpha = 0.05$ with the experimental value of $t = 6.3$.

The same interpretation may be formulated with the reference to p-values, which give the probability of obtaining a value as high as or higher than the test statistic assuming that H_0 is true. A quick method of obtaining the p-value is to perform a t-test in **R** using the t.test() function:

```
> RI_1 = read.table("RI_1.txt", header=TRUE)

> t.test(RI_1, mu=1.52214)

        One Sample t-test

data: RI_1
t = 6.3111, df = 39, p-value = 1.916e-07
```

```
alternative hypothesis: true mean is not equal to 1.52214
95 percent confidence interval:
 1.522174 1.522206
sample estimates:
mean of x
 1.52219
```

The calculated p-value is equal to 1.916×10^{-7} and is much lower than the significance level established at $\alpha = 0.05$. Therefore, the decision to reject H_0 should be taken.

3.4.4 Relation between tests and confidence intervals

Consider independent and identically distributed random variables $X_1, \ldots X_n \sim N(\mu, \sigma^2)$, with σ^2 unknown. The corresponding sample mean and variance are $\bar{x}$ and s^2.

A $100(1 - \alpha)\%$ confidence interval for μ is

$$\left(\bar{x} - \frac{t_{(n-1,\alpha/2)}s}{\sqrt{n}}, \quad \bar{x} + \frac{t_{(n-1,\alpha/2)}s}{\sqrt{n}} \right).$$

The critical region C of a level α test for $H_0 : \mu = \mu_0$ versus the two-sided alternative $H_1 : \mu \neq \mu_0$ is

$$C : \frac{|\bar{x} - \mu_0|}{s/\sqrt{n}} \geq t_{(n-1,\alpha/2)}.$$

Consider the complement $\bar{C}$ of the critical region:

$$\bar{C} : -t_{(n-1,\alpha/2)} < \frac{\bar{x} - \mu_0}{s/\sqrt{n}} < t_{(n-1,\alpha/2)},$$

which can be written as

$$\bar{C} : \bar{x} - \frac{t_{(n-1,\alpha/2)}s}{\sqrt{n}} < \mu_0 < \bar{x} + \frac{t_{(n-1,\alpha/2)}s}{\sqrt{n}}.$$

This last interval shows that any given value of μ, denoted as μ_0 for a (two-sided) test of $H_0 : \mu = \mu_0$ versus $H_1 : \mu \neq \mu_0$, will not be rejected at level α if μ_0 lies within the $100(1 - \alpha)\%$ confidence interval for μ. Once a $100(1 - \alpha)\%$ confidence interval for μ is established, it is known that all possible null hypothesis values μ_0 that lie outside this interval will be rejected at the $100\alpha\%$ level in a two-sided test of $H_0 : \mu = \mu_0$ versus $H_1 : \mu \neq \mu_0$ and all values of μ_0 lying inside the interval will not be rejected (Figure 3.22).

Figure 3.22 A $100(1 - \alpha)\%$ confidence interval for the mean of a normal distribution of unknown variance σ^2 given data $x_1, \ldots, x_n$ with sample mean $\bar{x}$ and sample standard deviation s.

3.4.5   Hypothesis test based on small samples for a difference in the means of two independent populations with unknown variances from normal distributions

Denote the random variables for the two independent samples by $X_{11}, \ldots, X_{1n_1}$ and $X_{21}, \ldots, X_{2n_2}$, with observations in each sample also independent. Assume $X_{1i} \sim N(\mu_1, \sigma_1^2)$, for $1 \leq i \leq n_1$, and $X_{2j} \sim N(\mu_2, \sigma_2^2)$, for $1 \leq j \leq n_2$. A confidence interval is required for $\mu_1 - \mu_2$.

Denote the random variables for the sample means and variances as $\bar{X}_1$, $\bar{X}_2$, S_1^2, and S_2^2. Then

$$\bar{X}_1 \sim N(\mu_1, \sigma_1^2/n_1), \quad \bar{X}_2 \sim N(\mu_2, \sigma_2^2/n_2).$$

The natural estimator of $\mu_1 - \mu_2$ is $\bar{X}_1 - \bar{X}_2$ with distribution $N(\mu_1 - \mu_2, \sigma_1^2/n_1 + \sigma_2^2/n_2)$.

An additional assumption is needed, and that is that the population variances are equal, thus $\sigma_1^2 = \sigma_2^2 = \sigma^2$. An analysis when this assumption cannot be made is given later.

Let the pooled estimate of the variance be

$$s^2 = \frac{(n_1 - 1)s_1^2 + (n_2 - 1)s_2^2}{n_1 + n_2 - 2}.$$

The standard error of $\bar{X}_1 - \bar{X}_2$, $\sqrt{\sigma^2/n_1 + \sigma^2/n_2}$, is estimated by $\sqrt{s^2/n_1 + s^2/n_2} = s\sqrt{\frac{1}{n_1} + \frac{1}{n_2}}$.

The $100(1 - \alpha)\%$ confidence interval for $\mu_1 - \mu_2$ is

$$\bar{x}_1 - \bar{x}_2 \pm t_{(n_1+n_2-2,\alpha/2)} \cdot s\sqrt{\frac{1}{n_1} + \frac{1}{n_2}}.$$

Consider a test of $H_0 : \mu_1 = \mu_2$ versus $H_1 : \mu_1 \neq \mu_2$. The test statistic is

$$t = \frac{|(\bar{x}_1 - \bar{x}_2) - (\mu_1 - \mu_2)|}{\sqrt{s^2\left(\frac{1}{n_1} + \frac{1}{n_2}\right)}},$$

and if H_0 is true this has a $t_{n_1+n_2-2}$ distribution. The test is known as a *two-sample t-test*.

Example

Elemental analysis of papers using ICP-MS (Kellner *et al.* 2004) has been reported on a wide range of papers from across the world (Adam 2010; Spence *et al.* 2000). Data are given for the results for five specimens for the manganese (Mn) concentrations (in $\mu g \cdot g^{-1}$) in Austria and Thailand. For Austria (group 1): $n_1 = 5$, $\bar{x}_1 = 13.3$, $s_1 = 0.6$. For Thailand (group 2): $n_2 = 5$, $\bar{x}_2 = 11.9$, $s_2 = 0.5$. The pooled estimate of the variance is $s^2 = (0.6^2 + 0.5^2)/2 = 0.305$, a simple average since $n_1 = n_2$. Hence $s = \sqrt{0.305} = 0.552$. There are $v = n_1 + n_2 - 2 = 8$ degrees of freedom.

Consider the example of the Mn content of paper. The form of the alternative hypothesis should be chosen before the data are collected. There are three possible questions we might ask: whether the Mn content of Austrian paper is

- different from the Mn content of Thai paper;
- less than the Mn content of Thai paper;
- greater than the Mn content of Thai paper.

If the first question is the one asked a two-sided test should be used, as below. If the second or third question is to be asked a one-sided test should be used.

Denote the true mean Mn concentration for paper from Austria by μ_1 and from Thailand by μ_2. The null hypothesis is that $\mu_1 - \mu_2$ equals zero.

The 95% confidence interval for $\mu_1 - \mu_2$ is

$$\bar{x}_1 - \bar{x}_2 \pm t_{n_1+n_2-2, 0.025} \cdot s \sqrt{\frac{1}{n_1} + \frac{1}{n_2}} = 1.4 \pm t_{8,0.025} \cdot 0.552\sqrt{\frac{1}{5} + \frac{1}{5}}$$

$$= 1.4 \pm 2.306 \cdot 0.552\sqrt{0.4}$$

$$= 1.4 \pm 0.81 = (0.59, 2.21)\,\mu g \cdot g^{-1}.$$

Consider a test of $H_0 : \mu_1 = \mu_2$ versus $H_1 : \mu_1 \neq \mu_2$. The test statistic is

$$t = \frac{|(\bar{x}_1 - \bar{x}_2) - (\mu_1 - \mu_2)|}{\sqrt{s^2\left(\frac{1}{n_1} + \frac{1}{n_2}\right)}} = \frac{(13.3 - 11.9) - 0}{0.552\sqrt{\left(\frac{1}{5} + \frac{1}{5}\right)}} = \frac{1.4\sqrt{5}}{0.552\sqrt{2}} = 4.01,$$

if H_0 is true (and hence $\mu_1 - \mu_2 = 0$).

The value of 4.01 corresponds to a p-value of 0.004 in a two-sided test ($2*(1-pt(4.01,8))$). The result is significant at the 0.5% level in a two-sided test. (The p-value of 0.004 is less than 0.005 or 0.5%.)

The conclusion is that there is a difference in Mn content between papers from the two countries. Further inspection of the data shows that Austrian paper has Mn content 1.4 $\mu g \cdot g^{-1}$ greater than the Mn content of Thai paper, with a 95% confidence interval of $(0.59, 2.21)\,\mu g \cdot g^{-1}$.

When the two-stage approach is considered for solving the comparison problem (Sections 2.2.1 and 2.2.3), the t-test could be applied at the first stage. Two examples of such analysis performed for glass evidence samples based on their RI values (Section 1.2.2) are presented below.

Example

The following refractive index values (Section 1.2.2) were determined for control (A) and recovered (B) samples:

- A: 1.51907, 1.51909, 1.51903, 1.51898, 1.51896,

- B: 1.51907, 1.51906, 1.51908, 1.51911, 1.51913.

So the hypotheses are as follows:

$$H_0 : \mu_A = \mu_B, \quad H_1 : \mu_A \neq \mu_B,$$

where μ_A and μ_B are the mean refractive indices of the populations from which the two samples of glass fragments were taken.

The sample sizes are $n_1 = n_2 = n = 5$ and the number of degrees of freedom is $v = n_1 + n_2 - 2 = 8$.

The test may be performed by using the following **R** code (Appendix D):

```
> RI_A = c(1.51907, 1.51909, 1.51903, 1.51898, 1.51896)
> RI_B = c(1.51907, 1.51906, 1.51908, 1.51911, 1.51913)

> t.test(RI_A, RI_B, var.equal = TRUE)
```

The outcome is as follows:

```
    Two Sample t-test

data: RI_A and RI_B
t = -2.2684, df = 8, p-value = 0.05302
alternative hypothesis: true difference in means is not equal to 0
95 percent confidence interval:
 -1.290604e-04 1.060383e-06
sample estimates:
mean of x mean of y
 1.519026 1.519090
```

The results suggest that the RI values of both samples are equal at the $\alpha = 0.05$ significance level. A two-stage approach to evidence evaluation would proceed to consideration of the rarity (Section 2.2.1).

When a very small change in the measurements is made, from 1.51911 to 1.51914 for one element from the RI_B data (which is lower than the precision of the method of analysis used),

```
> RI_A = c(1.51907, 1.51909, 1.51903, 1.51898, 1.51896)
> RI_C = c(1.51907, 1.51906, 1.51908, 1.51914, 1.51913)
```

the results of applying the **R** code (Appendix D), with the hypotheses $H_0 : \mu_A = \mu_C$ and $H_1 : \mu_A \neq \mu_C$, where μ_A and μ_C are the mean refractive indices of the populations from which the two samples of glass fragments were taken, are as follows:

```
> t.test(RI_A, RI_C, var.equal = TRUE)

        Two Sample t-test

data: RI_A and RI_C
t = -2.3438, df = 8, p-value = 0.04714
alternative hypothesis: true difference in means is not equal to 0
95 percent confidence interval:
 -1.388720e-04 -1.128029e-06
sample estimates:
mean of x mean of y
 1.519026 1.519096
```

Introducing only a small alteration in one of the measurements changes the significance probability from 0.053 to 0.047, so that H_0 is rejected and a rarity stage is not included.

There is a Welch test to use when $\sigma_1^2 \neq \sigma_2^2$. This involves a complicated expression for the associated degrees of freedom, the result of which may not be an integer:

$$\nu = \frac{\left(\frac{s_1^2}{n_1} + \frac{s_2^2}{n_2}\right)^2}{\left(\frac{s_1^4}{n_1^2(n_1-1)}\right) + \left(\frac{s_2^4}{n_2^2(n_2-1)}\right)}. \tag{3.6}$$

Example

Consider the previous example. Imagine two sets of RI values with $n_1 = 40$ and $n_2 = 30$ observations available in RI_1.txt and RI_2.txt. Both sets of values come from the normally distributed populations $N(\mu_1, \sigma_1^2)$ and $N(\mu_2, \sigma_2^2)$, where μ_1 and μ_2 are estimated as $\bar{x}_1 = 1.52219$ and $\bar{x}_2 = 1.52217$ and the standard deviations σ_1 and σ_2 are estimated as $s_1 = 5.036 \times 10^{-5}$ and $s_2 = 5.041 \times 10^{-5}$. The question of interest is whether both sets of observations come from populations with equal RI means, $\mu_1 = \mu_2$.

The hypotheses are

$$H_0 : \mu = 0, \quad H_1 : \mu \neq 0.$$

If the assumption of equality of variances is made (see the last example in Section 3.5), then

```
> RI_1 = read.table("RI_1.txt", header=TRUE)
> RI_2 = read.table("RI_2.txt", header=TRUE)

> t.test(RI_1, RI_2, var.equal=TRUE)

        Two Sample t-test

data: RI_1 and RI_2
t = 1.6368, df = 68, p-value = 0.1063
```

```
alternative hypothesis: true difference in means is not equal to 0
95 percent confidence interval:
 -4.364076e-06 4.419741e-05
sample estimates:
mean of x mean of y
 1.52219 1.52217
```

It can be stated that there is no reason to reject the H_0 at significance level 0.05 as 0.1063 > 0.05. This means that there exists insufficient evidence to indicate that the observed values come from populations with different means.

Typing t.test() without the restriction of equal variances (var.equal=TRUE), performs the Welch test. This test works under more rigid conditions when the equality of variances of observations from two datasets is not assumed. Then the t-value is equal to:

$$t = \frac{(\bar{x}_1 - \bar{x}_2) - (\mu_1 - \mu_2)}{\sqrt{s^2 \left(\frac{1}{n_1} + \frac{1}{n_2} \right)}}$$

$$= \frac{(1.52219 - 1.52217) - 0}{1.2168 \cdot 10^{-5}} = 1.64$$

as before, but $\nu = 62.6$ using equation (3.6).

Running **R** again:

```
> RI_1 = read.table("RI_1.txt", header=TRUE)
> RI_2 = read.table("RI_2.txt", header=TRUE)

> t.test(RI_1, RI_2)

        Welch Two Sample t-test

data: RI_1 and RI_2
t = 1.6366, df = 62.584, p-value = 0.1067
alternative hypothesis: true difference in means is not equal to 0
95 percent confidence interval:
 -4.405949e-06 4.423928e-05
sample estimates:
mean of x mean of y
 1.52219 1.52217
```

The Welch test changes the p-value slightly to 0.1067. Using the Welch test under more rigid conditions in this case does not lead to a change in the result of the hypothesis test.

3.4.6 Paired comparisons

A special form of the t-test is used when the data are paired. Data are paired when there is interest in the existence of a possible change in response between two treatments or over time, for example. There may be variation in response between different experimental units as well as between treatments or between different points in time. If it is possible to pair experimental units then the variation in response between experimental units can be separated from the variation between treatments or between different time points. This is best described by means of an example.

Example

In a comparison of two methods of chlorinating sewage, eight pairs of batches of sewage were treated. On each of eight days ($n = 8$) a pair of batches were taken. The two treatments were randomly assigned to the two batches in each pair (e.g. by tossing a fair coin). Treatment A involved an initial period of rapid mixing, while treatment B did not. The results in log coliform bacteria content per millilitre were as follows:

Day (i)	1	2	3	4	5	6	7	8
A	2.8	3.1	2.9	3.0	2.4	3.0	3.2	2.6
B	3.2	3.1	3.4	3.5	2.7	2.9	3.5	2.8
$D = B - A$	0.4	0.0	0.5	0.5	0.3	−0.1	0.3	0.2

The hypotheses are $H_0 : \mu = 0$ and $H_1 : \mu \neq 0$, where μ is the population mean difference in bacteria content.

It is also assumed that the distribution of D is normal with $D \sim N(\mu, \sigma^2)$. The test statistic is

$$t = \frac{\bar{d} - \mu}{s_d} \sqrt{n},$$

where $\bar{d}$ is the mean of the difference D between the paired variables ($D = B - A$) and s_d is its standard deviation. There are $v = n - 1 = 7$ degrees of freedom. We have

$$t = \frac{0.2625}{0.220} \sqrt{8} = 3.37,$$

when $\mu = 0$.

The corresponding p-value is equal to 0.012 (2*(1-pt(3.37,7), and is less than $\alpha = 0.05$). The conclusion is that there is evidence that the initial period of rapid mixing is associated with a reduction in mean coliform content.

The same results can also be obtained using **R** code (Appendix D):

```
> A = c(2.8,3.1,2.9,3.0,2.4,3.0,3.2,2.6)
> B = c(3.2,3.1,3.4,3.5,2.7,2.9,3.5,2.8)

> t.test(B, A, paired=TRUE)

        Paired t-test

data: B and A
t = 3.3751, df = 7, p-value = 0.01184
```

```
alternative hypothesis: true difference in means is not equal to 0
95 percent confidence interval:
 0.07858897 0.44641103
sample estimates:
mean of the differences
                0.2625
```

According to Zadora and Wilk (2009) it was experimentally shown that the differences between the RIs measured before and after the annealing process (Section 1.2.2), expressed as $RI_a - RI_b = \Delta RI$, for non-toughened glass fragments are insignificant ($\Delta RI = 0$ or $RI_a = RI_b$). In contrasting, for toughened glass fragments ΔRI is significant ($\Delta RI \neq 0$ or $RI_a \neq RI_b$).

Example

In order to establish whether the glass samples were toughened or not, the RI was measured before (RI_b) and after (RI_a) the annealing process for $n = 40$ ($n_{RIb} = n_{RIa} = n = 40$) glass samples (data are available in RI_annealing.txt). Both sets of RI values come from the normally distributed populations $N(\mu_{RIb}, \sigma^2_{RIb})$ and $N(\mu_{RIa}, \sigma^2_{RIa})$, where μ_{RIb} and μ_{RIa} are estimated as $\overline{RI}_b = 1.52219$ and $\overline{RI}_a = 1.522142$ and the standard deviations σ_{RIb} and σ_{RIa} are estimated as $s_{RIb} = 5.036 \times 10^{-5}$ and $s_{RIa} = 4.904 \times 10^{-5}$. The question arising is whether both sets of observations come from populations with equal RI means, $\mu_{RIb} = \mu_{RIa}$.

The hypotheses are as follows:

$$H_0 : \mu_{RIb} = \mu_{RIa}, \quad H_1 : \mu_{RIb} \neq \mu_{RIa}.$$

The test statistic is:

$$t = \frac{(\bar{d} - (\mu_{RIb} - \mu_{RIa}))}{s_d} \sqrt{n},$$

where $\bar{d}$ is the mean of the difference d between the paired variables ($d = RI_b - RI_a$) and s_d is its standard deviation. We have

$$t = \frac{(1.52219 - 1.522142) - 0}{1.847 \cdot 10^{-5}} \sqrt{40} = 16.6,$$

when $\mu_{RIb} - \mu_{RIa} = 0$. There are $\nu = n - 1 = 39$ degrees of freedom.

```
> RI_B = read.table("RI_annealing.txt", header=TRUE)$RI_before
> RI_A = read.table("RI_annealing.txt", header=TRUE)$RI_after

> t.test(RI_B, RI_A, paired=TRUE)

        Paired t-test
```

```
data: RI_B and RI_A
t = 16.6041, df = 39, p-value < 2.2e-16
alternative hypothesis: true difference in means is not equal to 0
95 percent confidence interval:
 4.259179e-05 5.440821e-05
sample estimates:
mean of the differences
             4.85e-05
```

The test result (p-value equal to 2.2×10^{-16}) suggests very strongly that the null hypothesis should be rejected. The significance probability indicates the probability of a result as large as this by chance alone is about 1 in 4.5×10^{15}. Therefore the difference in RI measured before and after the annealing process is statistically significant, which indicates that there is sufficient evidence to indicate that the glass fragments may come from toughened glass.

3.4.7 Hotelling's T^2 test

Hotelling's T^2 test is used to test hypotheses for multivariate normally distributed data. One of the most common applications of this test is to consider propositions about means. Consider p-variate data (where p refers to the number of variables and is not to be confused with the symbol p for the significance probability) from two populations with means $\mu_i = (\mu_{i1}, \ldots, \mu_{ip})^T$, $i = 1, 2$. The data are $\{x_{ijk}, i = 1, 2; j = 1, \ldots, n_i; k = 1, \ldots, p\}$, where n_i, $i = 1, 2$, are the sample sizes. The mean data vectors are $\bar{\mathbf{x}}_i = (\bar{x}_{i1}, \ldots, \bar{x}_{ip})^T$, $i = 1, 2$, where $\bar{x}_{ik} = \sum_{j=1}^{n_i} x_{ijk}/n_i$, for $i = 1, 2$ and $k = 1, \ldots, p$. The hypotheses under consideration are :

$$H_0 : (\mu_{11}, \ldots, \mu_{1p}) = (\mu_{21}, \ldots, \mu_{2p}),$$
$$H_1 : (\mu_{11}, \ldots, \mu_{1p}) \neq (\mu_{21}, \ldots, \mu_{2p}).$$

Hotelling's T^2 statistic is then

$$T^2 = \left(\frac{1}{n_1} + \frac{1}{n_2}\right)^{-1} (\bar{\mathbf{x}}_1 - \bar{\mathbf{x}}_2)^T \mathbf{S}^{-1} (\bar{\mathbf{x}}_1 - \bar{\mathbf{x}}_2).$$

where $\mathbf{S}$ is the pooled variance–covariance matrix

$$\mathbf{S} = \frac{(n_1 - 1)\mathbf{S}_1 + (n_2 - 1)\mathbf{S}_2}{n_1 + n_2 - 2}.$$

and $\mathbf{S}_1$ and $\mathbf{S}_2$ are the variance–covariance matrices (equation (3.5)) calculated for samples 1 and 2, respectively. Significance probabilities for this statistic may be determined with reference to the F distribution using the relationship

$$\frac{n_1 + n_2 - p - 1}{p(n_1 + n_2 - 2)} T^2 \sim F_{p, n_1 + n_2 - p - 1}.$$

Example

Imagine two glass fragments which were subjected to SEM-EDX analysis with the aim of determining the content of seven elements (O, Na, Mg, Al, Si, K, and Ca). Each element content was normalised by the oxygen content and transformed by taking the logarithm to base 10. Therefore, six variables described each of the glass fragments analysed. Each fragment was measured five times. The results of the analysis are presented in Table 3.2.

Table 3.2 SEM-EDX data for glass fragments used for performing Hotelling's T^2 test (data available in T2_Hotelling.txt).

Item	Na	Mg	Al	Si	K	Ca
1	−0.68	−1.30	−2.34	−0.16	−2.48	−0.93
1	−0.70	−1.30	−2.42	−0.26	−2.21	−0.74
1	−0.67	−1.29	−2.32	−0.16	−2.45	−0.93
1	−0.67	−1.39	−2.36	−0.21	−2.38	−0.86
1	−0.67	−1.28	−2.48	−0.11	−2.26	−0.85
$\bar{x}_1$	**−0.67**	**−1.29**	**−2.34**	**−0.14**	**−2.41**	**−0.90**
2	−0.78	−1.29	−2.34	−0.11	−2.36	−0.85
2	−0.66	−1.47	−2.49	−0.26	−2.34	−0.66
2	−0.76	−1.23	−2.43	−0.07	−2.33	−0.80
2	−0.68	−1.46	−2.44	−0.29	−2.40	−0.73
2	−0.56	−1.27	−2.49	−0.07	−2.24	−0.70
$\bar{x}_2$	**−0.66**	**−1.27**	**−2.40**	**−0.07**	**−2.32**	**−0.81**

The measurements are assumed to be independent and normally distributed. The within-group variances are assumed equal. The question of interest is whether the group mean log ratio (μ_{ij}, $i = 1, 2$; $j = 1, \ldots, 6$) elemental concentrations in the two fragments are equal. The fragments are described by vectors of length of six corresponding to content of six elements (Na, Mg, Al, Si, K, and Ca).

The hypotheses are

$$H_0 : (\mu_{11}, \ldots, \mu_{16}) = (\mu_{21}, \ldots, \mu_{26}),$$
$$H_1 : (\mu_{11}, \ldots, \mu_{16}) \neq (\mu_{21}, \ldots, \mu_{26}).$$

The data are $\bar{x}_1 = (-0.67, -1.29, -2.34, -0.14, -2.41, -0.90)$ and $\bar{x}_2 = (-0.66, -1.27, -2.40, -0.07, -2.32, -0.81)$.

Hotelling's test may be quickly computed in **R** software (Appendix D) using the package *Hotelling*.

```
> data = read.table("T2_Hotelling.txt", header=TRUE)
> install.packages("Hotelling")
> library(Hotelling)
```

```
> test = hotelling.test(.~Item, data = data)
> test

Test stat: 1.423
Numerator df: 6
Denominator df: 3
P-value: 0.4159
```

The p-value calculated in **R** as `1-pf(1.423,6,3)` is 0.4159. This is very large: a result as large as or larger than this is expected to occur by chance alone about four times in every ten when there is no difference in the mean elemental content between the objects of which the fragments were sampled. Thus, there is very little evidence to suggest differences in the mean elemental content of the objects.

3.4.8   Significance test for correlation coefficient

The t-test may be used to test for evidence of a non-zero correlation between a pair of normally distributed random variables. Given n independent pairs of observations $\{(x_{1k}, x_{2k}, \ k = 1, \ldots, n\}$, the sample correlation coefficient from Section 3.3.4 may be written as:

$$r(x_1, x_2) = \frac{1}{n-1} \sum_{k=1}^{n} \left(\frac{x_{1k} - \bar{x}_1}{s_1} \right) \left(\frac{x_{2k} - \bar{x}_2}{s_2} \right)$$

$$= \frac{\sum_{k=1}^{n}(x_{1k} - \bar{x}_1)(x_{2k} - \bar{x}_2)}{\sqrt{\sum_{k=1}^{n}(x_{1k} - \bar{x}_1)^2 \ \sum_{k=1}^{n}(x_{2k} - \bar{x}_2)^2}}.$$

Let ρ be the population correlation coefficient for a pair (X_1, X_2) of independent normally distributed random variables. Note that the observations are independent but the variables are not. For example, in a study of the variation of height with age amongst people aged 0 to 15, n unrelated individuals may be studied. The individuals can be considered independent but their age and height are related. The hypotheses to be tested are

$$H_0 : \rho = 0, \quad H_1 : \rho \neq 0.$$

Thus H_0 states there is no linear association between the two variables, and H_1 that there is a linear association between the two variables. The test statistic is

$$t = \frac{r}{\sqrt{1 - r^2}} \sqrt{n - 2},$$

and if H_0 is true this has a t distribution with $n - 2$ degrees of freedom.

Example

Imagine a set of 32 independent and normally distributed observations of a pair of variables for which the correlation coefficient was determined as 0.75. The null hypothesis is that there is no linear correlation between the two variables, and the alternative hypothesis is that there is a non-zero linear correlation but of unknown sign:

$$H_0 : \rho = 0, \quad H_1 : \rho \neq 0.$$

Thus,

$$t = \frac{r}{\sqrt{1 - r^2}} \sqrt{n - 2} = \frac{0.75}{\sqrt{1 - 0.75^2}} \sqrt{32 - 2} = 6.21.$$

The value, $t = 6.21$, of the test statistic is greater than the critical t-value at a significance level of 0.05, for a two-sided test. The critical value is equal to 2.04, obtainable from the **R** command qt(0.975,30) (1 - pt(2.04,30) = 0.025; Appendix D). Therefore the correlation coefficient is significantly different from 0. The significance probability p can be determined from 2*(1 - pt(6.21,30)), which is 7.8×10^{-7}. This p-value is extremely small. Thus, there is very strong evidence that the null hypothesis of zero linear correlation ($\rho = 0$) should be rejected in favour of the alternative hypothesis of non-zero linear correlation ($\rho \neq 0$). Further inspection of the data shows that the point estimate of ρ is 0.75 so one should act as if $\rho > 0$.

```
> install.packages("psych")
> library(psych)
> r.test(n = 32, r12 = 0.75)


Correlation tests
Call:r.test(n = 32, r12 = 0.75)
Test of significance of a correlation
 t value 6.21 with probability < 7.8e-07
 and confidence interval 0.54 0.87
```

Another example of testing the significance of the correlation coefficient is presented in Appendix D.7.6.

3.5   Analysis of variance

3.5.1   Principles of ANOVA

The analysis of variance (ANOVA) is a general method for studying linear statistical models by analysing the contribution of every source of uncertainty to the global variance observed in the sample. This can be done by subdividing the total sum of squares of a complex dataset into components due to various sources of variation. Here only the so-called *one-way classification* is considered, where it is investigated whether k samples of data from k normal populations with common unknown variance σ^2 might reasonably be supposed to have the same expectation. The essential information about variance is provided in Section 3.3.2.

Consider independent observations x_{ij}, for $i = 1, \ldots, k$ and $j = 1, \ldots, n_i$, where x_{ij} denotes the jth of n_i observations in group i. The total sum of squares of the data around the overall sample mean

$$\bar{x} = \frac{\sum_{i=1}^{k} \sum_{j=1}^{n_i} x_{ij}}{\sum_{i=1}^{k} n_i} \tag{3.7}$$

is defined to be

$$T = \sum_{i=1}^{k} \sum_{j=1}^{n_i} (x_{ij} - \bar{x})^2.$$

If there were no systematic differences between the k populations so that the $n = \sum_{i=1}^{k} n_i$ measurements could be regarded as one sample of data, then $T/(n-1)$ would be the usual sample estimate of the population variance σ^2. Here it is assumed that the measurements x_{ij} are realisations of independent random variables X_{ij}, where $X_{ij} \sim N(\mu_i, \sigma^2)$, and the null hypothesis to be tested is

$$H_0 : \mu_1 = \mu_2 = \cdots = \mu_k \, (= \mu, \text{ say})$$

against the general alternative that H_0 is false. Note that it is assumed that the population variance σ^2 is the same in every population. Also, if H_0 is false, some of the μ_i may be equal, but not all of them.

A procedure is also needed for testing the null hypothesis H_0, and it is to be expected that this reduces to something recognisable in the case of $k = 2$ groups (in which case the ANOVA is equivalent to a two-sample t-test).

Each observation x_{ij} can be split into three components:

$$x_{ij} = \bar{x} + (\bar{x}_i - \bar{x}) + (x_{ij} - \bar{x}_i), \quad i = 1, \ldots, k, \; j = 1, \ldots, n_i.$$

The first component is the overall mean of all n observations (equation (3.7)). The second component is the difference between the mean for the ith group,

$$\bar{x}_i = \frac{\sum_{j=1}^{n_i} x_{ij}}{n_i},$$

and the overall mean ($\bar{x}$). The third component is the difference between this particular observation (x_{ij}) and its own group mean ($\bar{x}_i$).

Therefore T may be written as

$$T = \sum_{i=1}^{k} \sum_{j=1}^{n_i} (x_{ij} - \bar{x})^2 = \sum_{i=1}^{k} \sum_{j=1}^{n_i} (x_{ij} - \bar{x}_i + \bar{x}_i - \bar{x})^2 \qquad (3.8)$$

$$= \sum_{i=1}^{k} \sum_{j=1}^{n_i} (x_{ij} - \bar{x}_i)^2 + \sum_{i=1}^{k} \sum_{j=1}^{n_i} (\bar{x}_i - \bar{x})^2 + 2 \sum_{i=1}^{k} (\bar{x}_i - \bar{x}) \sum_{j=1}^{n_i} (x_{ij} - \bar{x}_i)$$

$$= \sum_{i=1}^{k} \sum_{j=1}^{n_i} (x_{ij} - \bar{x}_i)^2 + \sum_{i=1}^{k} n_i (\bar{x}_i - \bar{x})^2 + 0,$$

since

$$\sum_{j=1}^{n_i} (x_{ij} - \bar{x}_i) = 0$$

Now $W_i = \sum_{j=1}^{n_i} (x_{ij} - \bar{x}_i)^2$ is the sum of squares within group i: it is the obvious measure of variability about the ith sample mean $\bar{x}_i$. Hence the within-groups sum of squares W is defined to be

$$W = \sum_{i=1}^{k} W_i = \sum_{i=1}^{k} \sum_{j=1}^{n_i} (x_{ij} - \bar{x}_i)^2. \qquad (3.9)$$

Similarly, the between-groups sum of squares B is defined to be

$$B = \sum_{i=1}^{k} n_i (\bar{x}_i - \bar{x})^2. \qquad (3.10)$$

This is a measure of the extent to which the group means $\bar{x}_i$ differ from the overall mean $\bar{x}$. Thus, $T = W + B$; note that expressions for T, B and W that are convenient for hand calculation may be easily derived (Table 3.3).

The entries for *degrees of freedom* (Section 3.3.2) need explanation. There are $k - 1$ degrees of freedom for group differences because although there are k groups, given the value of the overall mean, there is only freedom to choose the value of $k - 1$ of the group means,

Table 3.3 The procedure for ANOVA represented as a summary of the decomposition of the total sum of squares (T, W, and B are calculated according to equations (3.8)–(3.10)).

Source of variation	Sum of squares	Degrees of freedom	Mean square	F ratio
Between groups	B	$k - 1$	$B/(k - 1)$	$\frac{B(n-k)}{W(k-1)}$
Within groups	W	$n - k$	$W/(n - k)$	
Total (about mean)	T	$n - 1$		

as $\sum_{i=1}^{k} n_i \bar{x}_i = n\bar{x}$. Similarly, within group i there are $n_i - 1$ degrees of freedom to pick the n_i individual measurements x_{ij} subject to the constraint $\sum_{j=1}^{n_i} x_{ij} = n_i \bar{x}_i$, and so within the k groups there are $\sum_{i=1}^{k} (n_i - 1) = n - k$ degrees of freedom. Note that not only is it true that $B + W = T$, but also that $(k - 1) + (n - k) = n - 1$: the degrees of freedom for the two sums of squares of variation also sum to the total degrees of freedom about the overall mean $\bar{x}$.

It is also useful to think of B as the portion of the total sum of squares T which is explained by differences between the group means, and W as the portion not explained by differences between group means. The final two columns of Table 3.3 are explained as follows. In general, the mean square for any source of variation is the sum of squares divided by its degrees of freedom. The F-ratio

$$F = \frac{B/(k - 1)}{W/(n - k)} \tag{3.11}$$

is the ratio of the between-groups mean square to the within-groups mean square and it is used to test the null hypothesis H_0. The within-groups mean square $W/(n - k)$ is known as the *residual mean square* (RMS) (the residual after taking account of the variation between groups). It is denoted by s^2 and is an estimate of σ^2.

The question that arises is how ANOVA can be used to test for the equality of several means. The null hypothesis assumes that all means are equal ($H_0 : \mu_1 = \mu_2 = \cdots = \mu_k$) and the alternative hypothesis is that H_0 is false.

The between-groups sum of squares (B) is comparable to the within-groups sum of squares (W) in the case where the measurements do not differ significantly between groups. Conversely, when they differ between groups, then B is greater than W. Whether the difference between the sample means of the groups is statistically significant may be tested by means of the F-ratio with $k - 1$ degrees of freedom for the numerator and $n - k$ degrees of freedom for the denominator. This is illustrated in the following example.

Example

It was mentioned in Section 3.4.3 that in analytical chemistry it is very important to control whether the analytical procedure works properly (Kellner *et al.* 2004). One of the indicative parameters is the *intermediate precision*.

Imagine that the intermediate precision was controlled by having three analysts (named AR, GZ, and PA) determine the sodium content of a glass sample by SEM-EDX (Section 1.2.1). The measurements are shown in Figure 3.23. Even though the question is about testing the equality of three means, the problem can be solved by application of the analysis of variance (ANOVA).

ANOVA may be performed in **R** software (Appendix D) using the data available in ANOVA_data.txt:

```
> data=read.table("ANOVA_data.txt", header=TRUE)
> anova(lm(data$Sodium ~ data$Person))
```

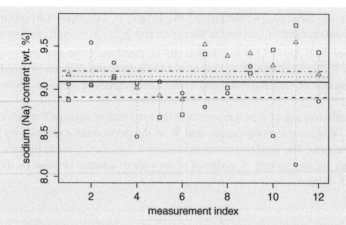

Figure 3.23 Measurements of sodium content determined by SEM-EDX and obtained by three analysts named AR, GZ, and PA (available in ANOVA_data.txt). Sample means are represented by dot-dashed lines for AR's measurements (triangles), dotted lines for GZ's measurements (squares), and dashed lines for AP's measurements (circles). The overall mean is represented by a solid line. Note that the data are not ordered. The three datasets are independent and only displayed as they are for convenience. The vertical lines have no significance.

The output is as follows:

```
Analysis of Variance Table

Response: Sodium
          Df Sum Sq Mean Sq F value  Pr(>F)
Person     2 0.6048 0.30242  2.9327 0.06725 .
Residuals 33 3.4029 0.10312
---
Signif. codes:  0    ***      0.001    **    0.01    *    0.05  .  0.1
                1
```

Here, the degrees of freedom (Df) describing the F statistic are equal to 2 (since $k = 3$) and 33 (since $n = 36$). The decision whether to reject or fail to reject the null hypothesis is taken by comparing the p-value with the desired level of significance $\alpha = 0.05$. Recall that the p-value is the probability that the differences occur by chance alone if H_0 holds. If $p > \alpha$, then there are no grounds for rejecting the null hypothesis. Since the p-value is 0.067, it is concluded that there is insufficient evidence to suggest there does not exist an intermediate precision of the measurements.

3.5.2   Feature selection with application of ANOVA

Comparison of the between- and within-group variation of data can also be employed to select the variables with the greatest classification abilities.

Objects are generally described by more than one feature (Chapter 1). Some of the features may be characteristic of a particular group of objects (e.g. a specific range of refractive

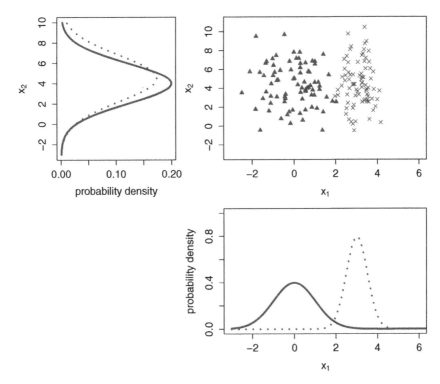

Figure 3.24 Selecting the most informative variable (x_1 or x_2) for classifying the objects forming two classes T (marked by triangles) and D (marked by crosses). Data are available in variables_selection.txt.

index values describing float glass). Such variables are highly desirable when considering a classification problem (Chapter 5), which focuses on determining the category to which the objects belong. The most appropriate variables then are those which enable the researcher to easily classify each object into the correct class. This means the best variables for a classification problem should have different ranges in each class with no overlap.

Suppose that a set of objects are described by variables x_1 and x_2 (Figure 3.24). The data distributions of each of the variables describing the objects from different classes are bell-shaped and in the case of x_1 are separate, whereas for x_2 they nearly completely overlap. Therefore, only the x_1 variable has an ability to classify objects since the ranges of this variable are separate (the areas of crosses and triangles overlap only slightly), whereas the classification cannot be carried out using x_2 as the values of this variable are independent of the classes.

In order to choose the best variables the F statistic and ANOVA may be employed. For this purpose, the k objects (groups) in the classical ANOVA are replaced by k classes, into which the objects are classified. In this example there are $k = 2$ classes marked by triangles and crosses in Figure 3.24. The n_i observations within the ith object ($i = 1, \ldots, k$) are substituted by n_i objects belonging to the ith class (each object is denoted by n_{ij}, where $j = 1, \ldots, n_i$). Therefore, the classical within-group variation (W) is calculated as the within-class variation of measurements. Consequently, the between-group variation (B) is given as the variation of measurements between classes. However, see Section 3.3.2 for estimation of the between- and within-group variances.

If there are no systematic differences in the values of a particular variable between observations from the k classes and the whole dataset could be treated as one sample then that variable is not a good classifier.

Large values of the F statistic imply that the dispersion between classes is greater than the dispersion within each class. The larger the value, the better the variable is at classification, since such a variable represents relatively separate sets of observed values.

Example

The F-test (and therefore ANOVA) was applied for the data illustrated in Figure 3.24. There are 80 objects belonging to category T (illustrated by triangles in Figure 3.24) and 80 belonging to category D (illustrated by crosses in Figure 3.24). Each object is described by two variables, x_1 and x_2 (variables_selection.txt). The aim was to check which variable would provide better classification. The ANOVA was done in **R** (Appendix D).

```
> data = read.table("variables_selection.txt", header = TRUE)
```

```
> anova(lm(data$x_1 ~ data$Factor)) ##ANOVA for the first variable x_1
Analysis of Variance Table
```

```
Response: data$x_1
              Df Sum Sq Mean Sq F value    Pr(>F)
data$Factor    1 389.01  389.01  604.61 < 2.2e-16   ***
Residuals    158 101.66    0.64
---
Signif. codes:   0    ***    0.001    **   0.01    *  0.05  .  0.1
                 1
```

The last line of the results provides a key to the asterisks against the p-value in the results (see, for example, < 2.2e-16 *** in the above results). The asterisks indicate the level at which the observed test value is significant.

```
> anova(lm(data$x_2 ~ data$Factor)) ##ANOVA for the second variable x_2
Analysis of Variance Table
```

```
Response: data$x_2
              Df   Sum Sq   Mean Sq  F value   Pr(>F)
data$Factor    1     0.42    0.4206   0.0792   0.7788
Residuals    158   839.34    5.3123
```

The F-values obtained are 604.61 for variable x_1 and 0.0792 for x_2. The F-value for the first variable is highly significant (the p-value is less than 2.2×10^{-16}), indicating a very strong separation between the two classes when using the first variable; this separation is easily visible in Figure 3.24. The F-value for the second variable is highly non-significant (the p-value is equal to 0.7788), indicating a very strong overlap of the two classes when using the second variable; this overlap is easily visible in Figure 3.24.

3.5.3   Testing of the equality of variances

Another example of the application of the F-test is in testing of the equality of variances of observations of two datasets. Such information may be important for the t-test assumptions (Section 3.4.5).

Example

The F-test for the datasets analysed in the last example in Section 3.4.5, in files `RI_1.txt` and `RI_2.txt`, is as follows:

```
> RI_1 = read.table("RI_1.txt", header = TRUE)
> RI_2 = read.table("RI_2.txt", header = TRUE)

> var.test(as.matrix(RI_1), as.matrix(RI_2))

        F test to compare two variances

data: as.matrix(RI_1) and as.matrix(RI_2)
F = 0.9979, num df = 39, denom df = 29, p-value = 0.9815
alternative hypothesis: true ratio of variances is not equal to 1
95 percent confidence interval:
 0.4909134 1.9576762
sample estimates:
ratio of variances
          0.9978629
```

The results suggest that there is insufficient evidence that the variances (equal to 2.5358×10^{-9} and 2.5413×10^{-9}) are not equal.

3.6   Cluster analysis

Cluster analysis (CA) is a data mining method especially designed for investigating data structure. Its main aim is to analyse the dataset and create clusters of objects similar in terms of the features considered. Clusters are defined by sets of objects that within a cluster are more similar than any others. Similarity is measured by the distance (in some sense) between objects: the smaller the distance, the greater the similarity.

When there is a lack of preliminary information in the context of data grouping, cluster analysis is a method belonging to the category of *unsupervised learning* methods. Sometimes it is mistakenly considered a classification method. However, it is only a method for data mining and visualisation of the data structure.

There are various algorithms for cluster analysis, differing in the way the clusters are found. Knowledge of clustering methods and their specific applications is key to the solution of data mining problems.

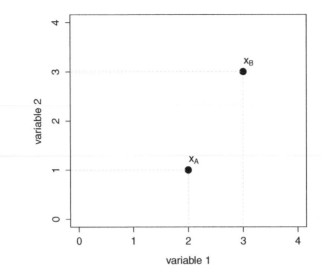

Figure 3.25   The location of objects $\mathbf{x}_A = (2, 1)$ and $\mathbf{x}_B = (3, 3)$ in a two-dimensional space $(p = 2)$.

3.6.1   Similarity measurements

In chemometrics, the similarity of objects or variables A and B is considered based on a measure of the distance between them in a multidimensional space. For measuring the distances between objects, the space is created by the variables and the objects become points (Figure 3.25). Even though the distance as a measure can be expressed in various ways, it must meet some obvious conditions for being an objective and reliable measure of similarity:

- the distance between A and B (d_{AB}) cannot be negative ($d_{AB} \geq 0$);
- the distance of A from itself is 0 ($d_{AA} = 0$);
- the distance between A and B is equal to the distance between B and A ($d_{AB} = d_{BA}$);
- the distance must satisfy the triangular inequality stating that if the distance between A and B is d_{AB}, between B and C d_{BC}, and between A and C d_{AC}, and d_{AB} is the longest of these, then $d_{AB} \leq d_{BC} + d_{AC}$.

Consequently, A and B are more similar than A and C if the distance between A and C is larger than between A and B: $d_{AB} < d_{AC}$.

Let $\mathbf{x}_A = (x_{A1}, \ldots, x_{Ap})$ and $\mathbf{x}_B = (x_{B1}, \ldots, x_{Bp})$ be two p-dimensional vectors describing the location of the object in the variable space. For example, in Figure 3.25, where $p = 2$, we have $(x_{A1}, x_{A2}) = (2, 1)$ and $(x_{B1}, x_{B2}) = (3, 3)$. The measures of distances between A and B frequently used in chemometrics are *Euclidean distance* and its square variation, *Manhattan distance*, and *Chebyshev distance* (Figure 3.26).

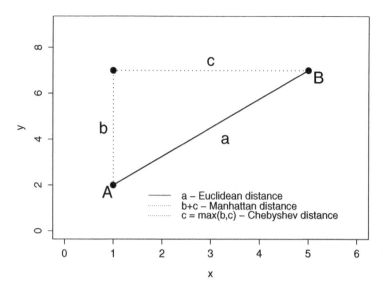

Figure 3.26 Geometrical definition of different measures of distance between A and B.

Euclidean distance is defined as a geometrical measure of distance between points in p-dimensional space:

$$d_{AB}^E = \sqrt{\sum_{i=1}^{p}(x_{Ai} - x_{Bi})^2}.$$

Consequently, the *square Euclidean distance* is given by

$$(d_{AB}^E)^2 = \sum_{i=1}^{p}(x_{Ai} - x_{Bi})^2.$$

The idea of the *Manhattan distance*, also referred to as *street distance* or *city-block distance*, stems from the grid arrangement of streets typical of the Manhattan borough of New York City. The feature characteristic of Manhattan streets is that they intersect at right angles. Thus, to move from point A to B a grid arrangement has to be followed so that the distance d_{AB}^M is given by

$$d_{AB}^M = \sum_{i=1}^{p}|x_{Ai} - x_{Bi}|.$$

The maximum value of two orthogonal (perpendicular; for orthogonal variables see also Section 3.7.1) distances composing the Manhattan distance between A and B is referred to as the *Chebyshev distance*:

$$d_{AB}^C = \max_{i=1}^{p}(|x_{Ai} - x_{Bi}|).$$

A commonly used (e.g. in SIMCA; Section 2.3.1) measure of distance between objects is the *Mahalanobis distance*. Its characteristic feature is that it takes into account the correlation between the variables describing the objects in p-dimensional space:

$$d_{AB} = \sqrt{(\mathbf{x}_A - \mathbf{x}_B)^T \mathbf{S}^{-1}(\mathbf{x}_A - \mathbf{x}_B)},$$

where $\mathbf{S}$ is the variance–covariance matrix (equation (3.5)), for the dataset under analysis.

Imagine a set of observations of p variables measured for m objects. The variables may be treated as vectors with length equal to the number of objects, m. In such a case vectors are viewed in a space determined by m objects. If the angle (a part of a plane that is formed by two rays with common endpoint) between the vectors is small, then the variables they represent are not distant and hence similar. The cosine of the angle between two vectors gives the correlation between them: $\cos \alpha = r$ (Section 3.3.4). Figure 3.27 shows the geometrical interpretation of the distance between variables in a two-dimensional space. If there are more than two objects, the relevant multidimensional spaces are created. However, the angle between two variables still lies on a plane.

One of the popular measures of distance between variables (e.g. X and Y) uses the tangent of the angle between the vectors of variables in a multidimensional space (Figure 3.27):

$$d_{XY} = \sqrt{\frac{1 - r_{XY}^2}{r_{XY}^2}}.$$

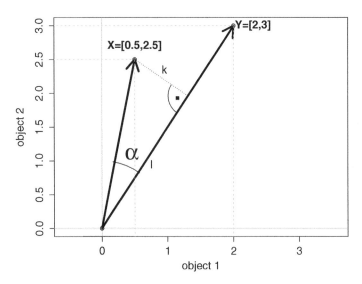

Figure 3.27 Geometrical presentation of the distance between variables in a space defined by two objects (α is the angle between the vectors of variables and k and l are introduced in equation (3.12)).

Because the correlation coefficient can be expressed by the cosine of the angle between the variables ($r = \cos(X, Y)$), we have (see also Figure 3.27):

$$d_{XY} = \sqrt{\frac{1 - \cos^2(X, Y)}{\cos^2(X, Y)}} = \sqrt{\frac{\sin^2(X, Y)}{\cos^2(X, Y)}}$$

$$= \sqrt{\tan^2(X, Y)} = |\tan(X, Y)| = \frac{k}{l}. \tag{3.12}$$

Other measures of distance between X and Y involving the correlation coefficient, r (Section 3.3.4), are:

- $\sqrt{2 - 2r_{XY}}$,

- $1 - r_{XY}$,

- $1 - |r_{XY}|$,

- $\sqrt{1 - r_{XY}^2}$.

The last two options must be used quite carefully as they do not take into account any distinction between positively and negatively correlated variables. For $d_{XY} = \sqrt{2 - 2r_{XY}}$ and $1 - r_{XY}$, it is true that variables are most distant when they are perfectly negatively correlated ($r = -1$) and most similar when they coincide ($r = 1$).

These measures of distances between variables are commonly used in a database search of drug profiling. More details in this topic can be found in Andersson *et al.* (2007) and Klemenc (2001) as well as in Section 2.3.1.

3.6.2 Hierarchical cluster analysis

One of the most popular ways of grouping objects is *hierarchical cluster analysis*, which presents the clusters in the form of a *dendrogram* (Figure 3.28). The vertical axis of a dendrogram illustrates the distances between objects and clusters (e.g. clusters of objects p_1 and p_3 or w_1 and w_3 in Figure 3.28), which are placed along the horizontal axis. The dendrogram is formed based on a distance matrix. As previously mentioned, there are many types of distances between objects (Section 3.6.1), as well as linkage methods designed for computing the distances between clusters:

- The *nearest neighbour method* uses the distance between the two nearest objects belonging to separate clusters (Figure 3.29 (left)).

- The *furthest neighbour method* uses the distance between the two furthest objects belonging to separate clusters (Figure 3.29 (right)).

- *Averaged linkage clustering method* calculates the mean distance between all possible pairs of objects belonging to separate clusters.

- *Ward's method* computes the distance between clusters as a minimum of within-cluster variance. The two clusters that are characterised by the smallest variance between their objects are merged together to build a new cluster.

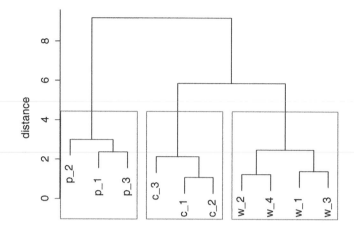

Figure 3.28 The dendrogram for glass data including car windows (c), building windows (w), and package glass (p). The distance measure is Euclidean, and the clustering method is Ward's algorithm. All data are available in `database_CA.txt`.

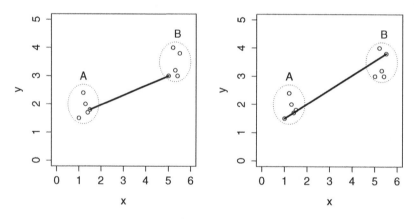

Figure 3.29 The nearest (left) and furthest neighbour (right) clustering methods (A and B are clusters comprising 5 objects each).

Example

Cluster analysis was employed for the visual presentation of glass objects of three different kinds such as car windows (c; 3 objects), building windows (w; 4 objects), and package glass (p; 3 objects). The data used for the cluster analysis are available in `database_CA.txt` and shown in Table 3.4. All objects were described by six variables referring to the Na, Mg, Al, Si, K, and Ca content (Table 3.4). The distance between objects was calculated using Euclidean measure after autoscaling the data (Section 3.2),

Table 3.4 SEM-EDX data on glass fragments used for performing the cluster analysis available in `database_CA.txt` (c, car windows; w, building windows; p, package glass).

Item	logNaO	logMgO	logAlO	logSiO	logKO	logCaO
c 1	−0.677	−1.302	−2.340	−0.159	−2.478	−0.929
c 2	−0.697	−1.225	−2.501	−0.144	−2.422	−0.928
c 3	−0.669	−1.195	−2.579	−0.109	−2.483	−0.880
p 1	−0.761	−1.734	−1.796	−0.250	−2.108	−0.915
p 2	−0.737	−2.266	−1.753	−0.178	−2.166	−0.887
p 3	−0.718	−1.665	−1.875	−0.228	−2.283	−0.965
w 1	−0.719	−1.349	−2.213	−0.171	−2.514	−0.916
w 2	−0.704	−1.355	−2.021	−0.228	−2.455	−1.001
w 3	−0.707	−1.364	−2.038	−0.202	−2.470	−0.953
w 4	−0.710	−1.340	−2.125	−0.208	−2.626	−0.993

Table 3.5 Euclidean distance matrix for glass objects belonging to three distinct categories (c, car windows; w, building windows; p, package glass). All data are available in `database_CA.txt`.

	c 1	c 2	c 3	p 1	p 2	p 3	w 1	w 2	w 3
c 2	1.08								
c 3	1.90	1.83							
p 1	4.99	4.90	6.22						
p 2	4.78	4.83	5.65	2.61					
p 3	3.31	3.54	4.98	2.37	3.08				
w 1	1.72	1.63	2.88	3.93	4.00	2.74			
w 2	2.82	3.17	4.70	3.97	4.73	1.83	2.64		
w 3	1.95	2.28	3.70	3.60	4.02	1.76	1.40	1.34	
w 4	2.59	2.89	4.32	4.59	5.08	2.64	2.22	1.24	1.42

and the distance matrix is presented in Table 3.5. For example, the distance between objects `p_1` and `w_1` is

$$d_{p_1, w_1} = [(-0.761 - (-0.719))^2 + (-1.734 - (-1.349))^2 + (-1.796 - (-2.213))^2$$
$$+ (-0.250 - (-0.171))^2 + (-2.108 - (-2.514))^2 + (-0.915 - (-0.916))^2]^{\frac{1}{2}}$$
$$= 3.93.$$

The linkage method for clustering is Ward's algorithm, which is based on minimal variance of distances between objects. The dendrogram obtained in the course of the cluster analysis performed using the **R** code (Appendix D) given below (available also in the file CA_code.R) is presented in Figure 3.28. It reveals the inner structure of the data, successfully dividing the glass objects into three distinct categories.

```
> data = read.table("database_CA.txt", header=TRUE)

> auto = scale(data[,2:7]) ## autoscaling the original data
> rownames(auto) = data[,1]

> distance.matrix = dist(auto, method = "euclidean") ## distance matrix

> CA=hclust(distance.matrix, method="ward") # Ward Hierarchical Clustering
> plot(CA, main="", sub="", xlab="", ylab="distance") ## displaying a dendrogram

> clusters = cutree(CA, k=3) ## cutting the dendrogram into 3 clusters
> rect.hclust(CA, k=3, border="darkviolet") ## drawing the dendrogram with violet
## borders around the 3 clusters
```

3.7 Dimensionality reduction

Dealing with multivariate data may be problematic, especially when the data are searched for their similarities or differences, since visualisation of the full dataset in many dimensions is not feasible. A limitation of multivariate modelling is the lack of background data from which to estimate the parameters of the distributions assumed such as means, variances, and covariances. For example, if objects are described by p variables, then it is necessary to reliably estimate up to p means, p variances, and $p \cdot (p - 1)/2$ covariances in the LR model (Chapters 4 and 5) if it is desired to use all p variables in one calculation run in the form:

$$LR = \frac{f(X_1, X_2, \ldots, X_p \mid H_1)}{f(X_1, X_2, \ldots, X_p \mid H_2)}.$$

Thus this process requires far more data than are usually available in many forensic databases. This effect has been dubbed the *curse of dimensionality*.

The simplest way around this problem is to assume independence of the variables considered. Then, if one wishes to use all p variables, the LR formula takes the form

$$LR = \frac{(X_1|H_1)}{(X_1|H_2)} \cdot \frac{(X_2|H_1)}{(X_2|H_2)} \cdot \cdots \cdot \frac{(X_p|H_1)}{(X_p|H_2)} = LR_{X_1} \cdot LR_{X_2} \cdot \cdots \cdot LR_{X_p}.$$

However, this is, in most of cases, a naïve assumption as by nature the features measured are correlated, especially those pertaining to the main components of the materials analysed. Where feasible, models should take this correlation into account.

3.7.1   Principal component analysis

All projection methods have been developed and especially designed for the purpose of finding patterns describing multivariate data. Their aim is to search for data projections which enable interesting data features to be revealed. At this point *principal component analysis* (PCA) is introduced as a method of projection which aims to maximise the variation of the data by means of a relevant rotation of the coordinate system and to reduce the dimensionality. Such an operation results in creating uncorrelated variables, called *principal components*, from the original set of correlated variables.

In general, in PCA the matrix containing the data for m objects described by p variables ($\mathbf{X}_{[m,p]}$) is presented as the sum of the product of matrices $\mathbf{T}_{[m,f]}$ and $\mathbf{P}_{[f,p]}$ and the residue matrix $\mathbf{E}_{[m,p]}$:

$$\mathbf{X}_{[m,p]} = \mathbf{T}_{[m,f]} \times \mathbf{P}_{[f,p]} + \mathbf{E}_{[m,p]}. \tag{3.13}$$

The elements of $\mathbf{T}_{[m,f]}$ give the coordinates of the objects (called *scores*), and $\mathbf{P}_{[f,p]}$ gives the variable vector coordinates (called *loadings*) in the newly created coordinate system of principal components.

PCA is the method of choice when the variables are highly correlated. Since one variable partly contains information that is simultaneously carried by other variables, some of them are to be treated as redundant. In such cases it is possible to use only a part of the data carrying all the relevant information. The higher the correlation coefficient (Section 3.3.4), the more effective the data dimensionality reduction is. PCA is concerned with the amount of information contained in the variables, with the aim of reducing the dimensionality. This information is expressed by variance, which may provide useful or insignificant information (e.g. measurement errors). The newly obtained variables (principal components) maximise the variation and are orthogonal (the vectors of the variables are perpendicular; Appendix B), which means they are uncorrelated and hence provide unique information about the whole dataset. So the reduction of data dimensionality, which is the most important advantage of PCA, may arise because of either a high correlation between variables or the fact that the data may contain some irrelevant information such as measurement errors.

Assume that the data are organised in the form of a matrix as presented in Figure 3.1, which contains values of p variables obtained in n measurements performed for each of m objects. Each row is related to a single measurement of p features describing the object analysed. Each column, therefore, contains information about a single variable determined in n measurements for m objects.

The first step for performing PCA is to transform the data by using an autoscaling procedure (Section 3.2) which produces variables centred at zero carrying an equal part of the overall variance (each has unit variance). Autoscaling (Figure 3.30) ensures that none of the variables dominates other variables with regard to the results of the data exploration, affecting principal component creation.

PCA involves a linear combination of the p variables, with the coefficients referred to as *weights* or *loadings*. The loadings are the elements of the eigenvectors of the variance–covariance (equation (3.5)) matrix obtained for the autoscaled data. Each eigenvector has a corresponding eigenvalue (Appendix B), which determines the variance explained by each principal component. Obtaining the coordinates of each of the data points in a newly

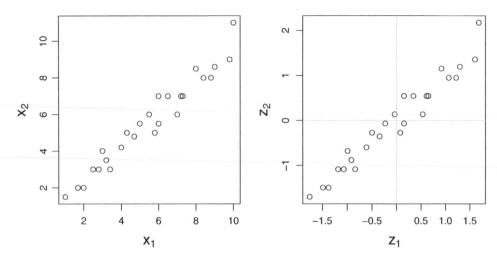

Figure 3.30 Data projection: (left) the original variable space x_1x_2 before autoscaling and (right) the autoscaled variable space z_1z_2 ($z_{ij} = (x_{ij} - \bar{x}_i)/s_i$). Note that two of the points overlap as they have the same coordinates. The data are available in `database_PCA.txt`.

formed coordinate system involves multiplying the matrix of autoscaled data by a matrix of loadings.

The crucial stage when performing the PCA is to choose the best number of principal components (PCs) carrying sufficient information (best explaining the variation) but at the same time enabling users maximally to reduce the dimensionality of the variable space. The idea of choosing only a few PCs for further analysis is based on the property of eigenvalues and eigenvectors, providing PCs in order of significance, which means each PC introduced in order describes less variation than the previous one. This enables the selection of the most informative PCs in terms of those with the greatest variance and a corresponding reduction of problem dimensionality arising from the removal of some repetitive or useless information contained in the PCs ignored.

Several criteria can be used to determine the appropriate number of PCs. The most widely used procedure takes into consideration only two or three components, sufficient for the visualisation of the data grouping in two- or three-dimensional space. However, such an approach in some cases may lead to false or incomplete interpretation, since two or three components may still not be enough to reliably describe the dataset.

A commonly used criterion assumes that the significant PCs are those characterised by a variance of at least 1. This approach is mainly due to the data transformation which produced variables characterised by unit variance. Therefore, this criterion favours PCs that provide more information about the data than the variables from which they are created.

Another widely used method for choosing the number of PCs is to ignore all the PCs that do not produce a substantial increase in the explained variance. This is a visual criterion. The meaning of *substantial* is specific to the data under analysis. This criterion is therefore very subjective.

Another method is to include those PCs which together contribute a given proportion (commonly 90% or 95%) of the cumulative variance.

Example

Let us consider an example of PCA carried out on a small dataset involving two variables, x_1 and x_2, describing some features of 27 objects (data available in database_PCA.txt). The two variables are highly correlated (Figure 3.30), with a correlation coefficient of 0.97. Such a high correlation indicates that nearly all the significant information in the data is accounted for one of the variables. Therefore, there is considerable redundancy of information and PCA will prove helpful in the effective reduction of data space dimensionality.

The variance–covariance matrix (equation (3.5)) for the autoscaled data (the mean is set to 0 and the standard deviation to 1; Figure 3.30) is given by

$$\begin{bmatrix} s_{x_1}^2 & cov(x_1, x_2) \\ cov(x_2, x_1) & s_{x_2}^2 \end{bmatrix} = \begin{bmatrix} 1 & 0.97 \\ 0.97 & 1 \end{bmatrix}.$$

The next step in PCA involves retrieving eigenvectors and eigenvalues from the variance–covariance matrix. The eigenvectors are $(0.707, 0.707)$ and $(-0.707, 0.707)$ and their corresponding eigenvalues are 1.97 and 0.025, respectively. Note that the eigenvalue associated with the first eigenvector is two orders of magnitude greater than that for the second vector. This means that first PC explains nearly all the variance (in fact the proportion explained is $\frac{1.97}{(1.97+0.025)} = 0.988$).

The PCA rotates the original coordinate system so that the new coordinates maximise the variance contained in the data. The vectors (PC1 and PC2) in Figure 3.31 determine the

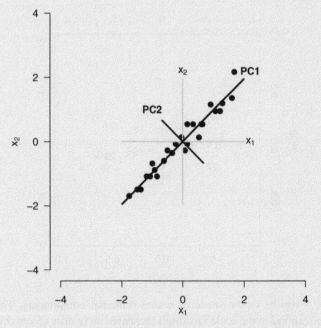

Figure 3.31   Rotating the coordinate system in PCA.

location of the variable vectors in the new two-dimensional space. A summary of the analysis is given in Table 3.6.

Table 3.6 Summary of PCA performed for the small dataset described by two variables. The data are available in database_PCA.txt.

	PC1	PC2
Standard deviation	1.405	0.158
Proportion of variance	0.988	0.012
Cumulative proportion	0.988	1.000

In order to obtain the coordinates of the objects analysed in the new space of PCs, it is necessary to multiply the matrix of autoscaled data by the matrix of loadings (elements of eigenvectors). The projection of all objects on new PCs (as new variables) is shown in Figure 3.32. The figure presents the scores of the data points on PC1 and PC2, obtained by multiplying the matrix of autoscaled variables by the matrix of loadings. The second set of information presented in the biplot illustrates the vectors of the original variables in the space of the first two PC.

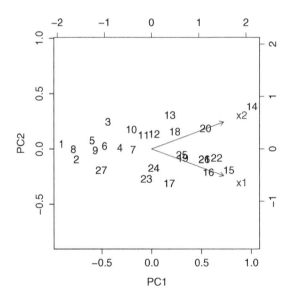

Figure 3.32 PCA results in the space of scaled principal components. The bottom and left-hand axes correspond to the scaled loadings presented as vectors of variables and the top and right-hand axes correspond to the scaled scores of data points denoted by the numbers corresponding to the object labels. The data are available in database_PCA.txt.

Example

This example gives the results of a PCA carried out on a set of 200 glass objects for which elemental composition (Section 1.2.1) was determined (see `glass_data.txt`) in order to illustrate how effectively data dimensionality can be reduced.

Seven variables were given in the form of log ratios of each element to oxygen concentration, denoted, for example, by *logNaO*. An inspection of Table 3.1 leads to the conclusion that most of the variables have significant correlation. Therefore, the use of PCA to effectively reduce the data dimensionality is clearly justified.

The PCA can be performed in **R** (Appendix D) by running the following code, available in the file `PCA_code.R`:

```
> data = read.table("glass_data.txt", header = TRUE)

> pca = prcomp(data[,2:8], scale.=TRUE) ## performing
    the PCA algorithm for the autoscaled variables

> sd = pca$sdev ## retrieving the standard deviations of PCs
> print(sd)

> loadings = pca$rotation ## retrieving the loadings (weights)
> print(loadings)

> scores = pca$x ## retrieving the scores for each data point
> print(scores)

> biplot(pca, cex=c(2,0.7), cex.lab=1, xlabs=rep(".",
    times=length(data [,1])), scale=0.5)

> summary(pca) ## providing the summary of PCA
```

The summary of the PCA for a set of 200 glass samples is given in Table 3.7 and graphically presented in Figure 3.33 in the two-dimensional space of the first two scaled principal components. Also presented are the scores of the data points on PC1 and PC2, obtained by multiplying the matrix of autoscaled variables by the matrix of loadings. The biplot also illustrates the vectors of the original variables in the space of the first two PCs. Running the **R** code also produces matrices of scores and loadings along with the PCA summary.

Table 3.7 Summary of PCA performed on a set of 200 glass samples. The data are available in `glass_data.txt`.

	PC1	PC2	PC3	PC4	PC5	PC6	PC7
Standard deviation	1.629	1.118	1.034	0.913	0.695	0.670	0.513
Proportion of variance	0.379	0.179	0.152	0.119	0.069	0.064	0.038
Cumulative proportion	0.379	0.558	0.710	0.829	0.898	0.962	1.000

Let us consider the methods mentioned above for choosing the appropriate number of principal components.

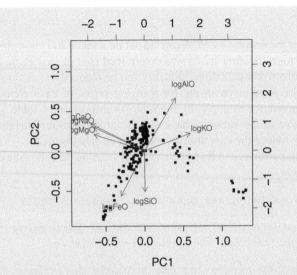

Figure 3.33 PCA performed on a set of 200 glass samples described by seven variables (data available in `glass_data.txt`). The bottom and left-hand axes correspond to the scaled loadings presented as vectors of variables and the top and right-hand axes correspond to the scaled scores of the data points.

The most widely used and least objective criterion is based on the assumption that PCs may be ignored as long as they do not produce a high increase in cumulative variance. Such an approach may be demonstrated in the proportion of variance plot (Figure 3.34). For the first four PCs there is a strong decrease in variance explained by each subsequent variable. The variance explained appears to stabilise for PCs starting from the fifth, thus PC5–PC7 may be ignored and the problem dimensionality reduces from 7 to 4.

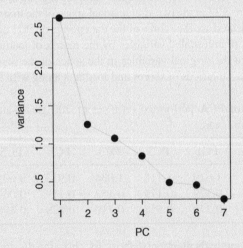

Figure 3.34 Variance explained by each PC for the data available in `glass_data.txt`.

One method involves comparing the variance that is explained by each PC to unity. In this case only the three first components PC1–PC3 complying with this rule (with standard deviations of 1.629, 1.118, and 1.034) are chosen. Thus the data dimensionality reduces from 7 to 3.

Another rule uses an arbitrarily set value of cumulative proportion of variance that the analyst deems sufficient. If this value is fixed at a commonly used level of 0.90 or 0.95, then 5 (or perhaps 6) PCs should be taken into account.

As can be seen from this example, the rules indicate various numbers of significant PCs. However, the final decision must be taken by the investigator with due regard to the type of data and what he or she wishes to find out by employing PCA for the given dataset.

Investigating the variables in the context of their discriminating abilities leads to the conclusion that $logAlO$, $logFeO$, and $logSiO$ are the variables with the greatest influence on the formation of PC2. This is because these vectors have the longest projections in the PC2 direction (> 0.5 in scale). However, $logKO$, $logCaO$, $logNaO$, and $logMgO$ mainly affect PC1 with scaled projections in the PC1 direction of length > 0.5 in scale. The $logSiO$ variable does not influence PC1, but has a very significant contribution to the formation of PC2 as this vector is parallel to the PC2 direction. An angle between variables close to $0°$ in a PCA plot indicates that they are highly positively correlated. An angle close to $180°$ means that the correlation is negative. In this example it can be stated that $logCaO$, $logNaO$, and $logMgO$ are highly positively correlated, while $logFeO$ and $logAlO$ are negatively correlated.

The application of PCA to obtain orthogonal variables, which are further used for likelihood ratio calculations based on the lead (Pb) isotope ratios for glass data collected as a result of multicollector inductively coupled mass spectrometry (MC-ICP-MS), is presented in Martyna et al. (2013).

3.7.2   Graphical models

Graphical models are a dimension reduction technique based on graph theory (Whittaker 1990). In a graph, each variable is represented as a node. Variables (nodes) which are directly associated are joined by a line (edge). The graphical model (GM) is constructed by the sequential addition of edges created by inspection of the partial correlation coefficient matrix, which is obtained by inverting the between-object variance–covariance matrix $\mathbf{C}$. The inverse matrix $(\mathbf{C}^{-1})$ is scaled in such a way that the diagonal terms are equal to 1, and the off-diagonal terms are the correlation coefficients. An **R** function for computing such a matrix is given below or is available in the relevant folder on the website (scaled_inverse_cov_matrix.R). The code returns the scaled inverse variance–covariance matrix for the data provided in the file SEMEDX_glass_database.txt.

Let a_{ij} be element in the ith row and jth column. The scaling is then implemented by setting the entry b_{ij} in the scaled matrix as

$$b_{ij} = \frac{a_{ij}}{\sqrt{a_{ii}a_{jj}}};$$

thus for $i = j$, $b_{ij} = 1$. In **R**:

```
population=read.table("SEMEDX_glass_database.txt", header=TRUE)
variables = c(4:10)
p = length(variables)
n = length(unique(population$Piece))

source("UC_comparison_calculations.R")
results.UC = UC(population, variables, p, n)
C = results.UC$C

scaled.matrix = matrix(0,nrow=p, ncol=p)
rownames(scaled.matrix) = rownames(C)
colnames(scaled.matrix) = colnames(C)
for (i in 1:p)
{
 var.i = solve(C)[i,i]
 sd.i = sqrt(var.i)

 for (j in 1:p)
 {

  cov = solve(C)[i,j]
  var.j = solve(C)[j,j]
  sd.j = sqrt(var.j)

  scaled.matrix[i,j] = cov/(sd.i*sd.j)

 }
}
print(round(scaled.matrix,3))
```

The output is as follows:

	logNaO	logMgO	logAlO	logSiO	logKO	logCaO	logFeO
logNaO	1.000	-0.134	-0.186	-0.238	0.216	-0.592	-0.026
logMgO	-0.134	1.000	-0.008	0.157	0.200	-0.297	-0.179
logAlO	-0.186	-0.008	1.000	0.115	-0.475	0.108	0.219
logSiO	-0.238	0.157	0.115	1.000	-0.091	0.131	-0.145
logKO	0.216	0.200	-0.475	-0.091	1.000	-0.002	-0.058
logCaO	-0.592	-0.297	0.108	0.131	-0.002	1.000	0.015
logFeO	-0.026	-0.179	0.219	-0.145	-0.058	0.015	1.000

By way of illustration, let us consider the three-variable problem for which the values of partial correlation coefficients were obtained as in Table 3.8. First, the partial correlation of

Table 3.8 Correlation coefficients for variables X_1, X_2, and X_3.

	X_1	X_2	X_3
X_1	1	−0.1	0.6
X_2		1	0.8
X_3			1

largest magnitude is selected, and an edge is added between the two nodes connected by this partial correlation. In this case this is $|0.8|$ for X_2 and X_3 (Figure 3.35(a)). This process is repeated until all nodes are part of the graphical model (Figure 3.35) or the model cannot be factorised (Whittaker 1990).

The factorisation of the p-dimensional density (here $p = 3$) is given by

$$f(C_i|S_i) = \frac{f(C_i)}{f(S_i)}, \tag{3.14}$$

where C_i is the ith clique in the model, that is, a subset of variables in which all the nodes are connected to each other and known as a complete subgraph. The set of all separators for the ith clique is denoted by S_i, that is, a node or a set of nodes at the intersection of two cliques. In

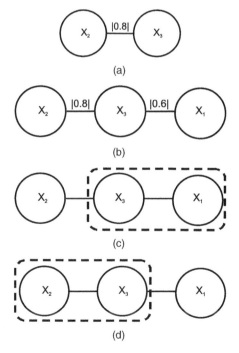

Figure 3.35 Creating a graphical model. Details are provided in the text.

the case of the simple model presented in Figures 3.35(c),(d) there are two cliques (X_1, X_3) and (X_2, X_3), and one separator, the node X_3. Thus, the following factorisation is obtained:

$$f(C_i|S_i) = \frac{f(X_1, X_3)f(X_2, X_3)}{f(X_3)}.$$

This factorisation of the density corresponds to the following factorisation for the LR models:

$$LR = \frac{LR(X_1, X_3)LR(X_2, X_3)}{LR(X_3)},$$

where, for example, $LR(X_1, X_3)$ is the value of the likelihood ratio calculated based on multivariate densities with two variables (X_1, X_3) by application of the LR models described in Chapters 4 and 5.

In the case of more sophisticated graphical models the following algorithm should be used in order to find a set chain and the factorisation of the model (Aitken *et al.* 2007):

1. Select a node arbitrarily from the model graph and denote this as the lowest numbered node.

2. Number each remaining node in turn ordered by the number of edges linking it to any other already numbered node; break ties arbitrarily.

3. Assign a rank to each clique based upon the highest numbered node in the clique; if two cliques share the highest numbered node then rank arbitrarily between the two nodes.

Given the cliques for the model and a suitable set chain, the set of separators (S_i in equation (3.14)) for each clique is found. The first clique in the set chain is always a complete subgraph and there are no separator sets. After that the next clique presented in the set chain is added to the model. The intersection of elements between these two cliques becomes the first separator set. The process is continued until all cliques are joined to the model.

An example (Zieba-Palus *et al.* 2008) is given of the construction of a GM for the data obtained during the analysis of 36 paint samples by Py-GC/MS (Section 1.4). The graphical model was selected by the sequential addition of edges determined by inspection of the negative partial correlation coefficients presented in Table 3.9.

First, the partial correlation of largest magnitude was selected. This was $|0.627|$ between $M2P$ and $M2E$, and an edge was added between these two nodes. Then, the second largest magnitude partial correlation was selected, $|0.607|$, and an edge added joining the corresponding nodes, $M2P$ and BMA, to the graph. This process was repeated until all nodes were part of the model (Figure 3.36).

The model presented in Figure 3.36(a) shows the final graph obtained after addition of all variables subjected to factorisation. First of all, the cliques were determined as described above from the graphical model. The clique $(M2P, BMA)$ has the highest numbered node BMA, so the clique $(i = 1)$ is given the same number as BMA, which is 7. A clique of variables TOL and MST $(i = 2)$ has the next highest numbered node TOL, and so has the number 6. Putting these into numerical order, the set chain is obtained in Table 3.10 (first column). The results of the application of the above algorithm are presented in Figure 3.36(b).

Table 3.9 Partial correlation matrix for the seven variables describing the car paint samples. Only the upper right triangle of the matrix is given as the lower left triangle is given by symmetry. The variables are the logarithm to base 10 of the ratio of the peak areas of seven organic compounds to the peak area of styrene: methylmethacrylate (*MMA*); toluene (*TOL*); butylmethacrylate (*BMA*); methylstyrene (*MST*); 2-hydroxyethylmethacrylate (*M2E*); 2-hydroxypropylmethacrylate (*M2P*); 1,6-diisocyanatehexane (*I*16).

	MMA	*TOL*	*BMA*	*MST*	*M2E*	*M2P*	*I*16
MMA	1.000	−0.085	−0.266	−0.091	−0.422	−0.216	0.106
TOL		1.000	−0.093	0.364	0.260	0.074	−0.164
BMA			1.000	0.130	0.105	0.607	−0.199
MST				1.000	−0.076	−0.139	0.295
M2E					1.000	−0.627	0.239
M2P						1.000	0.521
*I*16							1.000

Given the cliques for the model, and a suitable set chain, the set of separators (S_i for each clique) was found. The first clique in the set chain is (*M2P*, *M2E*). This is a complete subgraph and at the moment there are no other cliques added to the graph. Therefore, there can be no separator sets. The next clique in the set chain is (*M2E*, *MMA*) and it is added to the model. The intersection of elements between these two cliques is (*M2E*), and so this becomes the first separator set. The running union of the first two sets is now (*M2P*, *M2E*, *MMA*).

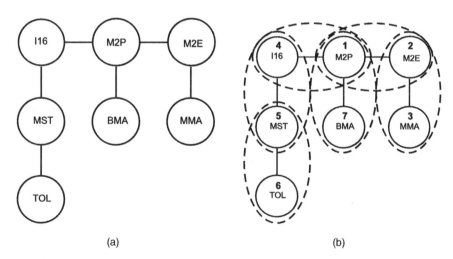

(a)   (b)

Figure 3.36 Results of construction of the undirected graphical model: (a) a model obtained on the basis of the rescaled inverse of the variance–covariance matrix for the seven variables from the 36 car paint objects (Table 3.9); (b) model (a) with marked cliques and numbers assigned to each node. (Reproduced by permission of Elsevier.)

Table 3.10   Cliques, separators, and clique ordering for the graphical model for paint analysis. The clique ordering for those cliques suggested by the model is based upon the description in the text. The separator sets S_i are composed of those elements of each clique which also appear in the running union along the set chain at position R_{i-1}. The running union R_i is composed of the set of all elements from each clique in the set up to and including clique C_i. $\emptyset$ denotes the empty set.

i	Clique (C_i)	Running union (R_i)	Separator set (S_i)
1	$(M2P, M2E)$	$(M2P, M2E)$	$\emptyset$
2	$(M2E, MMA)$	$(M2P, M2E, MMA)$	$(M2E)$
3	$(I16, M2P)$	$(M2P, M2E, MMA, I16)$	$(M2P)$
4	$(MST, I16)$	$(M2P, M2E, MMA, I16, MST)$	$(I16)$
5	(TOL, MST)	$(M2P, M2E, MMA, I16, MST, TOL)$	(MST)
6	$(M2P, BMA)$	$(M2P, M2E, MMA, I16, MST, TOL, BMA)$	$(M2P)$

Working through the entire set chain one arrives at the factorisation

$$f(C_i|S_i) = \frac{f(M2P, M2E)f(M2E, MMA)f(I16, M2P)}{f(M2P)f(M2E)}$$
$$\times \frac{f(MST, I16)f(TOL, MST)f(M2P, BMA)}{f(I16)f(M2P)f(MST)}, \qquad (3.15)$$

which could also be expressed as

$$f(C_i|S_i) = \frac{LR(M2P, M2E)LR(M2E, MMA)LR(I16, M2P)}{LR(M2P)LR(M2E)}$$
$$\times \frac{LR(MST, I16)LR(TOL, MST)LR(M2P, BMA)}{LR(I16)LR(M2P)LR(MST)}.$$

In general, it is suggested that graphical models should also make sense from a chemical point of view, as the graph presented in Figure 3.36 does. The nodes that create the (MST, TOL) clique represent compounds of polystyrene degradation, node $I16$ represents products of pyrolysis of urethane modifier, and other nodes represent products of acrylic components of the car paint.

Sometimes it is not possible to conclude which of several graphical models should be used. This is because they are based on a limited number of results. Therefore, a graphical model could be constructed taking into account general knowledge of relations between variables within a particular material (e.g. glass). For example, a graphical model was proposed for glass data obtained by application of SEM-EDX (Section 1.2.1). This so-called *standardised* graphical model was used with the aim of eliminating the influence of the application of various graphical models on the basis of the results obtained. It was constructed from the authors' experience regarding previous research and knowledge of the chemical composition of glass. Furthermore, it is a simplified version of a previously obtained graphical model for

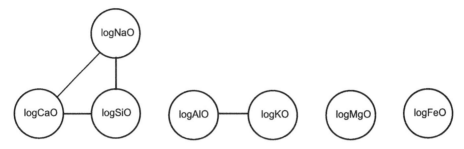

Figure 3.37 A so-called standardised graphical model used for evaluation of the evidential value of glass chemical composition obtained by SEM-EDX.

glass data (Aitken *et al.* 2007; Zadora *et al.* 2010; Zadora and Neocleous 2009). The model proposed in Figure 3.37 and calculated using the equation

$$LR = LR(logNaO, logSiO, logCaO)LR(logAlO, logKO)LR(logMgO)LR(logFeO) \qquad (3.16)$$

accounts for the main relationships between the most important chemical compounds in glass. The seven variables were divided into four cliques (*logNaO, logSiO, logCaO*), (*logAlO, logKO*), (*logMgO*), and (*logFeO*). In general, a strong correlation between variables (*logSiO, logCaO, logNaO*) corresponding to silicone, calcium, and sodium was observed. This is not surprising from a chemical point of view, as glass is formed by oxides of these elements. Also, a relatively high correlation between variables based on potassium and aluminium concentrations was observed. These variables represent compounds used in glass as additives, whose role is to improve the physical properties of glass samples (optical properties in the case of K_2O, and to avoid re-crystallisation in the case of Al_2O_3). A relatively low correlation of variables based on concentration of magnesium (*logMgO*) or iron (*logFeO*) with other variables was observed. Therefore, these variables were treated like independent variables. Magnesium oxide is used as a substitute for calcium oxide in glass and avoids the re-crystallisation process. Localisation of the *logFeO* node makes sense, too. Iron oxides are additives used in glass to produce colour (green or brown). For example, ordinary non-coloured window panes reveal a green shade in a cross-section. It is a cheap additive, being present in sand. Because a small quantity of iron oxides is sufficient to give glass a suitable colour, iron oxides are commonly used as colouring agents. Nevertheless, the concentration of iron in non-coloured glass objects is below the detection limits of the SEM-EDX method. Iron is only detected by this analytical method for coloured (green or brown) glass objects.

Other examples of the application of the graphical model approach for dimension reduction of physicochemical data can be found in Zadora and Neocleous (2009) and Zadora *et al.* (2010), and in Section 4.4.6.

References

Adam C 2010 *Essential Mathematics and Statistics for Forensic Science*. John Wiley & Sons, Ltd., Chichester.

Aitken CGG and Taroni F 2004 *Statistics and the Evaluation of Evidence for Forensic Scientists*. John Wiley & Sons, Ltd, Chichester.

Aitken CGG, Zadora G, Lucy D 2007 A two-level model for evidence evaluation. *Journal of Forensic Sciences* **52**, 412–419.

Andersson K, Lock E, Jalava K, Huizer H, Jonson S, Kaa E, Lopes A, Poortman-van der Meer A, Sippola E, Dujourdy L, Dahlen J 2007 Development of a harmonized method for the profiling of amphetamines VI. Evaluation of methods for comparison of amphetamine. *Forensic Science International* **169**, 86–99.

Bozza S, Taroni F, Marquis R, Schmittbuhl M 2008 Probabilistic evaluation of handwriting evidence: Likelihood ratio for authorship. *Applied Statistics* **57**, 329–341.

Curran JM 2011 *Introduction to Data Analysis with R for Forensic Scientists*. CRC Press, Boca Raton, FL.

Evans M, Hastings N, Peacock B 2000 *Statistical Distributions*, 3rd edn. John Wiley & Sons, Inc., New York.

Kellner R, Mermet JM, Otto M, Valcarcel M, Widmer HM (eds) 2004 *Analytical Chemistry: A Modern Approach to Analytical Science*. 2nd edn. Wiley VCH, Wienheim, Germany.

Klemenc S 2001 In common batch searching of illicit heroin samples – evaluation of data by chemometric methods. *Forensic Science International* **115**, 43–52.

Lucy D 2005 *Introduction to Statistics for Forensic Scientists*. John Wiley & Sons, Ltd, Chichester.

Martyna A, Sjastad K-E, Zadora G, Ramos D 2013 Analysis of lead isotopic ratios of glass objects with the aim of comparing them for forensic purposes. *Talanta* **105**, 158–166.

Otto M 2007 *Statistics and Computer Application in Analytical Chemistry*. Wiley VCH, Weinheim, Germany.

Silverman BW 1986 *Density Estimation for Statistics and Data Analysis*. Chapman & Hall, London.

Spence LD, Baker AT, Byrne JP 2000 Characterization of document paper using elemental compositions determined by inductively coupled plasma mass spectrometry. *Journal of Analytical Atomic Spectroscopy* **15**, 813–819.

Stephenson G 1973 *Mathematical Methods for Science Students* (2nd edn). Longman, London.

Whittaker J 1990 *Graphical Models in Applied Multivariate Statistics*. John Wiley & Sons, Ltd, Chichester.

Zadora G, Neocleous T 2009 Likelihood ratio model for classification of forensic evidence. *Analytica Chimica Acta* **64**, 266–278.

Zadora G, Wilk D, 2009 Evaluation of evidence value of refractive index measured before and after annealing of container and float glass fragments. *Problems of Forensic Sciences* **80**, 365–377.

Zadora G, Neocleous T, Aitken CGG 2010 A two-level model for evidence evaluation in the presence of zeros. *Journal of Forensic Sciences* **55**, 371–384.

Zieba-Palus J, Zadora G, Milczarek JM 2008 Differentiation and evaluation of evidence value of styrene acrylic urethane topcoat car paints analysed by pyrolysis-gas chromatography. *Journal of Chromatography A* **1179**, 47–58.

4

Likelihood ratio models for comparison problems

4.1 Introduction

The comparison of physicochemical data (mostly of continuous type; Chapter 1) describing objects subjected to forensic analysis (e.g. glass fragments) is one of the most commonly encountered problems in forensic science. Data obtained from analysis of material obtained, for example, from a suspect's clothes (the recovered sample) and from a crime scene (the control sample) may be compared. Recovered material is so called because its source is not known and it has been recovered from somewhere. Control material is so called because its source is known.

The value of evidence (E) in the form of physicochemical data can be assessed by means of the likelihood ratio (LR) approach (Section 2.2.2). This makes it possible to compare data obtained from two fragments in the context of two contrasting hypotheses. The first, termed the prosecutor's hypothesis (H_1) is the proposition that the two fragments come from the same source, while the second, termed the defence hypothesis, (H_2), is the proposition that the fragments have different sources. The general LR expressions are as follows: $LR = \frac{P(E|H_1)}{P(E|H_2)}$ in the case of discrete data and $LR = \frac{f(E|H_1)}{f(E|H_2)}$ in the case of continuous data (where $P(\cdot)$ denotes probability and $f(\cdot)$ the probability density function; Appendix A).

It is worth recalling that the evaluation of the evidence in the form of physicochemical data, within a comparison problem, requires taking into account the following aspects:

- similarities between the data for the objects compared (e.g. glass fragments);

- possible sources of uncertainty, which include variations in the measurements of characteristics within recovered and control objects, and variations in the measurements of characteristics between various objects in the relevant population;

Statistical Analysis in Forensic Science: Evidential Value of Multivariate Physicochemical Data, First Edition.
Grzegorz Zadora, Agnieszka Martyna, Daniel Ramos and Colin Aitken.
© 2014 John Wiley & Sons, Ltd. Published 2014 by John Wiley & Sons, Ltd.
Companion website: www.wiley.com/go/physicochemical

- information about the rarity of the physicochemical data;

- existing correlation between variables in the case of multidimensional data.

In this chapter *LR* models meeting all the aforementioned requirements are introduced.

Information on the within- (**U**) and between-object (**C**) variability can be assessed on the basis of background information obtained from a suitable database. Details of how to calculate these matrices can be found in Section 3.3.2. With the LR models presented in this chapter the within-object distribution is in all cases assumed normal. The between-object distribution is assumed normal or is estimated by the kernel density estimation procedure using Gaussian kernels (Section 3.3.5). Information on the analysis of correlation between variables (in the case of multidimensional data and related to an appropriate reduction in dimensionality) can be found in Section 3.7.

Data obtained from the analysis of m objects, each measured n times, will take the form of p-dimensional vectors, where p is the number of variables of interest. Such a vector may be written as $\mathbf{x}_{ij} = \left(x_{ij1}, \ldots, x_{ijp}\right)^T$, where $i = 1, \ldots, m$ and $j = 1, \ldots, n$ (see, for example, the file `glass_database.txt`;[1] Figure 4.20; Appendix B).

Moreover, suppose that there are two datasets (control and recovered) with n_1 and n_2 measurements and a comparison between them is required. In this case an appropriate notation is

$$\mathbf{y}_{lj} = \left(y_{lj1}, \ldots, y_{ljp}\right)^T,$$

where $l = 1, 2$ (for the control and recovered samples, respectively) and $j = 1, \ldots, n_l$, with n_l being number of measurements performed on the control or recovered objects.

The expressions for uni- and multivariate LR computations are given in Sections 4.2 and 4.3 depending on whether the between-object data distribution is assumed normal or is estimated by the KDE procedure (Section 3.3.5).

4.2   Normal between-object distribution

Denote the mean vector for the ith object by $\boldsymbol{\theta}_i$ (estimated by $\bar{\mathbf{x}}_i = \frac{1}{n}\sum_{j=1}^{n}\mathbf{x}_{ij}$), the within-object covariance matrix by $\mathbf{U}$, and the between-object covariance matrix by $\mathbf{C}$. Estimates of $\mathbf{U}$ and $\mathbf{C}$ are given by equations (3.3) and (3.4).

Then, given $\boldsymbol{\theta}_i$ and $\mathbf{U}$, the distribution of the observations for the ith object $\mathbf{X}_{ij}$ is

$$\left(\mathbf{X}_{ij} \mid \boldsymbol{\theta}_i, \mathbf{U}\right) \sim N_p(\boldsymbol{\theta}_i, \mathbf{U}), \quad i = 1, \ldots, m, \; j = 1, \ldots, n.$$

It is also assumed that

$$(\boldsymbol{\theta}_i \mid \boldsymbol{\mu}, \mathbf{C}) \sim N_p(\boldsymbol{\mu}, \mathbf{C}), \quad i = 1, \ldots, m,$$

where $\boldsymbol{\mu}$ is a vector of the overall means of p variables estimated using n measurements for m objects from the database and is estimated by

$$\bar{\mathbf{x}} = \frac{1}{mn}\sum_{i=1}^{m}\sum_{j=1}^{n}\mathbf{x}_{ij}.$$

---

[1] All the files are available from www.wiley.com/go/physicochemical

4.2.1   Multivariate data

The LR model, in general, can be expressed as (Aitken and Lucy 2004; Neocleous *et al.* 2011; Zadora 2010)

$$LR = \frac{f(\bar{\mathbf{y}}_1, \bar{\mathbf{y}}_2 | \mathbf{U}, \mathbf{C}, \bar{\mathbf{x}}, H_1)}{f(\bar{\mathbf{y}}_1, \bar{\mathbf{y}}_2 | \mathbf{U}, \mathbf{C}, \bar{\mathbf{x}}, H_2)},$$

where $\bar{\mathbf{y}}_1$ and $\bar{\mathbf{y}}_2$ are vectors of the means of p variables calculated using n_1 and n_2 measurements for control and recovered objects, respectively:

$$\bar{\mathbf{y}}_l = \frac{1}{n_l} \sum_{j=1}^{n_l} \mathbf{y}_{lj}, \quad l = 1, 2.$$

In the numerator of the likelihood rato it is assumed that the materials compared originate from the same object (H_1). Thus, it can be shown (Aitken and Lucy 2004) that the numerator can be expressed as explicitly taking into account the evaluation of the difference in the physicochemical data between objects $\bar{\mathbf{y}}_1 - \bar{\mathbf{y}}_2$, as well as their rarity, the latter being expressed by the distance of the μ from the weighted mean $\bar{\mathbf{y}}^*$:

$$\bar{\mathbf{y}}^* = \frac{n_1 \bar{\mathbf{y}}_1 + n_2 \bar{\mathbf{y}}_2}{n_1 + n_2}.$$

Therefore, the numerator can be expressed as

$$f(\bar{\mathbf{y}}_1, \bar{\mathbf{y}}_2 | \mathbf{U}, \mathbf{C}, \bar{\mathbf{x}}, H_1) = f(\bar{\mathbf{y}}_1 - \bar{\mathbf{y}}_2, \bar{\mathbf{y}}^* | \mathbf{U}, \mathbf{C}, \bar{\mathbf{x}}, H_1)$$

$$= (2\pi)^{-p} \left| \frac{\mathbf{U}}{n_1} + \frac{\mathbf{U}}{n_2} \right|^{-\frac{1}{2}} \exp\left\{ -\frac{1}{2} (\bar{\mathbf{y}}_1 - \bar{\mathbf{y}}_2)^T \left(\frac{\mathbf{U}}{n_1} + \frac{\mathbf{U}}{n_2} \right)^{-1} (\bar{\mathbf{y}}_1 - \bar{\mathbf{y}}_2) \right\}$$

$$\times \left| \frac{\mathbf{U}}{n_1 + n_2} + \mathbf{C} \right|^{-\frac{1}{2}} \exp\left\{ -\frac{1}{2} (\bar{\mathbf{y}}^* - \bar{\mathbf{x}})^T \left(\frac{\mathbf{U}}{n_1 + n_2} + \mathbf{C} \right)^{-1} (\bar{\mathbf{y}}^* - \bar{\mathbf{x}}) \right\}.$$

$$(4.1)$$

In the denominator of the likelihood ratio, $\bar{\mathbf{y}}_1$ and $\bar{\mathbf{y}}_2$ are taken to be independent as the data are assumed to originate from different objects (H_2). Thus

$$f(\bar{\mathbf{y}}_1, \bar{\mathbf{y}}_2 | \mathbf{U}, \mathbf{C}, \bar{\mathbf{x}}, H_2) = f(\bar{\mathbf{y}}_1 | \mathbf{U}, \mathbf{C}, \bar{\mathbf{x}}, H_2) f(\bar{\mathbf{y}}_2 | \mathbf{U}, \mathbf{C}, \bar{\mathbf{x}}, H_2)$$

$$= (2\pi)^{-p} \left| \frac{\mathbf{U}}{n_1} + \mathbf{C} \right|^{-\frac{1}{2}} \exp\left\{ -\frac{1}{2} (\bar{\mathbf{y}}_1 - \bar{\mathbf{x}})^T \left(\frac{\mathbf{U}}{n_1} + \mathbf{C} \right)^{-1} (\bar{\mathbf{y}}_1 - \bar{\mathbf{x}}) \right\}$$

$$\times \left| \frac{\mathbf{U}}{n_2} + \mathbf{C} \right|^{-\frac{1}{2}} \exp\left\{ -\frac{1}{2} (\bar{\mathbf{y}}_2 - \bar{\mathbf{x}})^T \left(\frac{\mathbf{U}}{n_2} + \mathbf{C} \right)^{-1} (\bar{\mathbf{y}}_2 - \bar{\mathbf{x}}) \right\}.$$

$$(4.2)$$

This analysis assumes **U** and **C** fixed for all objects. The model proposed by Bozza *et al.* (2008) removes this restriction and is suitable for unbalanced data. Other models can also be found in Aitken and Lucy (2004).

4.2.2 Univariate data

In case of univariate data ($p = 1$), the covariance matrices become scalar variances (e.g. **U** becomes u^2 and **C** becomes c^2) and the vectors become scalars (e.g. $\bar{\mathbf{x}}$ becomes $\bar{x}$). Thus, the following equations should be used (Aitken and Lucy 2004; Neocleous *et al.* 2011; Zadora 2010): for the numerator,

$$f(\bar{y}_1, \bar{y}_2 | u^2, c^2, \bar{x}, H_1) = f(\bar{y}_1 - \bar{y}_2, \bar{y}^* | u^2, c^2, \bar{x}, H_1)$$

$$= \frac{1}{\sqrt{2\pi u_0^2}} \exp\left\{-\frac{(\bar{y}_1 - \bar{y}_2)^2}{2u_0^2}\right\} \times \frac{1}{\sqrt{2\pi \left(\frac{u^2}{n_1+n_2} + c^2\right)}} \exp\left\{-\frac{(\bar{y}^* - \bar{x})^2}{2\left(\frac{u^2}{n_1+n_2} + c^2\right)}\right\},$$

(4.3)

where

$$u_0^2 = u^2 \left(\frac{1}{n_1} + \frac{1}{n_2}\right);$$

and for the denominator,

$$f(\bar{y}_1, \bar{y}_2 | u^2, c^2, \bar{x}, H_2) = f(\bar{y}_1 | u^2, c^2, \bar{x}, H_2) f(\bar{y}_2 | u^2, c^2, \bar{x}, H_2)$$

$$= \frac{1}{\sqrt{2\pi \left(\frac{u^2}{n_1} + c^2\right)}} \exp\left\{-\frac{(\bar{y}_1 - \bar{x})^2}{2\left(\frac{u^2}{n_1} + c^2\right)}\right\} \times \frac{1}{\sqrt{2\pi \left(\frac{u^2}{n_2} + c^2\right)}} \exp\left\{-\frac{(\bar{y}_2 - \bar{x})^2}{2\left(\frac{u^2}{n_2} + c^2\right)}\right\}.$$

(4.4)

4.3 Between-object distribution modelled by kernel density estimation

The distribution of θ may not always be normal (either in the univariate or multivariate case). The assumption of normality can be replaced by considering a kernel density estimate (KDE) for the between-object distribution (Section 3.3.5). Earlier examples of the use of the KDE in forensic science are available in Aitken (1986), Aitken and Lucy (2004), Aitken and Taroni (2004), Berry (1991), Berry *et al.* (1992), Chan and Aitken (1989), and Evett *et al.* (1987).

The kernel density function is expressed as follows. For multivariate data $\boldsymbol{\theta}$,

$$K(\boldsymbol{\theta} \mid \bar{\mathbf{x}}_i, \mathbf{H}) = (2\pi)^{-p/2}|\mathbf{H}|^{-1/2}\exp\left\{-\frac{1}{2}(\boldsymbol{\theta} - \bar{\mathbf{x}}_i)^T\mathbf{H}^{-1}(\boldsymbol{\theta} - \bar{\mathbf{x}}_i)\right\},$$

where $\bar{\mathbf{x}}_i = \frac{1}{n}\sum_{j=1}^n \mathbf{x}_{ij}$ is the mean of the n measurements for the ith item, and $\mathbf{H} = h^2\mathbf{C}$ is the kernel bandwidth matrix, in which h is a smoothing parameter (Section 3.3.5) commonly calculated as proposed in Aitken and Lucy (2004):

$$h = h_{\text{opt}} = \left(\frac{4}{m(2p+1)}\right)^{\frac{1}{p+4}}.$$

For univariate data,

$$K(\theta \mid \bar{x}_i, h) = (2\pi h^2 c^2)^{-1/2}\exp\left\{-\frac{1}{2}(\theta - \bar{x}_i)^2(h^2c^2)^{-1}\right\}.$$

4.3.1 Multivariate data

In the case of multivariate data (when the between-object distribution is modelled by KDE) the following equations should be used (Aitken and Lucy 2004; Neocleous *et al.* 2011; Zadora 2010): for the numerator,

$$f(\bar{\mathbf{y}}_1, \bar{\mathbf{y}}_2|\mathbf{U}, \mathbf{C}, \bar{\mathbf{x}}_1, \dots, \bar{\mathbf{x}}_m, h, H_1) = f(\bar{\mathbf{y}}_1 - \bar{\mathbf{y}}_2, \bar{\mathbf{y}}^*|\mathbf{U}, \mathbf{C}, \bar{\mathbf{x}}_1, \dots, \bar{\mathbf{x}}_m, h, H_1)$$

$$= (2\pi)^{-p/2}\left|\frac{\mathbf{U}}{n_1} + \frac{\mathbf{U}}{n_2}\right|^{-\frac{1}{2}}\exp\left\{-\frac{1}{2}(\bar{\mathbf{y}}_1 - \bar{\mathbf{y}}_2)^T\left(\frac{\mathbf{U}}{n_1} + \frac{\mathbf{U}}{n_2}\right)^{-1}(\bar{\mathbf{y}}_1 - \bar{\mathbf{y}}_2)\right\}$$

$$\times\frac{1}{m}\sum_{i=1}^m\left\{(2\pi)^{-p/2}\left|\frac{\mathbf{U}}{n_1 + n_2} + h^2\mathbf{C}\right|^{-\frac{1}{2}}\exp\left\{-\frac{1}{2}(\bar{\mathbf{y}}^* - \bar{\mathbf{x}}_i)^T\left(\frac{\mathbf{U}}{n_1 + n_2} + h^2\mathbf{C}\right)^{-1}(\bar{\mathbf{y}}^* - \bar{\mathbf{x}}_i)\right\}\right\};$$

$$(4.5)$$

for the denominator,

$$f(\bar{\mathbf{y}}_1, \bar{\mathbf{y}}_2|\mathbf{U}, \mathbf{C}, \bar{\mathbf{x}}_1, \dots, \bar{\mathbf{x}}_m, h, H_2) = f(\bar{\mathbf{y}}_1|\mathbf{U}, \mathbf{C}, \bar{\mathbf{x}}_1, \dots, \bar{\mathbf{x}}_m, h, H_2)f(\bar{\mathbf{y}}_2|\mathbf{U}, \mathbf{C}, \bar{\mathbf{x}}_1, \dots, \bar{\mathbf{x}}_m, h, H_2)$$

$$= (2\pi)^{-p/2}\left|\frac{\mathbf{U}}{n_1} + h^2\mathbf{C}\right|^{-1/2}\times\frac{1}{m}\sum_{i=1}^m\exp\left\{-\frac{1}{2}(\bar{\mathbf{y}}_1 - \bar{\mathbf{x}}_i)^T\left(\frac{\mathbf{U}}{n_1} + h^2\mathbf{C}\right)^{-1}(\bar{\mathbf{y}}_1 - \bar{\mathbf{x}}_i)\right\}$$

$$\times(2\pi)^{-p/2}\left|\frac{\mathbf{U}}{n_2} + h^2\mathbf{C}\right|^{-1/2}\times\frac{1}{m}\sum_{i=1}^m\exp\left\{-\frac{1}{2}(\bar{\mathbf{y}}_2 - \bar{\mathbf{x}}_i)^T\left(\frac{\mathbf{U}}{n_2} + h^2\mathbf{C}\right)^{-1}(\bar{\mathbf{y}}_2 - \bar{\mathbf{x}}_i)\right\}.$$

$$(4.6)$$

4.3.2 Univariate data

In case of univariate data ($p = 1$), the covariance matrices become scalar variances, and the vectors become scalars. The the following equations are used (Aitken and Lucy 2004;

Neocleous *et al.* 2011; Zadora 2010): for the numerator,

$$f(\bar{y}_1, \bar{y}_2 | u^2, c^2, \bar{x}_1, \ldots, \bar{x}_m, h, H_1) = f(\bar{y}_1 - \bar{y}_2, \bar{y}^* | u^2, c^2, h, H_1)$$

$$= \frac{1}{\sqrt{2\pi u_0^2}} \exp\left\{-\frac{(\bar{y}_1 - \bar{y}_2)^2}{2u_0^2}\right\}$$

$$\times \frac{1}{m} \sum_{i=1}^{m} \frac{1}{\sqrt{2\pi \left(\frac{u^2}{n_1+n_2} + h^2c^2\right)}} \exp\left\{-\frac{(\bar{y}^* - \bar{x}_i)^2}{2\left(\frac{u^2}{n_1+n_2} + h^2c^2\right)}\right\}, \qquad (4.7)$$

where

$$u_0^2 = u^2 \left(\frac{1}{n_1} + \frac{1}{n_2}\right);$$

for the denominator,

$$f(\bar{y}_1, \bar{y}_2 | u^2, c^2, \bar{x}_1, \ldots, \bar{x}_m, h, H_2)$$
$$= f(\bar{y}_1 | u^2, c^2, \bar{x}_1, \ldots, \bar{x}_m, h, H_2) f(\bar{y}_2 | u^2, c^2, \bar{x}_1, \ldots, \bar{x}_m, h, H_2)$$

$$= \frac{1}{\sqrt{2\pi \left(\frac{u^2}{n_1} + h^2c^2\right)}} \times \frac{1}{m} \sum_{i=1}^{m} \exp\left\{-\frac{(\bar{y}_1 - \bar{x}_i)^2}{2\left(\frac{u^2}{n_1} + h^2c^2\right)}\right\}$$

$$\times \frac{1}{\sqrt{2\pi \left(\frac{u^2}{n_2} + h^2c^2\right)}} \times \frac{1}{m} \sum_{i=1}^{m} \exp\left\{-\frac{(\bar{y}_2 - \bar{x}_i)^2}{2\left(\frac{u^2}{n_2} + h^2c^2\right)}\right\}. \qquad (4.8)$$

4.4 Examples

There are numerous software packages available for performing all the calculations in this book. The recommended package is **R** (Appendix D). Suitable **R** routines which allow the user to calculate *LR* according to the equations given can be found in Appendix D and are available from the book's website. They have the advantage that they can be modified as required by the user. A special **R** package for carrying out *LR* calculations, called *comparison* (version 1.0–2), removes the need for the user to create his or her own programs. For analysing univariate data, a program called **calcuLatoR** is also available. Details, including an example of its application, can be found in Appendix F. For applications involving Bayesian networks, **Hugin Researcher**[TM] software may be useful (see Appendix E).

4.4.1 Univariate research data – normal between-object distribution – R software

In this example **R** code (given in Section D.9.2) is applied for research purposes to data obtained during the analysis of the colour of blue inks by microspectrophotometry with diode array detector (MSP-DAD; Section 1.5). The aim of the research was to select the variable which created a model giving the lowest rates of false positive and false negative answers (Martyna *et al.* 2013, Section 2.4).

4.4.1.1   Database creation

Forty blue inks were subjected to MSP-DAD analysis (Martyna *et al.* 2013). Lines were made by writing instruments (ballpens) on white printing paper (80 g/m^2, A4). A fragment of paper containing the writing was cut and fixed to the microscope base slide and placed on the stage of the microscope. The area for measurements was chosen using the diaphragm. For each ink ten measurements in reflection mode were made.

The blue inks were analysed by a Zeiss Axioplan 2 microspectrophotometer with J&M Tidas DAD (MCS/16 1024/100-1, Germany), which was configured for VIS range (380–800 nm) analyses. The ink spectra were obtained in reflection mode using an integration time of 2.5 seconds and at a magnification of 400×. The software used for the analysis was Spectralys v1.82 from J&M Tidas, which provided the possibility of calculating the chromaticity coordinates ($x, y, z; x + y + z = 1$; Section 1.5). For the results of the analysis, including the chromaticity coordinates for the 40 inks, see MSP_inks_database.txt. The columns of the matrix are as follows:

- column 1 (Name), unique ink name;

- column 2 (Item), serial number of ink sample;

- column 3 (Piece), measurement index;

- columns 4–6: respectively x, y, and z variables.

There are 400 rows (10 measurements performed for 40 inks), each referring to a single measurement.

4.4.1.2   Graphical summary

Graphical summaries of the data available in the file MSP_inks_database.txt are presented in the form of box-plots (Figure 4.1; Section 3.3.3) for each of the variables considered. In order to check whether the data are normally distributed, Q-Q plots as well as KDE plots (Figure 4.2; Section 3.3.5) were drawn. Even though the Q-Q plots reveal that the distribution deviates from normality at the left and right ends, it appears normal for the data in the middle and therefore the assumption that the data are normally distributed was made.

4.4.1.3   LR modelling

In this example the authors focused on three univariate LR models (Martyna *et al.* 2013), each referring to one of the variables x, y, z, using the expressions derived under the assumption that the data are normally distributed in the database (equations (4.3) and (4.4)).

The **R** function calculating the LR for each univariate problem is available in MSP_inks_comparison_code.R. The code uses the functions given in Section D.9.1:

- UC_comparison_calculations.R, which contains the UC function to calculate the within- and between-object variability (**U** and **C** in equations (3.3) and (3.4), respectively),

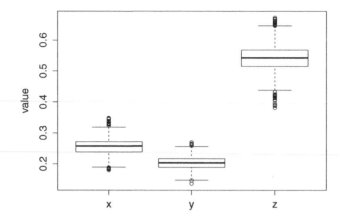

Figure 4.1 Box-plots for chromaticity coordinates x, y, z (data available in `MSP_inks_database.txt`).

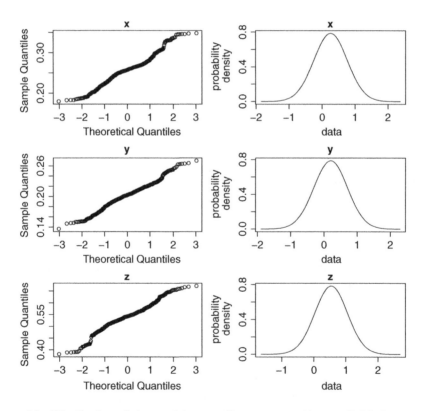

Figure 4.2 Distribution of chromaticity coordinates x, y, z (data available in `MSP_inks_database.txt`) in the form of (left) Q-Q plots and (right) kernel density estimation curves.

- LR_comparison_Nor.R, which contains the LR.Nor.function to calculate the LR assuming a normal between-object data distribution (equations (4.3) and (4.4)).

4.4.1.4 Results

LR calculations are used in research within a comparison problem to obtain information about the performance of LR models by estimating the levels of false positive and false negative answers. In order to choose the model with the best discriminating power (with low rates of false positive and false negative answers; Section 2.4), the following experiments were performed. Firstly, experiments for estimating the false positive and false negative answers were carried out (Section 2.4). The rates of false positive answers were obtained by performing pairwise comparisons of ink samples, one treated as the recovered sample and the other as the control sample (a total of $(40 \cdot 39)/2 = 780$ comparisons). The false negative rates were estimated by comparing two samples both created from each of the 40 inks. This involved dividing the samples of ten measurements into two equal halves (the control sample (measurements 1:5) and the recovered sample (measurements 6:10)) for a particular ink sample.

In order to obtain the results for all three univariate LR models, the user needs to run the code MSP_inks_comparison_code.R three times, by changing the index of the column in which the variable considered is located (4 for calculating the LR model based on x; 5 for the LR model based on y; and 6 for the LR model based on z). This can be done in the code file MSP_inks_comparison_code.R in the line:

```
variables = c(4) ##4 for x, 5 for y, 6 for z
```

The code MSP_inks_comparison_code.R can be run in the **R** Console after setting the proper working directory (Appendix D) and by inserting the command:

```
source(file = "MSP_inks_comparison_code.R")
```

Running this code provides the user with the following files as an outcome:

- x_comparison_research_Nor.txt, which contains a matrix of size $m \times m$ (where m is the number of objects in the database; here $m = 40$ inks) of the *LR* values (calculated under assumption of a normal between-object distribution; equations (4.3) and (4.4)) for compared objects. For example when ink_1 and ink_2 are compared the LR equals to 10.74. Only half of the matrix is presented, as the other half is given by symmetry – for example, comparison of ink_1 (control sample) and ink_20 (recovered sample) gives the same *LR* value as ink_20 (control sample) and ink_1 (recovered sample). The diagonal elements of the matrix provide the *LR* results obtained from experiments performed in order to estimate the rates of false negative answers. The remaining elements present the *LR* values obtained from experiments performed in order to estimate the rates of false positive answers.

- x_comparison_research_Nor_different.txt, which presents only the *LR* values obtained from the pairwise comparisons performed for estimating the rates of false positive answers, organised in one column (under the assumption of a normal between-object distribution; equations (4.3) and (4.4)). This type of data presentation is suitable for model performance evaluation, for example by empirical cross entropy (Section 6.6.1).

- `x_comparison_research_Nor_same.txt`, which presents only the *LR* values obtained from the comparisons performed for estimating the rates of false negative answers, organised in one column (under the assumption of a normal between-object distribution; equations (4.3) and (4.4)). This type of data presentation is suitable for model performance evaluation, for example by empirical cross entropy (Section 6.6.1).

- `x_research_descriptive_statistics.eps`, which is a graphical file presenting the descriptive statistics (box-plots, Q-Q plots, and KDE functions).

- `x_research_LR_distribution.eps`, which is a graphical file presenting the distribution of $\log_{10} LR$ for experiments performed to estimate the false negative and false positive rates.

For the remaining variables (y and z), the same set of outcome files is obtained, but with appropriate file name changes, for example `x_comparison_research_Nor.txt` is replaced by `y_comparison_research_Nor.txt`.

There is one additional file, `comparison_research_error_rate.txt`, which provides information on the rates of false negative (`fn`) and false positive (`fp`) answers for each univariate problem.

For x, y, and z respectively, 12.9%, 20.8%, and 18.7% false positive and 2.5%, 5.0%, and 2.5% false negative answers were obtained. Therefore, x seems to be the best choice if one wishes to evaluate the evidential value of the chromaticity coordinates determined for blue inks in casework.

Nevertheless, as described in Chapter 2, it is important to control not only the rates of false answers, but also the discriminating power and the calibration of the *LR* values obtained. This could be achieved by performance measures such as empirical cross entropy. Therefore, the analysis of this example is continued in Section 6.6.1.

See also Section 4.4.5 in which multivariate LR models (equations (4.1) and (4.2)) are applied to MSP data.

4.4.2   Univariate casework data – normal between-object distribution – Bayesian network

In this example a Bayesian network (BN) model (Appendix E) using **Hugin Researcher**™ is applied for casework purposes to data obtained during the analysis of colour variation of green fibres by MSP-DAD (Section 1.5). The results of the research and modelling are described in detail in Zadora (2010), where it was concluded that chromaticity coordinates (x, y, z; Section 1.5) should be used to help solve a comparison problem.

4.4.2.1   Database creation

The microspectrophotometer used in this study was a Zeiss Axioplan 2 with J&M Tidas DAD (MCS/16 1024/100-1, Germany) configured for the VIS region (380–800 nm). Transmission spectra of fibres were recorded using an integration time of 200 ms. Measurements were

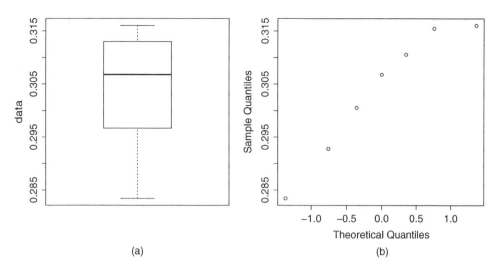

(a)   (b)

Figure 4.3 Graphical summaries for x describing green fibres in the form of (a) a box-plot, (b) a Q-Q plot (data available in MSP_fibres_database.txt).

performed on fibres at a magnification of 400×. Seven fibre samples made of cotton, polyester, polyamide, viscose, and wool were analysed. They were put on a microscope slide (76 × 26 mm, SuperFrost, Menzel-Gläser, Germany) immersed in an oil (Nikon 50, type A, $n_d^{23} = 1.515$) and covered with a cover slip (20 × 20 mm, SuperFrost, Menzel-Gläser, Germany).

The measurements were taken from ten different locations on each fibre sample. The same procedure was used for a control sample. A single recovered fibre was measured ten times at different points. The software used for analysis was Spectralys v1.82 from J&M Tidas, which allowed for the calculation of the chromaticity coordinates ($x + y + z = 1$).

For the results of the analysis see MSP_fibres_database.txt, which contains information about the x variable describing the seven fibres, one in each row of the matrix. Column 1 (Name) provides the unique fibre names, while column 2 (x) provides the mean of each set of ten measurements performed on the fibre samples.

4.4.2.2 Graphical summary

In this example only information relating to the chromaticity coordinate x is used. The descriptive statistics for this variable are presented in the form of a box-plot (Figure 4.3(a); Section 3.3.3) and Q-Q plot (Figure 4.3(b); Section 3.3.5). Based on the analysis, the distribution of x is determined as $N(0.3037, 1 \times 10^{-4})$; that is, the mean μ is estimated by $\bar{x} = 0.3037$ and the variance τ^2 is set equal to 1×10^{-4}. Values for the control sample and for the recovered sample for the chromaticity coordinate x are used to estimate the within-object means and variances as follows: $\bar{y}_c = 0.3067$ and $\sigma_c^2 = 5.6 \cdot 10^{-6}$ for the control sample; and $\bar{y}_r = 0.2919$ and $\sigma_r^2 = 3.4 \cdot 10^{-7}$ for the recovered sample.

4.4.2.3 LR modelling

LR values may be calculated using the BN model shown in Figure 4.4 (file green_fibres.oobn). A detailed description of LR calculations using BN models can

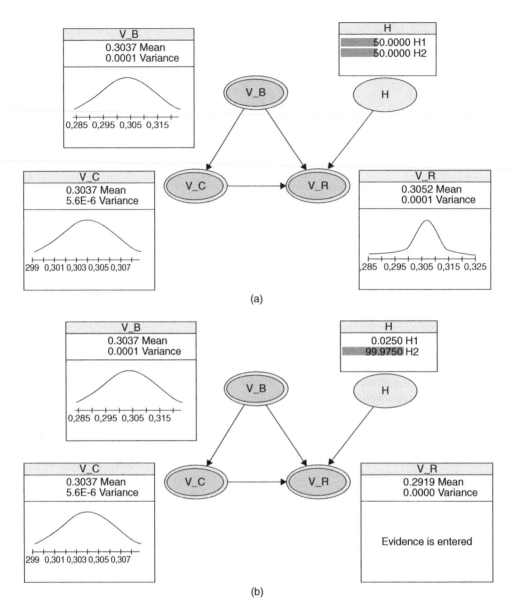

Figure 4.4 Bayesian Network model prepared in **Hugin Researcher**[TM] to solve a comparison problem on the basis of continuous type data; (a) the BN model after entering historical data; (b) the results of entering a piece of evidence into the V_R node ($\mu = 0.2919$). The H node represents hypotheses; V_B represents background information; V_C represents measurements made on the control sample; V_R represents measurements made on the recovered sample. (Reproduced by permission of Wiley.)

be found in Appendix E. H is a node of the type for discrete data with two hypotheses, H_1 (the fibres compared originated from the same object) and H_2 (the fibres compared originated from different objects). The values of the prior probabilities $P(H_1)$ and $P(H_2)$ were taken to be 0.5. Nodes V_B, V_C, and V_R are continuous data type nodes which represent information about the distribution of a particular variable in a general population (V_B), control samples (V_C), or recovered samples (V_R).

The continuous nodes V_B, V_C, and V_R are assumed to be represented by normal distributions, the parameters (means and variances) of which were estimated taking into account rules of propagation of information in a BN (Taroni *et al.* 2010; Zadora 2010).

(a) V_B represents background information with $N(\mu, \tau^2)$, where μ is the overall mean of all objects in the database of interest and τ^2 is the between-object variability. In this example V_B is characterised by an overall mean of $\mu = 0.3037$ and a between-object variance of $\tau^2 = 1 \cdot 10^{-4}$ (equation (3.2)).

(b) V_C represents measurements made on the control sample with a within-group distribution $N(\theta, \sigma^2)$, where $\theta \sim N(\mu, \tau^2)$ is the variation of the mean across groups, σ^2 represents the within-group variability and τ^2 represents the between-group variability. In this example V_C is characterised by the overall mean ($\mu = 0.3037$) and a within-object variability of $\sigma_c^2 = 5.6 \cdot 10^{-6}$ (equation (3.2)).

(c) V_R represents measurements made on the recovered sample. The node V_R depends on nodes V_B, V_C, and H (Appendix E). Therefore, appropriate normal distribution parameters should be calculated as follows:

- $N(\bar{y}_c, \sigma_c^2 + \frac{\sigma_c^2 \cdot \tau^2}{\sigma_c^2 + \tau^2})$ when H_1 is true, and where $\bar{y}_c$ is the control sample mean. In this example we have $\bar{y}_c = 0.3067$ and $\sigma_c^2 + \frac{\sigma_c^2 \cdot \tau^2}{\sigma_c^2 + \tau^2} = 1.1 \cdot 10^{-5}$;

- $N(\mu, \sigma_c^2 + \tau^2)$ when H_2 is true. In this example we have $\mu = 0.3037$ and $\sigma_c^2 + \tau^2 = 1 \times 10^{-4}$.

4.4.2.4   Results

The posterior probabilities, $P(H_1|E) = 0.00029$ and $P(H_2|E) = 0.99971$ (equation (2.1)) were obtained in node H after entering information on the evidence value of the measurement of x on the recovered fibre into the node V_R, that is, $\bar{y}_r = 0.2919$ (Figure 4.4(b)). It was assumed that the prior probabilities in node H were equal for each of the two states (Figure 4.4(a)). Therefore, the ratio of posterior probabilities could be treated as LR (equation (2.1); details in Taroni *et al.* (2010)). The value obtained, $LR = 2.9 \times 10^{-4}$, suggests that the evidence is about 3450 times more likely to support the hypothesis that the samples originated from different objects (H_2) than from the same object (H_1).

4.4.3   Univariate research data – kernel density estimation – R software

This example illustrates the application of the **R** code presented in Section D.9.2 for research purposes to data obtained during the analysis of car paint samples by Py-GC/MS (Section 1.4). The aim of the research was to select the model which gave the lowest rates of false positive and false negative answers (Section 2.4).

4.4.3.1 Database creation

Py-GC/MS (TurboMass Gold, Perkin Elmer, USA) was used to analyse 36 samples (each with three replications) of acrylic topcoat paints from 36 different cars that were indistinguishable on the basis of their infrared spectra and elemental composition (Section 1.4; Zieba-Palus *et al.* 2008). The top layer, the so-called clear coat, was isolated from solid and metallic paints under an optical microscope (SMXX Carl Zeiss, Jena, Germany; 40× magnification) using a scalpel, and was placed in a quartz tube (Analytix Ltd., UK) in the platinum coil of a pyrolyser. The sample size was 50–100 μg. The GC programme was: 40°C held for 2 min; increased by 10°C· min^{-1} to 300°C; 300°C maintained for 2 min; increased by 30°C·min^{-1} to 320°C; 320°C maintained for 3 min. An RTx-35MS capillary column (30 m × 0.25 mm × 0.25 μm) was used. Pyrolysis was performed at 750°C for 20 s.

The compounds of interest were represented by the main peaks shown on the pyrogram in Figure 4.5. Where peaks were not observed, zero was substituted by a relatively small value of 1000 (a regular peak area was of the order of magnitude 10^6–10^{12}). The $\log_{10}$ values of the ratios of the peak areas of the following seven organic compounds to the peak area of styrene were calculated: 2-hydroxyethylmethacrylate (*M2E*), α-methylstyrene (*MST*), toluene (*TOL*), butylacrylate (*BMA*), 2-hydroxypropylmethacrylate (*M2P*), methylmethacrylate (*MMA*), 1,6-diisocyanatehexane (*I16*). The normalisation by logarithmic transformation of the proposed ratios effectively removed stochastic fluctuations in the instrumental measurement. Three replicate measurements were made for each of 36 car paint objects. For the results of the analysis see GCMS_car_paints_database.txt. The columns of the matrix in the file are as follows:

- column 1 (Name), unique car paint name;

- column 2 (Item), car paint sample serial numbers;

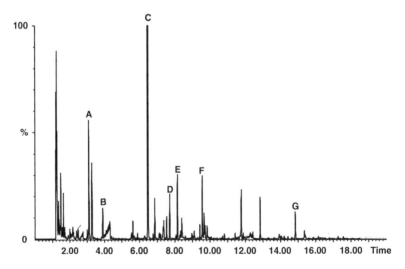

Figure 4.5 Pyrograms obtained during analysis of a paint sample: A, methyl-methacrylate; B, toluene; C, styrene; D, butylmethacrylate; E, α-methylstyrene; F, 2-hydroxypropylmethacrylate; G, 1,6-diisocyanatehexane.

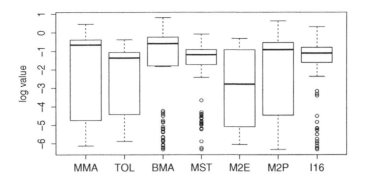

Figure 4.6 Box-plots for car paint data available in GCMS_car_paints_database.txt.

- column 3 (Piece), measurement index;

- columns 4–10: MMA, TOL, BMA, MST, M2E, M2P, I16 variables, respectively.

There are 108 rows (three measurements performed for 36 car paint samples), each referring to a single measurement.

4.4.3.2   Graphical summary

Box-plots show the descriptive statistics for the data available in the file GCMS_car_ paints_database.txt for each variable describing the car paint samples (Figure 4.6; Section 3.3.3). Q-Q plots (Figure 4.7; Section 3.3.5) and KDE curves were plotted (Figure 4.8; Section 3.3.5) to check assumptions of normality. The Q-Q plots clearly demonstrate that the distributions for all the seven variables are not normal and the KDE curves show that the distributions are bimodal for each variable.

4.4.3.3   LR modelling

The analysis focused first on univariate *LR* computations for single variables using the expressions for the KDE procedure (equations (4.7) and (4.8)). However, univariate models cannot use all the evidence collected, which consists of seven variables. On the other hand the multivariate model based on all seven variables is not relevant in this case, as the parameters needed for LR computation would not be reliably estimated because of the small size of the database (comprising only 36 samples). Therefore, an alternative LR model was proposed: based on the naïve assumption that the variables are independent, the *LR* values obtained for the seven univariate problems representing each individual variable were multiplied together to give

$$LR = LR_{MMA} \cdot LR_{TOL} \cdot LR_{BMA} \cdot LR_{MST} \cdot LR_{M2P} \cdot LR_{M2E} \cdot LR_{I16}. \qquad (4.9)$$

The **R** function calculating the *LR* for each univariate problem as well as the alternative *LR* model is available in GCMS_car_paints_comparison_code.R. The code uses the functions given in Section D.9.1:

- UC_comparison_calculations.R, which contains the UC function to calculate the within- and between-object variability (**U** and **C**; equations (3.3) and (3.4), respectively),

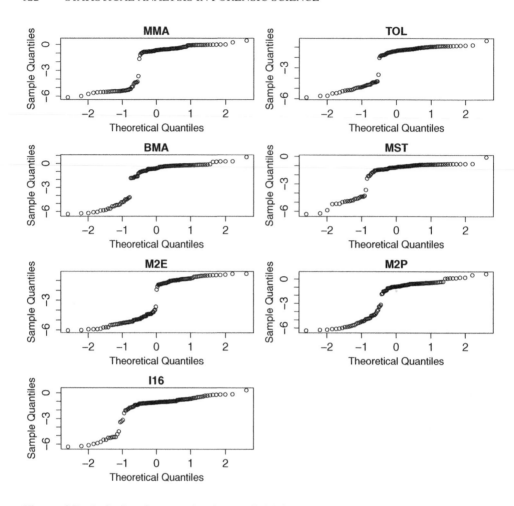

Figure 4.7   Q-Q plots for car paint data available in GCMS_car_paints_database.txt.

- LR_comparison_KDE.R, which contains the LR.KDE.function to calculate *LR* using KDE for estimating the between-object data distribution (equations (4.7) and (4.8)).

4.4.3.4   Results

LR calculations are used in research within a comparison problem to obtain information about the performance of LR models by estimating the levels of false positive and false negative answers. In order to choose the model with the best discriminating power (with low rates of false positive and false negative answers; Section 2.4), the following experiments were performed. Firstly, experiments for estimating the false positive and false negative answers were carried out (Section 2.4). The rates of false positive answers were obtained by performing pairwise comparisons of car paint samples, one treated as the recovered sample and the other as the control sample (a total of $(36 \cdot 35)/2 = 630$ comparisons). The false negative rates

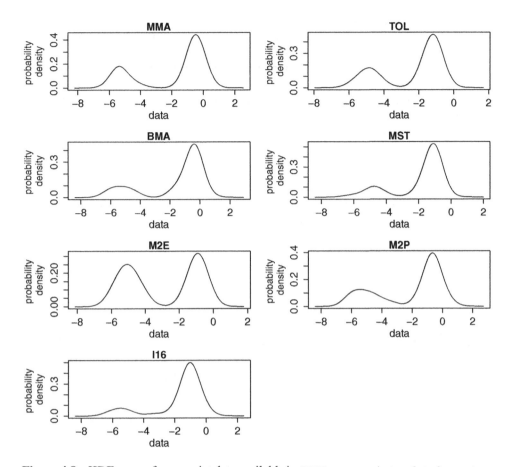

Figure 4.8   KDE curves for car paint data available in GCMS_car_paints_database.txt.

were estimated by comparing two samples both created from each of the 36 car paints. This involved dividing the samples of three measurements into a control sample (measurements 1:2) and the recovered sample (measurement 3) for a particular car paint sample.

All the results (for seven univariate models and the alternative naïve model) are obtained in one code run (GCMS_car_paints_comparison_code.R). The code can be run in the **R** Console after setting the proper working directory (Appendix D) and by inserting the command:

```
source(file = "GCMS_car_paints_comparison_code.R")
```

The code provides the user with the following files as an outcome:

- MMA_comparison_research_KDE.txt, which contains a matrix of size $m \times m$ (where m is the number of objects in the database; here $m = 36$ car paint samples) of the *LR* values (calculated when the between-object variability is estimated by KDE;

equations (4.7) and (4.8)) for compared objects. For example when a_lak210 and a_lak212 are compared the LR equals to 11.53. Only half of the matrix is presented, as the other half is given by symmetry – for example, comparison of a_lak033 (control sample) and b_lak151 (recovered sample) gives the same *LR* value as b_lak151 (control sample) and a_lak033 (recovered sample). The diagonal elements of the matrix provide the *LR* results obtained from experiments performed in order to estimate the rates of false negative answers. The remaining elements present the *LR* values obtained from experiments performed in order to estimate the rates of false positive answers.

- MMA_comparison_research_KDE_different.txt, which presents only the *LR* values obtained from the pairwise comparisons performed for estimating the rates of false positive answers, organised in one column (calculated when between-object variability is estimated by KDE; equations (4.7) and (4.8)). This type of data presentation is suitable for model performance evaluation, for example by empirical cross entropy (Section 6.6.2).

- MMA_comparison_research_KDE_same.txt, which presents only the *LR* values obtained from the comparisons performed for estimating the rates of false negative answers, organised in one column (calculated when between-object variability is estimated by KDE; equations (4.7) and (4.8)). This type of data presentation is suitable for model performance evaluation, for example by empirical cross entropy (Section 6.6.2).

- MMA_research_descriptive_statistics.eps, which is a graphical file presenting the descriptive statistics (box-plots, Q-Q plots, and KDE functions).

- MMA_research_LR_distribution.eps, which is a graphical file presenting the $\log_{10} LR$ distribution for experiments performed for estimating the false negative and false positive rates.

For the remaining variables (*TOL*, *BMA*, *MST*, *M2E*, *M2P*, and *I*16), the same set of outcome files is obtained, but with appropriate file name changes, for example MMA_comparison_research_KDE.txt is replaced by MST_comparison_research_KDE.txt. When the multivariate naïve *LR* model is considered (model_comparison_research_KDE.txt), there is one file fewer, since no descriptive statistics are available.

There is also one additional file, comparison_research_error_rate.txt, which provides information of the rates of false negative (fn) and false positive (fp) answers for each univariate problem and the alternative model.

In the first step of the analysis (concerning the univariate models) false positive answer rates of 29.7% for *MMA*, 32.9% for *TOL*, 23.3% for *BMA*, 36.3% for *MST*, 24.6% for *M2P*, and 28.1% for *I*16 were obtained. The lowest rate was 18.7% for *M2E*. The lowest false negative answers rate occurred with *TOL*, *MST*, *M2E*, *I*16 (2.8%). There are 36 same source comparisons and a false negative rate of 2.8% corresponds to one false negative in 36 comparisons ($1/36 = 0.028$). Therefore the best results are provided by the *M2E* variable.

The rates of false positive and false negative answers obtained for the proposed alternative LR model (equation (4.9)) were effectively lowered and equal to 2.8% for both. However, the quality of the *LR* values should be checked, for example, by the empirical cross entropy approach, as shown in Section 6.6.2.

For the data under analysis, LR models could be used together with graphical models as described in Section 3.7.2 and Zieba-Palus *et al.* (2008).

4.4.4 Univariate casework data – kernel density estimation – calcuLatoR **software**

This subsection provides an example of the application of the **calcuLatoR** software (Appendix F; Zadora and Surma 2010) for casework purposes to data obtained during the analysis of window glass samples by the GRIM technique (Martyna *et al.* 2013, Section 1.2.2).

4.4.4.1 **Database creation**

The evidence analysed in this example consists of five recovered items of window glass (RI_recovered.txt) and one control item of window glass (RI_control.txt). The background database used for the calculations consists of 23 window glass samples (RI_w_database.txt). The analysis of glass fragments was performed according to the procedure described in Section 1.2.2. The control glass fragment was measured 16 times, while the recovered items were measured eight times each. Additionally, the objects from the database were measured four times.

The columns of the data matrices are as follows:

- column 1 of database file (Category), glass category (w standing for "windows");

- column 1 of control and recovered objects data files (Name), unique glass fragment name;

- column 2 (Item), glass fragment serial numbers;

- column 3 (Piece), measurement index;

- column 4 (RI), RI data.

Each row refers to a single measurement.

4.4.4.2 **Graphical summary**

A box-plot (Section 3.3.3) and Q-Q plot (Section 3.3.5) for the population data available in RI_w_database.txt are shown in Figure 4.9. The box-plots for the control and recovered objects are displayed in Figure 4.10.

4.4.4.3 **LR modelling**

Casework calculations involve pairwise LR computations for control and recovered objects to provide information on whether they may or may not share a common origin. Univariate LR calculations were performed for each comparison between control and recovered glass fragments using the expressions for the KDE procedure for proper parameter estimation (equations (4.7) and (4.8)) as the between-object distribution cannot be modelled by the normal distribution (Figure 4.9(b)).

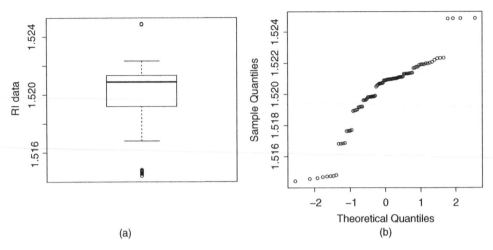

(a)   (b)

Figure 4.9   Graphical summaries for population RI data available in `RI_w_database.txt`: (a) box-plot; (b) Q-Q plot.

4.4.4.4   Results

In order to perform the calculations in the **calcuLatoR** software (when between-object variability is estimated by KDE; equations (4.7) and (4.8)), refer to Appendix F. The results of the calculations are presented in Figure 4.11 and are available in a file generated by **calcuLatoR** named `RI_comparison_casework_KDE.txt`.

The results suggest that for comparisons between the control object (`c1`) and four recovered glass fragments (`e1-e4`) the evidence more likely supports the prosecutor's hypothesis (H_1) rather than the defence hypothesis (H_2). When comparing objects `c1` and `e5`, the evidence $1/2.29 \times 10^{-16} = 4.4 \times 10^{15}$ times more likely supports H_2 than H_1.

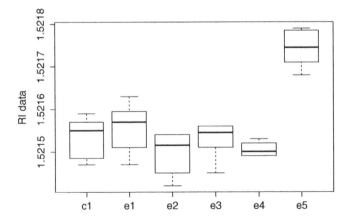

Figure 4.10   Box-plots for the control (`c1`) and recovered (`e1-e5`) sample RI data available in `RI_control.txt` and `RI_recovered.txt`.

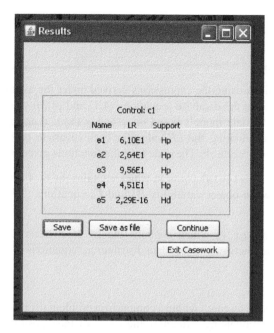

Figure 4.11 Results of the calculations provided by the **calcuLatoR** software (Appendix F) for data available in RI_control.txt and RI_recovered.txt.

4.4.5 Multivariate research data – normal between-object distribution – R software

In this example **R** code (given in Section D.9.2) is applied for research purposes to data obtained during the analysis of the colour of blue inks by MSP-DAD (Sections 1.5 and 4.4.1). The aim of the research was to select the set of variables which created a multivariate model giving the lowest rates of false positive and false negative answers (Section 2.4).

4.4.5.1 Database creation

Data obtained as a result of the colour analysis of 40 blue inks by MSP-DAD were used (MSP_inks_database.txt). Details of the MSP-DAD analysis can be found in Section 4.4.1.

4.4.5.2 Graphical summary

The descriptive statistics for the data can be found in Section 4.4.1. Recall from Section 1.5 the mathematical expressions combining all the chromaticity coordinates together. Their mutual dependence results in high correlation coefficients: 0.72 between x and y, -0.95 between x and z, and -0.90 between y and z. That is the reason for the bivariate calculations in the present section. The trivariate LR model was not analysed due to the fact that such calculations require nine parameters to be reliably estimated (three means, three variances, and three covariances) which is problematic for a database of only 40 objects. Therefore, it

was decided to restrict the dimensionality to univariate (Section 4.4.1) and bivariate cases
(this section).

4.4.5.3   LR modelling

Bivariate *LR* computations using the expressions derived under the assumption that the data
are normally distributed in the database (equations (4.1) and (4.2)) were performed. There
were three proposed bivariate models based on *x* and *y* (*xy*), *x* and *z* (*xz*), and *y* and *z*
(*yz*). The **R** function calculating the *LR* value for each bivariate problem is available in
MSP_inks_comparison_code.R. The code uses the functions given in Section D.9.1:

- UC_comparison_calculations.R, which contains the UC function to calculate the
  within- and between-object variability (**U** and **C**; equations (3.3) and (3.4), respectively),

- LR_comparison_Nor.R, which contains the LR.Nor.function to calculate *LR*
  assuming a normal between-object data distribution (equations (4.1) and (4.2)).

4.4.5.4   Results

LR calculations are used in research within a comparison problem to obtain information about
the performance of LR models by estimating the levels of false positive and false negative
answers. In order to choose the model with the lowest rates of false positive and false negative
answers (Section 2.4), the following experiments were performed. Firstly, experiments for
estimating the false positive and false negative answers were carried out (Section 2.4). The
rates of false positive answers were obtained by performing pairwise comparisons of ink
samples, one treated as the recovered sample and the other as the control sample (a total
of $(40 \cdot 39)/2 = 780$ comparisons). The false negative rates were estimated by comparing
two samples both created from each of the 40 inks. This involved dividing the samples of
ten measurements into two equal halves (the control sample (measurements 1:5) and the
recovered sample (measurements 6:10)) for a particular ink sample.

In order to obtain the results for all three bivariate *LR* models (*xy*, *xz*, and *yz*) the user needs
to run the code MSP_inks_comparison_code.R three times, by changing the indices of
the columns in which the variables considered are located (4,5 for calculating the LR model
based on *xy*; 5,6 for the LR model based on *yz*; and 4,6 for the LR model based on *xz*). This
can be done in the code file MSP_inks_comparison_code.R in the line:

```
variables = c(4,5) ##4 for x, 5 for y, 6 for z
```

The code MSP_inks_comparison_code.R can be run in the **R** Console after setting the
proper working directory (Appendix D) and by inserting the command:

```
source(file = "MSP_inks_comparison_code.R")
```

The code provides the user with the following files as an outcome:

- x_y_comparison_research_Nor.txt, which contains a matrix of size $m \times m$
  (where *m* is the number of objects in the database; here $m = 40$ inks) of the *LR*

values (calculated when between-object variability is assumed normal; equations (4.1) and (4.2)) for compared objects. For example when `ink_1` and `ink_2` are compared the LR equals to 57.24. Only half of the matrix is presented, as the other half is given by symmetry – for example, the comparison of `ink_1` (control sample) and `ink_20` (recovered sample) gives the same *LR* value as `ink_20` (control sample) and `ink_1` (recovered sample). The diagonal elements of the matrix provide the *LR* results obtained from experiments performed in order to estimate the rates of false negative answers. The remaining elements present the *LR* values obtained from experiments performed in order to estimate the rates of false positive answers.

- `x_y_comparison_research_Nor_different.txt`, which presents only the *LR* values obtained from the pairwise comparisons performed for estimating the rates of false positive answers, organised in one column (calculated when between-object variability is assumed normal; equations (4.1) and (4.2)). This type of data presentation is suitable for model performance evaluation, for example by empirical cross entropy (Chapter 6).

- `x_y_comparison_research_Nor_same.txt`, which presents only the *LR* values obtained from the comparisons performed for estimating the rates of false negative answers, organised in one column (calculated when between-object variability is assumed normal; equations (4.1) and (4.2)). This type of data presentation is suitable for model performance evaluation e.g. by empirical cross entropy (Chapter 6).

- `x_y_research_descriptive_statistics.eps`, which is a graphical file presenting the descriptive statistics (box-plots, Q-Q plots, and KDE functions).

- `x_y_research_LR_distribution.eps`, which is a graphical file presenting the $\log_{10} LR$ distribution for experiments performed to estimate the false negative (`fn`) and false positive rates (`fp`).

For the remaining models (*xz* and *yz*), the same set of outcome files is obtained, but with appropriate file name changes, for example `x_y_comparison_research_Nor.txt` is replaced by `y_z_comparison_research_Nor.txt`.

There is also one additional file `comparison_research_error_rate.txt`, which delivers information of the rates of false negative and false positive answers for each bivariate problem.

Each bivariate combination of variables *xy*, *yz*, and *xz* gave equal rates of false positive answers (4.6% for each model) and 2.5% of false negative answers.

Nevertheless, as described in Chapter 2, it is important to control not only the rates of false answers, which in this case do not point to any set of variables giving the best results, but also the quality of *LR* values obtained. This could be done using the empirical cross entropy approach (Chapter 6).

4.4.6 Multivariate research data – kernel density estimation procedure – R software

In this example **R** code (given in Section D.9.2) is applied for research purposes to data obtained during the analysis of glass fragments by SEM-EDX (Section 1.2.1). The aim of the research was to create and evaluate the usefulness of a graphical LR model involving all the variables, by estimating the rates of false positive and false negative answers (Section 2.4).

4.4.6.1   Database creation

Fragments of 200 glass objects with surfaces as smooth and flat as possible were placed on self-adhesive carbon tabs on an aluminium stub and then carbon-coated using a sputter (Bal-Tech, Switzerland). Analysis of the elemental content of each glass fragment was carried out using a scanning electron microscope (JSM-5800 Jeol, Japan), with an energy dispersive X-ray detector (Link ISIS 300, Oxford Instruments Ltd., UK). An example of the spectra obtained is shown in Figure 4.12. The measurement conditions were: accelerating voltage 20 kV, lifetime 50 s, magnification 1000–2000×, and the calibration element was cobalt. The SEMQuant option (part of the LINK ISIS software, Oxford Instruments Ltd, UK) was used in the process of determining the percentages of particular elements in a fragment. A report on the quantitative analysis only contained data on selected elements, which in this case were O, Na, Mg, Al, Si, K, Ca, and Fe, so that the sum of concentrations of these elements was equal to 100 wt.%. From the eight elements measured, seven different variables were derived by taking the $\log_{10}$ of the ratio of each element concentration to the oxygen concentration, for example $\log_{10}(Na/O) \equiv logNaO$. If 0 was obtained, that is, an element was not present or its concentration was below the SEM-EDX detection limit (0.1 wt. % for most of the elements considered), then zero was substituted by a very small value (0.0001 wt. %).

For the results of the analysis see SEMEDX_glass_database.txt which gives the logarithms of the ratios of each element's concentration to the oxygen concentration (*logNaO*, *logMgO*, *logAlO*, *logSiO*, *logKO*, *logCaO*, *logFeO*) for the 200 glass samples. The columns of the matrix are as follows:

- column 1 (Name), unique glass fragment name;

- column 2 (Item), glass fragment serial number;

- column 3 (Piece), measurement index;

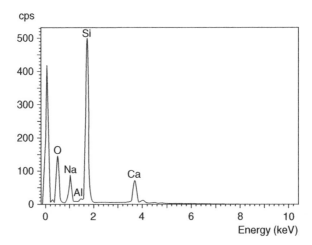

Figure 4.12   An SEM-EDX spectrum of glass (Section 1.2.1), in which only sodium (Na), aluminium (Al), silicone (Si), calcium (Ca), and oxygen (O) content was determined. Magnesium (Mg), potassium (K), and iron (Fe) were not detected.

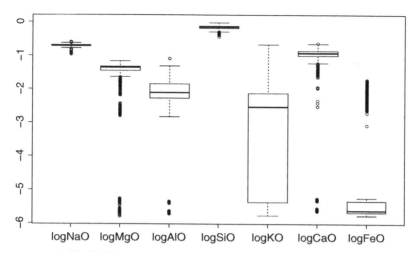

Figure 4.13 Box-plots presenting distributions of seven variables observed for 200 glass objects (data available in SEMEDX_glass_database.txt).

- columns 4–10, logNaO, logMgO, logAlO, logSiO, logKO, logCaO, logFeO variables.

There are 2400 rows (12 measurements performed for 200 glass fragments), each referring to a single measurement.

4.4.6.2 Graphical summary

Box-plots were used to illustrate the descriptive statistics for the data under analysis (Figure 4.13; Section 3.3.3). Q-Q plots and KDE plots (Section 3.3.5) of the data distribution are shown in Figures 4.14 and 4.15. It is evident that, except for the *logSiO* variable, none of the distributions can be assumed normal.

4.4.6.3 LR modelling

In this example multivariate *LR* computations for creating the LR model given by equation (4.11) were performed using the expressions derived under the assumption that the data are not normally distributed in the database. The KDE procedure (equations (4.5)–(4.8)) was used to estimate the distributional parameters.

4.4.6.4 Dimensionality reduction using graphical models

In many cases the number of objects in a database with multivariate data is far too low for reliable estimation of parameters such as means, variances, and covariances. For example, for a database consisting of 200 glass objects characterised by 7 variables, it is necessary to evaluate 7 means, 7 variances, and 21 covariances. It is a difficult task to estimate these reliably using a scarce database of 200 objects. This is referred to the *curse of dimensionality*. It can be addressed by means of the graphical model approach (Section 3.7.2). Graphical

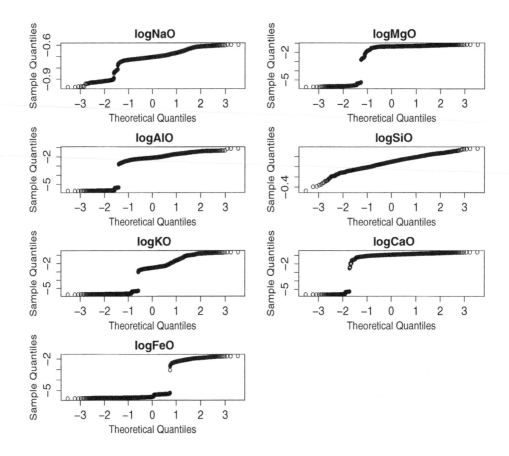

Figure 4.14 Q-Q plots for seven variables observed for 200 glass objects (data available in `SEMEDX_glass_database.txt`).

models are a dimension reduction technique based on graph theory (Whittaker 1990). The graphical model (GM) is selected by the sequential addition of edges determined by inspection of the rescaled inverse of the variance–covariance matrix (partial correlation matrix obtained using the **R** code provided in Section 3.7.2), for example as obtained for the seven variables observed for 200 glass objects from the background population (Table 4.1).

Based on Table 4.1 a graphical model was created for the glass data following the procedure described in Section 3.7.2. Firstly, the largest partial correlation coefficient was selected. This was 0.592 between *logCaO* and *logNaO*, and an edge was added between these two nodes. Then the second largest partial correlation coefficient was selected (0.475) and an edge was added joining the corresponding nodes, *logKO* and *logAlO*, to the graph. This process was repeated until all the nodes were part of the model (Figure 4.16).

The model presented in Figure 4.16(a) represents the final graph obtained after addition of all the variables which were subjected to factorisation. First of all, the cliques were determined and numbered as described in Section 3.7.2. The results of this process are presented in Figure 4.16(b). The clique (*logNaO*, *logSiO*) has the highest numbered node,

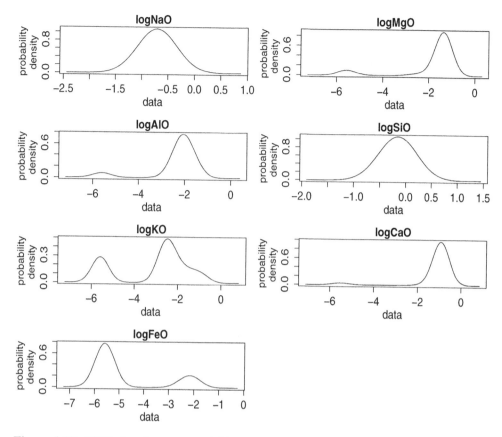

Figure 4.15   KDE plots for seven variables observed for 200 glass objects (data available in SEMEDX_glass_database.txt).

$logSiO$, so the clique is given the same number as $logSiO$, which is 7. The clique ($logAlO$, $logFeO$) has the next highest numbered node, $logFeO$, and so has the number 6. Putting these into numerical order, the set chain was obtained using procedure described in Section 3.7.2 (results are provided in Table 4.2).

Given the cliques for the model and a suitable set chain, the sets of separators (S_i) for each clique were found. The first clique (Table 4.2) in the set chain is ($logNaO$, $logCaO$). This is a complete subgraph, and at the moment there are no other cliques added to the graph, so there are no separator sets. The next clique in the set chain is ($logMgO$, $logCaO$) and it is added to the model. The intersection of elements between these two cliques is ($logCaO$), and so this becomes the first separator set. The running union of the first two sets is now ($logNaO$, $logCaO$, $logMgO$). Working through the entire set chain one arrives at the factorisation

$$f\,(C_i|S_i) = \frac{f\,(logNaO, logCaO)\,f\,(logMgO, logCaO)\,f\,(logKO, logNaO)}{[f\,(logNaO)]^2\,f\,(logCaO)}$$
$$\times \frac{f\,(logAlO, logKO)\,f\,(logFeO, logAlO)\,f\,(logNaO, logSiO)}{f\,(logKO)\,f\,(logAlO)}. \qquad (4.10)$$

Table 4.1 The partial correlation matrix (Section 3.7.2) for seven variables observed for 200 glass objects (only the upper right triangle of the matrix is given; the lower left triangle is given by symmetry).

	logNaO	logMgO	logAlO	logSiO	logKO	logCaO	logFeO
logNaO	1.000	0.134	0.186	0.238	−0.216	0.592	0.026
logMgO		1.000	−0.008	−0.157	−0.200	0.297	0.179
logAlO			1.000	−0.115	0.475	−0.108	−0.219
logSiO				1.000	0.091	−0.131	0.145
logKO					1.000	0.002	0.058
logCaO						1.000	−0.015
logFeO							1.000

The probability density functions presented in equation (4.10) can be expressed by equations (4.5)-(4.8), and then equation (4.10) takes the form of

$$LR = \frac{LR(logNaO, logCaO)LR(logMgO, logCaO)LR(logKO, logNaO)}{[LR(logNaO)]^2 LR(logCaO)}$$
$$\times \frac{LR(logAlO, logKO)LR(logFeO, logAlO)LR(logNaO, logSiO)}{LR(logKO)LR(logAlO)}, \tag{4.11}$$

which is represented as sets of one or two variables (columns C_i and S_i in Table 4.2). Parameters (means, variances, and covariances) of the bivariate distributions can then be estimated on the basis of information represented by a relatively small database.

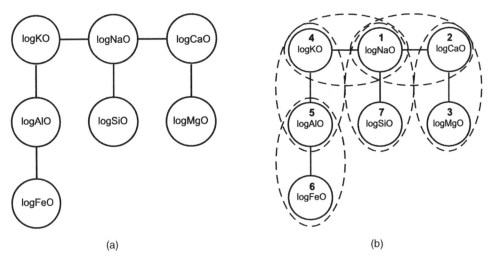

(a)   (b)

Figure 4.16 Decomposable undirected graphical model for the glass data calculated from Table 4.1: (a) model obtained on the basis of the rescaled inverse of the variance–covariance matrix for the seven variables from the 200 glass objects; (b) model (a) with marked cliques and numbers assigned to each node.

Table 4.2 Cliques, separators, and clique ordering for the glass graphical model. The clique ordering for those cliques suggested by the model is based upon the algorithm featured in Section 3.7.2. The separator sets S_i are composed of those elements of each clique which also appear in the running union along the set chain at position R_{i-1}. The running union R_i is composed of the set of all elements from each clique in the set up to and including clique C_i. $\emptyset$ denotes the empty set.

i	Clique (C_i)	Running union (R_i)	Separator set (S_i)
1	$(logNaO, logCaO)$	$(logNaO, logCaO)$	$\emptyset$
2	$(logMgO, logCaO)$	$(logNaO, logCaO, logMgO)$	$(logCaO)$
3	$(logKO, logNaO)$	$(logNaO, logCaO, logMgO, logKO)$	$(logNaO)$
4	$(logAlO, logKO)$	$(logNaO, logCaO, logMgO, logKO, logAlO)$	$(logKO)$
5	$(logFeO, logAlO)$	$(logNaO, logCaO, logMgO, logKO, logAlO, logFeO)$	$(logAlO)$
6	$(logNaO, logSiO)$	$(logNaO, logCaO, logMgO, logKO, logAlO, logFeO, logSiO)$	$(logNaO)$

Using the graphical model, LR calculations were reduced to six bivariate and four univariate problems. The proposed LR model involved the KDE procedure for probability density function estimation (equations (4.5)–(4.8)). The **R** function calculating the LR for the proposed model is available in SEMEDX_glass_comparison_code.R. The code uses the functions given in Section D.9.1:

- UC_comparison_calculations.R, which contains the UC function to calculate the within- and between-object variability (**U** and **C**; equations (3.3) and (3.4), respectively);

- LR_comparison_KDE.R, which contains the LR.KDE.function to calculate the LR using KDE for estimating the between-object data distribution (equations (4.5)–(4.8)).

4.4.6.5 Results

LR calculations are used in research within a comparison problem to obtain information about the performance of LR models by estimating the levels of false positive and false negative answers (Section 2.4). The rates of false positive answers were obtained by performing pairwise comparisons of samples of glass fragments, one treated as recovered sample and the second as the control sample (a total of $(200 \cdot 199)/2 = 19\,900$ comparisons). The false negative rates were estimated by comparing two samples both created from each of the 200 glass fragments. This involved dividing the samples of 12 measurements into two equal halves (the control sample (measurements $1:6$) and the recovered sample (measurements $7:12$)) for a particular glass fragment.

The code SEMEDX_glass_comparison_code.R can be run in the **R** Console after setting the proper working directory (Appendix D) and by inserting the command:

```
source(file = "SEMEDX_glass_comparison_code.R")
```

It may take some time (up to a few hours, depending on the compter power available) to run the code and obtain the results.

Running the code provides the user with the following files as an outcome:

- model_comparison_research_KDE.txt, which contains a matrix of size $m \times m$ (where m is the number of objects in the database; here $m = 200$ glass samples) of the *LR* values (calculated when KDE is used for between-object variability estimation; equations (4.5)–(4.8)) for compared objects. For example when s1 and s2 are compared the LR equals to 3736. Only half of the matrix is presented, as the other half is given by symmetry – for example, the comparison of s1 (control sample) and s20 (recovered sample) gives the same *LR* value as s20 (control sample) and s1 (recovered sample). The diagonal elements of the matrix provide the *LR* results obtained from experiments performed in order to estimate the rates of false negative answers. The remaining elements present the *LR* values obtained from experiments performed in order to estimate the rates of false positive answers.

- model_comparison_research_KDE_different.txt, which presents only the *LR* values obtained from the pairwise comparisons performed for estimating the rates of false positive answers, organised in one column (calculated when KDE is used for between-object variability estimation; equations (4.5)–(4.8)). This type of data presentation is suitable for model performance evaluation, for example by empirical cross entropy (Chapter 6).

- `model_comparison_research_KDE_same.txt`, which presents only the *LR* values obtained from the comparisons performed for estimating the rates of false negative answers, organised in one column (calculated when KDE is used for between-object variability estimation; equations (4.5)–(4.8)). This type of data presentation is suitable for model performance evaluation, for example by empirical cross entropy (Chapter 6).

- `logNaO_research_descriptive_statistics.eps`, which is a graphical file presenting the descriptive statistics (box-plots, Q-Q plots, and KDE functions). For the remaining variables (*logMgO*, *logAlO*, *logSiO*, *logKO*, *logCaO*, *logFeO*) similar graphical files are generated.

- `model_research_LR_distribution.eps`, which is a graphical file presenting the $\log_{10} LR$ distribution for experiments performed to estimate the false negative and false positive rates.

There is also one additional file `comparison_research_error_rate.txt`, which provides information of the rates of false negative (`fn`) and false positive (`fp`) answers for the proposed LR model.

The *LR* results showed that the level of false positive answers estimated using the graphical model was 4.7% and the level of false negative answers was 4.0%. A complete assessment of the model performance can be performed using the empirical cross entropy approach (Chapter 6; Zadora and Ramos 2010).

4.4.7 Multivariate casework data – kernel density estimation – R software

In this example **R** code (given in Section D.9.1) is applied for casework purposes to data obtained during the analysis of elemental composition of glass samples by the SEM-EDX technique (Section 1.2.1).

4.4.7.1 Database creation

This examination involved ten recovered (`r1-r10`) and two control (`c1-c2`) glass fragments subjected to comparison analysis. The recovered glass fragments (`SEMEDX_recovered.txt`) were found on clothing belonging to a suspect. The control glass fragments (`SEMEDX_control.txt`) were found at the crime scene and collected from the broken window. The elemental concentration of all the glass fragments was determined by SEM-EDX according to the procedure described in Sections 1.2.1 and 4.4.6, which produces $\log_{10}$ ratios of each elemental concentration to the oxygen concentration. The control glass fragments were measured 12 times, and the recovered items were measured 3 times each. Additionally, the objects from the database were measured 12 times.

The columns of the data matrices are as follows:

- column 1 (`Name`), unique glass fragment name;
- column 2 (`Item`), glass fragment serial number;

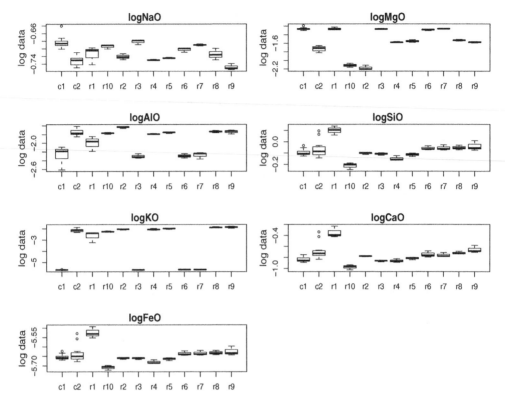

Figure 4.17 Box-plots of the seven variables observed for each of the control (c1-c2) and recovered (r1-r10) glass fragments. Data are available in the files SEMEDX_control.txt and SEMEDX_recovered.txt.

- column 3 (Piece), measurement index;

- columns 4–10, logNaO, logMgO, logAlO, logSiO, logKO, logCaO, logFeO variables.

Each row refers to a single measurement.

4.4.7.2 Graphical summary

To illustrate the data for the control and recovered fragments, box-plots were drawn (Figure 4.17). The box-plots (Section 3.3.3), Q-Q plots (Section 3.3.5) and KDE curves (Section 3.3.5) for the population data are presented in Figures 4.13–4.15.

4.4.7.3 LR modelling

LR calculations were performed according to equation (4.11). Computing *LR* for univariate and bivariate problems involved the KDE procedure for probability density function estimation (equations (4.5)–(4.8)). The **R** code used for the calculations is available in

SEMEDX_glass_comparison_casework_code.R. The code uses the functions given in Section D.9.1:

- UC_comparison_calculations.R, which contains the UC function to calculate the within- and between-object variability (**U** and **C**; equations (3.3) and (3.4), respectively);

- LR_comparison_KDE.R, which contains the LR.KDE.function to calculate the LR using KDE for estimating the between-object data distribution (equations (4.5)–(4.8)).

4.4.7.4   Results

Casework calculations involve pairwise LR computations for control and recovered objects to provide information on whether they may or may not share a common origin. The code SEMEDX_glass_comparison_casework_code.R can be run in the **R** Console after setting the proper working directory (Appendix D) and by inserting the command:

```
source(file = "SEMEDX_glass_comparison_casework_code.R")
```

Running the code provides the user with the following files as an outcome:

- model_comparison_casework_KDE.txt, which contains a matrix with the *LR* values (calculated when KDE is used for between-object variability estimation; equations (4.5)–(4.8)) for pairwise comparisons of the control and recovered fragments (Table 4.3).

- logNaO_casework_descriptive_statistics.eps (similar graphical files are obtained for the remaining variables), which is a graphical file presenting the descriptive statistics (box-plots, Q-Q plots, and KDE functions).

The results shown in Table 4.3 suggest that when comparing the elemental composition of the control and recovered objects, the evidence more likely supports the prosecutor's hypothesis (H_1, assuming their common origin) than the alternative hypothesis (H_2) in the case of pairwise comparisons between c1 and r3, r6, or r7 and between c2 and r2, r5, r8, or r9. The evidence more likely supports the H_2 hypothesis (stating that samples originated from different objects) only for fragments r1, r4, and r10 when compared with c1 and c2.

Table 4.3   Results of the LR pairwise comparisons between control (c1-c2) and recovered (r1-r10) glass fragments (available in model_comparison_casework_KDE.txt). In the data files the results are given in the form, for example, 2.097e-25 rather than $2.10 \cdot 10^{-25}$.

	c1	c2		c1	c2
r1	$2.10 \cdot 10^{-25}$	0.0035	r6	476.3	$2.85 \cdot 10^{-17}$
r2	$2.15 \cdot 10^{-23}$	335.1	r7	3815	$4.46 \cdot 10^{-20}$
r3	6785	$3.77 \cdot 10^{-25}$	r8	$2.61 \cdot 10^{-21}$	7648
r4	$1.02 \cdot 10^{-23}$	0.58	r9	$9.12 \cdot 10^{-32}$	57190
r5	$5.97 \cdot 10^{-22}$	160.2	r10	$1.29 \cdot 10^{-23}$	$1.55 \cdot 10^{-14}$

4.5   R Software

4.5.1   Routines for casework applications

Casework calculations involve pairwise LR computations for control and recovered objects to provide information on whether they may or may not share a common origin (see the examples in Section 4.4). A sample whose origin is known (e.g. collected from the crime scene), is referred to as a *control sample*. A sample whose origin is not known (e.g. recovered from the suspect's clothing), is referred to as a *recovered sample*.

The routines described in this section enable users to calculate likelihood ratios concerning two possible sources of data variability: within-object (U) and between-object (C) variability assuming a normal between-object data distribution or using kernel density estimation (Section 3.3.5) for computing a probability density function. The within-object distribution is always assumed normal. For full **R** routines refer to Section D.9.1 in Appendix D.

In order to work out the routines, an example is given which concerns a univariate comparison problem based on sodium content (column 4 in the data files), measured in two recovered and one control glass fragments. The data describing each of these objects are available in files named data_control.txt and data_recovered.txt and presented in Figures 4.18 and 4.19. A database of 200 glass objects (glass_database.txt) was used to estimate the parameters. Part of it is presented in Figure 4.20.

Data files and **R** code files must be stored in the same file folder. Data file names should have the *.txt file extension and **R** code file names the *.R file extension. The rows in the

Name	Item	Piece	logNaO	logMgO	logAlO	logSiO	logKO	logCaO	logFeO
recovered_1	1	1	-0.7038	-1.3388	-2.1741	-0.1559	-5.3782	-0.8532	-5.6792
recovered_1	1	2	-0.6941	-1.3347	-2.2167	-0.1535	-5.6791	-0.853	-5.378
recovered_1	1	3	-0.6952	-1.3366	-2.1719	-0.1532	-5.6771	-0.8478	-5.6771
recovered_1	1	4	-0.688	-1.2973	-2.0658	-0.0899	-5.3446	-0.7442	-5.6456
recovered_1	1	5	-0.694	-1.3145	-2.1114	-0.0836	-5.6429	-0.7243	-5.3418
recovered_1	1	6	-0.689	-1.3162	-2.0314	-0.1059	-5.6547	-0.7796	-5.6547
recovered_1	1	7	-0.6905	-1.3489	-2.0759	-0.2039	-5.6991	-0.9395	-5.6991
recovered_1	1	8	-0.7059	-1.3267	-2.0692	-0.1006	-5.3479	-0.7084	-5.3479
recovered_1	1	9	-0.6911	-1.347	-2.1212	-0.1838	-5.6894	-0.8798	-5.6894
recovered_1	1	10	-0.6767	-1.3062	-2.1569	-0.151	-5.6754	-0.8638	-5.6754
recovered_1	1	11	-0.6824	-1.3365	-2.1107	-0.1585	-5.3779	-0.8666	-5.3779
recovered_1	1	12	-0.6819	-1.3164	-2.0852	-0.1526	-5.6762	-0.864	-5.6762
recovered_2	2	1	-0.7254	-1.3489	-2.1627	-0.2089	-2.7068	-0.9825	-5.7068
recovered_2	2	2	-0.7244	-1.3664	-2.262	-0.2517	-2.683	-1.0607	-5.4233
recovered_2	2	3	-0.7336	-1.3765	-2.3247	-0.2003	-2.7049	-0.9622	-5.7049
recovered_2	2	4	-0.7398	-1.3723	-2.308	-0.1625	-2.433	-0.8896	-5.6882
recovered_2	2	5	-0.7294	-1.3574	-2.3888	-0.167	-2.4857	-0.9059	-5.3888
recovered_2	2	6	-0.7339	-1.3717	-2.2727	-0.162	-2.409	-0.8939	-5.6877
recovered_2	2	7	-0.7304	-1.3711	-2.2946	-0.2127	-2.5956	-0.9869	-5.7095
recovered_2	2	8	-0.7243	-1.3667	-2.2902	-0.2026	-2.5291	-0.9793	-5.4041
recovered_2	2	9	-0.7232	-1.3679	-2.2914	-0.2059	-2.5023	-0.9854	-5.7064
recovered_2	2	10	-0.7214	-1.3499	-2.3558	-0.1863	-2.4941	-0.9578	-5.6982
recovered_2	2	11	-0.7235	-1.3502	-2.327	-0.1651	-2.4582	-0.9134	-5.3877
recovered_2	2	12	-0.7228	-1.3554	-2.4733	-0.1994	-2.5276	-0.9721	-5.7037

Figure 4.18   Data on recovered glass objects available in data_recovered.txt.

Name	Item	Piece	logNaO	logMgO	logAlO	logSiO	logKO	logCaO	logFeO
control_1	1	1	-0.7231	-1.3352	-2.1433	-0.1633	-2.7331	-0.9099	-5.6874
control_1	1	2	-0.7287	-1.3588	-2.2121	-0.1655	-2.4851	-0.9125	-5.3882
control_1	1	3	-0.7305	-1.3379	-2.1696	-0.1647	-2.5742	-0.905	-5.6882
control_1	1	4	-0.7245	-1.3369	-2.1634	-0.1797	-2.7406	-0.9413	-5.6949
control_1	1	5	-0.7231	-1.353	-2.2163	-0.1755	-2.6143	-0.9353	-5.3924
control_1	1	6	-0.7255	-1.3721	-2.2814	-0.1819	-2.5824	-0.9352	-5.6964
control_1	1	7	-0.7306	-1.3747	-2.2737	-0.2018	-2.559	-0.9703	-5.7051
control_1	1	8	-0.7257	-1.3651	-2.2603	-0.2092	-2.5614	-0.984	-5.4065
control_1	1	9	-0.722	-1.36	-2.2612	-0.2082	-2.7083	-0.989	-5.7083
control_1	1	10	-0.7409	-1.373	-2.3095	-0.1992	-5.7075	-0.9865	-5.7075
control_1	1	11	-0.7385	-1.3607	-2.2901	-0.196	-5.7051	-0.9776	-5.4041
control_1	1	12	-0.7429	-1.3652	-2.1742	-0.1973	-5.4047	-0.9757	-5.7057

Figure 4.19 Data on control glass objects available in `data_control.txt`.

data files correspond to n measurements carried out on each of m objects (so that there are nm rows in total). The **R** routines presented in this book require that the number of measurements (n) carried out for each object is the same throughout the database. As set out repeatedly in Section 4.4, the columns of the matrix are as follows:

- column 1 (`Name`), unique object name;

- column 2 (`Item`), object serial number;

- column 3 (`Piece`), measurement index from 1 to n;

- from column 4, variables (e.g. columns 4–10: `logNaO`, `logMgO`, `logAlO`, `logSiO`, `logKO`, `logCaO`, `logFeO` variables).

The code for casework purposes (`comparison_casework_code.R`, Section D.9.1) uses the following functions given in Section D.9.1 in Appendix D:

- `UC_comparison_calculations.R`, which contains the UC function to calculate the distributional parameters for LR computation when a comparison problem is considered. To estimate such parameters, it is necessary to use a database of similar objects described by the same variables, gathered during analyses of the same type. The function calculates the within-object variability (denoted by **U** and given by equation (3.3))

Name	Item	Piece	logNaO	logMgO	logAlO	logSiO	logKO	logCaO	logFeO
s1	1	1	-0.6603	-1.4683	-1.4683	-0.1463	-1.7047	-1.1096	-5.6778
s1	1	2	-0.6658	-1.4705	-1.4814	-0.1429	-1.7183	-1.1115	-5.3763
s1	1	3	-0.656	-1.4523	-1.4789	-0.1477	-1.6864	-1.1118	-5.6776
s1	1	4	-0.6309	-1.4707	-1.5121	-0.1823	-1.7743	-1.1306	-2.609
s1	1	5	-0.6332	-1.4516	-1.4996	-0.1792	-1.7577	-1.1332	-5.6871
s1	1	6	-0.6315	-1.4641	-1.4883	-0.171	-1.7548	-1.1291	-5.6842
s1	1	7	-0.642	-1.4437	-1.4708	-0.1165	-1.6789	-1.0358	-2.4307
s1	1	8	-0.6431	-1.4692	-1.4664	-0.1048	-1.6568	-1.0094	-5.656
s1	1	9	-0.658	-1.4698	-1.4612	-0.0843	-1.5881	-0.9888	-5.6488
s1	1	10	-0.6477	-1.463	-1.4956	-0.1653	-1.7289	-1.1138	-5.6831
s1	1	11	-0.633	-1.4563	-1.4944	-0.1579	-1.7297	-1.1074	-5.6791
s1	1	12	-0.6424	-1.4607	-1.4987	-0.1669	-1.7591	-1.1199	-2.6042

Figure 4.20 Part of the database consisting of 200 glass objects (entire database available in `glass_database.txt`).

and between-object variability (denoted by **C** and given by equation (3.4)) of the variables of interest.

- `LR_comparison_Nor.R`, which contains the `LR.Nor.function` to compute the LR when the between-object distribution is assumed normal, in accordance with equations (4.1)–(4.4).

- `LR_comparison_KDE.R`, which contains the `LR.KDE.function` to compute the LR when the between-object distribution is not assumed normal and KDE is used for its estimation, in accordance with equations (4.5)–(4.8).

In order to run the code file `comparison_casework_code.R` to obtain LR results, the user must include file names for the database, recovered object and control object data:

```
population = read.table("ENTER THE FILE NAME.txt", header = TRUE)
data.recovered = read.table("ENTER THE FILE NAME.txt", header = TRUE)
data.control = read.table("ENTER THE FILE NAME.txt", header = TRUE)
```

Also required are the the index (indices) of the column(s) containing the data for the relevant variable(s):

```
variables = c(ENTER THE INDEX (INDICES) OF THE COLUMNS, COMMA SEPARATED)
```

For *logNaO* and *logSiO*, for example, the correct input would be:

```
variables = c(4,7)
```

The code `comparison_casework_code.R` can be run in the **R** Console after setting the proper working directory (Appendix D) and by inserting the command:

```
source(file = "comparison_casework_code.R")
```

Running the code for the variable introduced in the example (*logNaO*) results in the following files as outcomes:

- `logNaO_comparison_casework_Nor.txt` containing *LR* values calculated for pairwise comparisons between control and recovered fragments when the between-object distribution is assumed normal (equations (4.1)–(4.4)). This example involves only one control and two recovered items (Figure 4.21(a)). If there are more control items, a matrix is created with a column for each control item and a row for each recovered item. In general, if other models (univariate or multivariate) are considered, the beginning of the file name will change according to the new variable name(s) specified in the column names of the data files.

- `logNaO_comparison_casework_KDE.txt` with *LR* values calculated for pairwise comparisons between the control and recovered fragments when KDE is used for

```
logNaO                          logNaO
control_1                       control_1
recovered_1 3.484e-10           recovered_1 2.954e-10
recovered_2 10.66               recovered_2 8.192
```

(a)   (b)

Figure 4.21 Results of LR calculations (for the *logNaO* variable) for comparison problem of glass fragments when the between-object distribution: (a) is assumed normal (available in `logNaO_comparison_casework_Nor.txt`); (b) estimated by KDE (available in `logNaO_comparison_casework_KDE.txt`).

estimation of distributional parameters (equations (4.5)–(4.8)). This example involves only one control and two recovered items (Figure 4.21(b)). If there are more control items, a matrix is created with a column for each control item and a row for each recovered item. In general, if other models (univariate or multivariate) are considered, the beginning of the file name will change according to the new variable name(s) specified in the column names of the data files.

- `logNaO_casework.eps`, which provides plots of descriptive statistics (in the form of box-plots) and data distributions (in the form of Q-Q plots and KDE curves), as shown in Figure 4.22, saving them in the relevant files denoted by the variable name. In general, if other models (univariate or multivariate) are considered, the beginning of the file name will change according to the new variable name(s) specified in the column names of the data files.

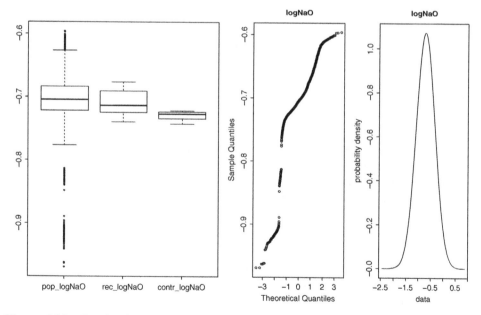

Figure 4.22 Graphs obtained by running the code `comparison_casework_code.R` for the *logNaO* variable (available in `logNaO_casework.eps`): (left) box-plots; (middle) Q-Q plot; (right) KDE plot.

4.5.2   Routines for research applications

LR calculations are used in research within a comparison problem to obtain information about the performance of LR models by estimating the levels of false positive and false negative answers (Section 2.4). This is done by pairwise comparisons of each pair of objects from different sources (to estimate false positives) and comparisons of control and recovered samples from the same object (to estimate false negatives).

The routines described in this section enable users to calculate likelihood ratios concerning two possible sources of data variability: within-object (**U**) and between-object (**C**) variability assuming a normal between-object data distribution or using KDE (Section 3.3.5) to compute the probability density function for the between-object variation. For full **R** routines refer to Section D.9.2 in Appendix D.

In order to work out the routines, an example is given which concerns a univariate comparison problem based on sodium content (column 4 in the data file), measured in 200 glass fragments (`glass_database.txt`; Figure 4.20).

Data files and **R** code files must be stored in the same file folder. Data file names should have the `*.txt` file extension and **R** code file names the `*.R` file extension. The rows in the data files correspond to n measurements carried out on each of m objects (so that there are nm rows in total). The **R** routines presented in this book require that the number of measurements (n) carried out for each object is the same throughout the database. As set out repeatedly in Section 4.4, the columns of the matrix are as follows:

- column 1 (`Name`), unique object name;

- column 2 (`Item`), object serial number;

- column 3 (`Piece`), measurement index from 1 to n;

- from column 4, variables (e.g. columns 4–10: `logNaO, logMgO, logAlO, logSiO, logKO, logCaO, logFeO` variables).

The code for research purposes (`comparison_research_code.R`, Section D.9.2) uses the following functions given in Section D.9.1:

- `UC_comparison_calculations.R`, which contains the `UC` function calculate the distributional parameters for LR computation when a comparison problem is considered. To estimate such parameters, it is necessary to use a database of similar objects described by the same variables, gathered during analyses of the same type. The function calculates the within-object variability (denoted by **U** and given by equation (3.3)) and between-object variability (denoted by **C** and given by equation (3.4)) of the variables of interest.

- `LR_comparison_Nor.R`, which contains the `LR.Nor.function` to compute the LR when the between-object distribution is assumed normal, in accordance with equations (4.1)–(4.4).

- `LR_comparison_KDE.R`, which contains the `LR.KDE.function` to compute the LR when the between-object distribution is not assumed normal and KDE is used for its estimation, in accordance with equations (4.5)–(4.8).

The code compares each pair of objects. There are two different types of experiments that provide information on the rates of false positive and false negative answers. For false positive rates, pairwise comparisons between different objects are performed. Therefore, there are $m(m-1)/2$ comparisons within a database of m objects. Each pair of objects constitutes a control and a recovered sample. The database used as background information for parameter estimation purposes is then diminished by 2 objects (a control and a recovered sample) and comprises $m-2$ objects. This is known as a jackknife procedure. For false negative rates, pairwise comparisons between samples from the same object are performed. This involves dividing a single sample into two parts that form the control and recovered items. For such comparisons, the database created according to the jackknife procedure consists of $m-1$ objects.

In order to run the code file `comparison_research_code.R` to obtain LR results, the user must include file names for the database data and the data to be analysed (usually the same as the database in research practice):

```
population = read.table("ENTER THE FILE NAME.txt", header = TRUE)
data.analysed = read.table("ENTER THE FILE NAME.txt", header = TRUE)
```

Also required are the the index (indices) of the column(s) containing the data for the relevant variable(s):

```
variables = c(ENTER THE INDEX (INDICES) OF THE COLUMNS, COMMA SEPARATED)
```

For *logNaO* and *logSiO*, for example, the correct input would be:

```
variables = c(4,7)
```

Finally, the user must include the index (indices) of the row(s) forming control and recovered samples for the estimation of false negative rates:

```
y.1 = data.frame(y.1.2[ENTER THE INDEX (INDICES) OF THE ROW(S) CREATING
        THE CONTROL SAMPLE, E.G. 1:6,])
y.2 = data.frame(y.1.2[ENTER THE INDEX (INDICES) OF THE ROW(S) CREATING
        THE RECOVERED SAMPLE, E.G. 7:12,])
```

For example:

```
y.1 = data.frame(y.1.2[1:6,])
y.2 = data.frame(y.1.2[7:12,])
```

In order to become familiar with the routines, the database of 200 glass objects available in `glass_database.txt` can be downloaded in order to perform the calculations. Our example concerns a univariate comparison problem based on sodium content (column 4 in the data file) measured in 200 glass samples.

The code comparison_research_code.R can be run in the **R** Console after setting the proper working directory (Appendix D) and by inserting the command:

```
source(file = "comparison_research_code.R")
```

Running the code for the variable *logNaO* provides the user with the following files as an outcome (Figures 4.23–4.25):

- logNaO_comparison_research_Nor.txt, which contains a matrix of size $m \times m$ (where m is the number of objects in the database; here $m = 200$ glass fragments) of the *LR* values (calculated under the assumption of a normal between-object distribution; equations (4.1)–(4.4)) for compared objects. For example when s1 and s2 are compared the LR equals to 0.0002919. Only half of the matrix is presented, as the other half is given by symmetry – for example, the comparison of s1 (control sample) and s20 (recovered sample) gives the same *LR* value as s20 (control sample) and s1 (recovered sample). The diagonal elements of the matrix provide the *LR* results obtained from experiments performed in order to estimate the rates of false negative answers. The remaining elements present the *LR* values obtained from experiments performed in order to estimate the rates of false positive answers (Figure 4.23(a)). In general, if other models (univariate or multivariate) are considered, the beginning of the file name will change according to the new variable(s) name(s) specified in the column names of the data files.

- logNaO_comparison_research_KDE.txt, which contains a matrix of size $m \times m$ (where m is the number of objects in the database; here $m = 200$ glass fragments) of the *LR* values (calculated using KDE; equations (4.5)–(4.8)) for compared objects. For example when s1 and s2 are compared the LR equals to 0.000362. Only half of the matrix is presented, as the other half is given by symmetry – for example, the

	s1	s2	s3		s1	s2	s3
s1	13.36	0	0	s1	16.2	0	0
s2	0.0002919	21.61	0	s2	0.000362	29.07	0
s3	1.66E-06	23.3	27.36	s3	2.04E-06	32.68	36.92

(a)   (b)

```
logNaO
fp_Nor 23.3
fp_KDE 21.6
fn_Nor 12
fn_KDE 12
```

(c)

Figure 4.23 Part of the results of LR calculations for the *logNaO* variable: (a) when a normal between-object distribution is assumed; (b) when KDE is used for between-object distribution estimation; (c) false positive and false negative rates. Complete results are available in files logNaO_comparison_research_Nor.txt, logNaO_comparison_research_KDE.txt and comparison_research_error_rate.txt.

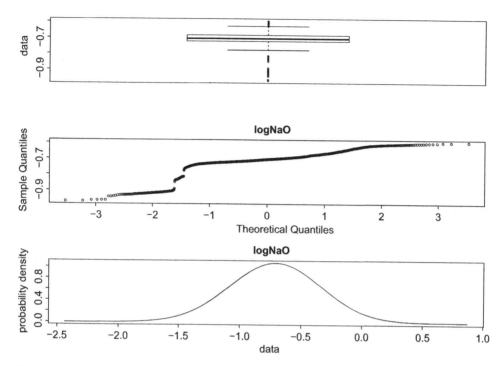

Figure 4.24 Graphs (from top to bottom: box-plot, Q-Q plot and KDE plot) obtained by running the code comparison_research_code.R, presenting descriptive statistics for the *logNaO* variable (available in logNaO_research_descriptive_statistics.eps).

comparison of s1 (control sample) and s20 (recovered sample) gives the same *LR* value as s20 (control sample) and s1 (recovered sample). The diagonal elements of the matrix provide the *LR* results obtained from experiments performed in order to estimate the rates of false negative answers (Figure 4.23(b)). The remaining elements present the *LR* values obtained from experiments performed in order to estimate the rates of false positive answers. In general, if other models (univariate or multivariate) are considered, the beginning of the file name will change according to the new variable(s) name(s) specified in the column names of the data files.

- logNaO_comparison_research_Nor_different.txt, which presents only the *LR* values obtained from the pairwise comparisons performed to estimate the rates of false positive answers, organised in one column (under the assumption of a normal between-object distribution, equations (4.1)–(4.4)). This type of data presentation is suitable for model performance evaluation, for example by empirical cross entropy (Chapter 6). In general, if other models (univariate or multivariate) are considered, the beginning of the file name will change according to the new variable(s) name(s) specified in the column names of the data files.

- logNaO_comparison_research_Nor_same.txt, which presents only the *LR* values obtained from the comparisons performed to estimate the rates of false negative answers, organised in one column (under the assumption of a normal between-object

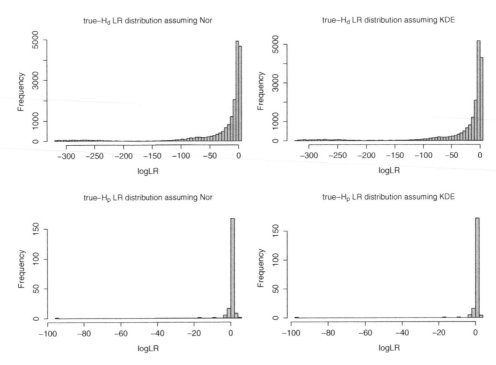

Figure 4.25 Graphs obtained by running the code `comparison_research_code.R`, showing the $\log_{10} LR$ distribution for the *logNaO* variable (available in `logNaO_research_LR_distribution.eps`).

distribution, equations (4.1)–(4.4)). This type of data presentation is suitable for model performance evaluation, for example by empirical cross entropy (Chapter 6). In general, if other models (univariate or multivariate) are considered, the beginning of the file name will change according to the new variable(s) name(s) specified in the column names of the data files.

- `logNaO_comparison_research_KDE_different.txt`, which presents only the *LR* values obtained from the pairwise comparisons performed to estimate the rates of false positive answers, organised in one column (when KDE is used, equations (4.5)–(4.8)). This type of data presentation is suitable for model performance evaluation, for example by empirical cross entropy (Chapter 6). In general, if other models (univariate or multivariate) are considered, the beginning of the file name will change according to the new variable(s) name(s) specified in the column names of the data files.

- `logNaO_comparison_research_KDE_same.txt`, which presents only the *LR* values obtained from the comparisons performed to estimate the rates of false negative answers, organised in one column (when KDE is used, equations (4.5)–(4.8)). This type of data presentation is suitable for model performance evaluation, for example by empirical cross entropy (Chapter 6). In general, if other models (univariate or multivariate) are considered, the beginning of the file name will change according to the new variable(s) name(s) specified in the column names of the data files.

- `logNaO_research_descriptive_statistics.eps`, which is a graphical file presenting the descriptive statistics (box-plots, Q-Q plots, and KDE functions) for variables of interest (Figure 4.24). For the remaining variables the same graphs are generated. In general, if other models (univariate or multivariate) are considered, the beginning of the file name will change according to the new variable(s) name(s) specified in the column names of the data files.

- `logNaO_research_LR_distribution.eps`, which is a graphical file presenting the $\log_{10} LR$ distribution for experiments performed to estimate the false negative and false positive rates (Figure 4.25). In general, if other models (univariate or multivariate) are considered, the beginning of the file name will change according to the new variable(s) name(s) specified in the column names of the data files.

There is also one additional file `comparison_research_error_rate.txt`, which provides information on the rates of false negative (`fn`) and false positive (`fp`) answers for the proposed LR model.

All the necessary functions (`UC_comparison_calculations.R`, `LR_comparison_Nor.R` and `LR_comparison_KDE.R`; Section D.9.1) must be enclosed in the relevant file folder.

References

Aitken CGG 1986 Statistical discriminant analysis in forensic science. *Journal of Forensic Sciences* **26**, 237–247.

Aitken CGG, Lucy D 2004 Evaluation of trace evidence in the form of multivariate data. *Applied Statistics* **53**, 109–122.

Aitken CGG, Taroni F 2004 *Statistics and the Evaluation of Evidence for Forensic Scientists*, 2nd edn. John Wiley & Sons, Ltd, Chichester.

Berry DA 1991 Inferences in forensic identification and paternity cases using hypervariable DNA sequences (with discussion). *Statistical Science* **6**, 175–205.

Berry DA, Evett IW, Pinchin R 1992 Statistical inference in crime investigations using deoxyribonucleic acid profiling (with discussion). *Applied Statistics* **41**, 499–531.

Bozza S, Taroni F, Marquis R, Schmittbuhl M 2008 Probabilistic evaluation of handwriting evidence: Likelihood ratio for authorship. *Applied Statistics* **57**, 329–341.

Chan KPS, Aitken CGG 1989 Estimation of the Bayes' factor in a forensic science problem. *Journal of Statistical Computation and Simulation* **33**, 249–264.

Evett IW, Cage PE, Aitken CGG 1987 Evaluation of the likelihood ratio for fibre transfer evidence in criminal cases. *Applied Statistics* **36**, 174–180.

Martyna A, Lucy D, Zadora G, Trzcinska BM, Ramos D, Parczewski A 2013 The evidential value of microspectrophotometry measurements made for pen inks. *Analytical Methods*, doi: 10.1039/C3AY41622D.

Neocleous T, Aitken CGG, Zadora G 2011 Transformations for compositional data with zeros with an application to forensic evidence evaluation. *Chemometrics and Intelligent Laboratory Systems* **109**, 77–85.

Taroni F, Bozza S, Biedermann A, Garbolino P and Aitken CGG 2010 *Data Analysis in Forensic Science: A Bayesian Decision Perspective*. John Wiley & Sons, Ltd, Chichester.

Whittaker J 1990 *Graphical Models in Applied Multivariate Statistics*. John Wiley & Sons, Ltd, Chichester.

Zadora G 2010 Evaluation of evidential value of physicochemical data by a Bayesian network approach. *Journal of Chemometrics* **24**, 346–366.

Zadora G and Ramos D 2010 Evaluation of glass samples for forensic purposes – an application of likelihood ratio model and information-theoretical approach. *Chemometrics and Intelligent Laboratory Systems* **102**, 62–63.

Zadora G, Surma M 2010 The calcuLatoR – software for evaluation of evidence value of univariate continuous type data. *Chemometrics and Intelligent Laboratory Systems* **103**, 43–52.

Zieba-Palus J, Zadora G, Milczarek JM 2008 Differentiation and evaluation of evidence value of styrene acrylic urethane topcoat car paints analysed by pyrolysis-gas chromatography. *Journal of Chromatography A* **1179**, 47–58.

5

Likelihood ratio models for classification problems

5.1 Introduction

In forensic cases, it is important to determine which category the object belongs to. This is a so-called *classification problem*, which can be solved in the context of two contrasting hypotheses: the first, H_1, the proposition that the object of interest comes from category 1, while the second, H_2, is the proposition that the object comes from category 2. To address this problem the likelihood ratio approach could be used in the form of $LR = \frac{P(E|H_1)}{P(E|H_2)}$ in the case of discrete data, and $LR = \frac{f(E|H_1)}{f(E|H_2)}$ in the case of continuous data (where $P(\cdot)$ stands for probability and $f(\cdot)$ stands for probability density function; Appendix A).

The classification problem is especially important when the objects of interest are recovered from, for example, the victim's clothes, and there is no control sample, for example a glass fragment collected on the scene of crime. A classification of glass fragments could help investigators (policemen, prosecutors) focus their search for appropriate control materials.

It should be explained that in the forensic sphere the word *classification* is used to solve a problem which in statistics and chemometrics is usually referred to as *discrimination*. This is because discrimination in the forensic sphere is related to the problem of *comparison*.

Undertaking the classification task using evidence evaluation methods requires taking into account the following aspects:

- sources of uncertainty, which include the variation of measurements within the object of interest, and the variation of measurements between objects in the relevant population;

- information about the rarity of the physicochemical data (e.g. elemental and/or chemical composition) of the sample of interest;

- existing correlation between variables in the case of multidimensional data.

Statistical Analysis in Forensic Science: Evidential Value of Multivariate Physicochemical Data, First Edition.
Grzegorz Zadora, Agnieszka Martyna, Daniel Ramos and Colin Aitken.
© 2014 John Wiley & Sons, Ltd. Published 2014 by John Wiley & Sons, Ltd.
Companion website: www.wiley.com/go/physicochemical

In this chapter, LR models which allow users to include all the above-mentioned factors are presented. More information on the application of the LR in forensic sciences can be found in Chapter 2.

The information about the within- ($\mathbf{U}_g$) and between-object ($\mathbf{C}_g$) variability within each of the two categories ($g = 1, 2$) can be assessed from background information obtained for each object from the appropriate database. Details of the relevant calculations can be found in Section 3.3.2. In the LR models presented in this chapter the within-object variance is always assumed normal, while between-object variance is assumed normal or is estimated by a kernel density estimation procedure using Gaussian kernels (Section 3.3.5).

The data obtained as a result of the analysis performed on m_g objects from category g ($g = 1, 2$), each measured n times, constitute a database in the form of p-vectors (where p is the number of variables considered): $\mathbf{x}_{gij} = (x_{gij1}, \ldots, x_{gijp})^T$, where $i = 1, \ldots, m_g$ and $j = 1, \ldots, n$.

Suppose that the features of an object whose category g ($g = 1, 2$) is to be determined are measured n_y times, such that

$$\mathbf{y}_j = (y_{j1}, \ldots, y_{jp})^T,$$

where $j = 1, \ldots, n_y$.

The expressions for one-level model (only between-object variability is considered, expressed by the between-object covariance matrix $\mathbf{C}$) and two-level model (both within- and between-object variabilities are considered, expressed by the within- and between-object covariance matrices $\mathbf{U}$ and $\mathbf{C}$) LR computations, for univariate or multivariate data, are given in Sections 5.2 and 5.3, depending on whether the between-object data distribution is assumed normal or is estimated by the KDE procedure.

5.2   Normal between-object distribution

Denote the mean vector for the ith object from the gth category ($g = 1, 2$) by $\boldsymbol{\theta}_{gi}$ (estimated by $\bar{\mathbf{x}}_{gi} = \frac{1}{n} \sum_{j=1}^{n} \mathbf{x}_{gij}$), the within-object covariance matrix by $\mathbf{U}_g$, and the between-object covariance matrix by $\mathbf{C}_g$. Estimates of $\mathbf{U}$ and $\mathbf{C}$ are given by equations (3.3) and (3.4).

Then, given $\boldsymbol{\theta}_{gi}$ and $\mathbf{U}_g$, the distribution of the observations for the ith object $\mathbf{X}_{gij}$ is

$$\left(\mathbf{X}_{gij} \mid \boldsymbol{\theta}_{gi}, \mathbf{U}_g\right) \sim N_p(\boldsymbol{\theta}_{gi}, \mathbf{U}_g), \quad i = 1, \ldots, m_g, \ j = 1, \ldots, n.$$

It is also assumed that

$$\left(\boldsymbol{\theta}_{gi} \mid \boldsymbol{\mu}_g, \mathbf{C}_g\right) \sim N_p(\boldsymbol{\mu}_g, \mathbf{C}_g), \quad i = 1, \ldots, m_g,$$

where $\boldsymbol{\mu}_g$ are the vectors of the overall means of p variables. The $\boldsymbol{\mu}_g$ are estimated by the means $\bar{\mathbf{x}}_g$ from the n measurements for m_g objects from the database for the gth category ($g = 1, 2$) with

$$\bar{\mathbf{x}}_g = \frac{1}{m_g n} \sum_{i=1}^{m_g} \sum_{j=1}^{n} \mathbf{x}_{gij}.$$

5.2.1 Multivariate data

The LR model based on ideas presented for solving a comparison problem (Chapter 4) by Aitken and Lucy (2004), Zadora and Neocleous (2009), and Zadora (2010) is expressed as

$$LR = \frac{f(\bar{\mathbf{y}}|\mathbf{U}_1, \mathbf{C}_1, \bar{\mathbf{x}}_1, H_1)}{f(\bar{\mathbf{y}}|\mathbf{U}_2, \mathbf{C}_2, \bar{\mathbf{x}}_2, H_2)},$$

where $\bar{\mathbf{y}}$ is the vector of the means of p variables calculated using n_y measurements,

$$\bar{\mathbf{y}} = \frac{1}{n_y} \sum_{j=1}^{n_y} \mathbf{y}_j.$$

For multivariate data ($p > 1$) the following equations should be used: for the numerator,

$$f(\bar{\mathbf{y}}|\mathbf{U}_1, \mathbf{C}_1, \bar{\mathbf{x}}_1, H_1)$$

$$= (2\pi)^{-p/2} \left| \frac{\mathbf{U}_1}{n} + \mathbf{C}_1 \right|^{-1/2} \exp\left\{ -\frac{1}{2}(\bar{\mathbf{y}} - \bar{\mathbf{x}}_1)^T \left(\frac{\mathbf{U}_1}{n} + \mathbf{C}_1 \right)^{-1} (\bar{\mathbf{y}} - \bar{\mathbf{x}}_1) \right\};$$

and for the denominator,

$$f(\bar{\mathbf{y}}|\mathbf{U}_2, \mathbf{C}_2, \bar{\mathbf{x}}_2, H_2)$$

$$= (2\pi)^{-p/2} \left| \frac{\mathbf{U}_2}{n} + \mathbf{C}_2 \right|^{-1/2} \exp\left\{ -\frac{1}{2}(\bar{\mathbf{y}} - \bar{\mathbf{x}}_2)^T \left(\frac{\mathbf{U}_2}{n} + \mathbf{C}_2 \right)^{-1} (\bar{\mathbf{y}} - \bar{\mathbf{x}}_2) \right\}.$$

Then

$$LR = \frac{\left| \frac{\mathbf{U}_1}{n} + \mathbf{C}_1 \right|^{-1/2} \exp\left\{ -\frac{1}{2}(\bar{\mathbf{y}} - \bar{\mathbf{x}}_1)^T \left(\frac{\mathbf{U}_1}{n} + \mathbf{C}_1 \right)^{-1} (\bar{\mathbf{y}} - \bar{\mathbf{x}}_1) \right\}}{\left| \frac{\mathbf{U}_2}{n} + \mathbf{C}_2 \right|^{-1/2} \exp\left\{ -\frac{1}{2}(\bar{\mathbf{y}} - \bar{\mathbf{x}}_2)^T \left(\frac{\mathbf{U}_2}{n} + \mathbf{C}_2 \right)^{-1} (\bar{\mathbf{y}} - \bar{\mathbf{x}}_2) \right\}}. \tag{5.1}$$

5.2.2 Univariate data

For univariate data ($p = 1$), the covariance matrices become scalar variances (e.g. $\mathbf{U}_g$ becomes u_g^2, and $\mathbf{C}_g$ becomes c_g^2), and the vectors of means become scalar means (e.g. $\bar{\mathbf{x}}_g$ becomes $\bar{x}_g$). Thus, the following equations should be used: for the numerator,

$$f(\bar{y}|u_1^2, c_1^2, \bar{x}_1, H_1)$$

$$= \left(2\pi \left(\frac{u_1^2}{n} + c_1^2 \right) \right)^{-1/2} \exp\left\{ -\left(2\left(\frac{u_1^2}{n} + c_1^2 \right) \right)^{-1} (\bar{y} - \bar{x}_1)^2 \right\};$$

and for the denominator,

$$f(\bar{y}|u_2^2, c_2^2, \bar{x}_2, H_2)$$

$$= \left(2\pi \left(\frac{u_2^2}{n} + c_2^2\right)\right)^{-1/2} \exp\left\{-\left(2\left(\frac{u_2^2}{n} + c_2^2\right)\right)^{-1}(\bar{y} - \bar{x}_2)^2\right\}.$$

Then

$$LR = \frac{\left(\frac{u_1^2}{n} + c_1^2\right)^{-1/2} \exp\left\{-\left(2\left(\frac{u_1^2}{n} + c_1^2\right)\right)^{-1}(\bar{y} - \bar{x}_1)^2\right\}}{\left(\frac{u_2^2}{n} + c_2^2\right)^{-1/2} \exp\left\{-\left(2\left(\frac{u_2^2}{n} + c_2^2\right)\right)^{-1}(\bar{y} - \bar{x}_2)^2\right\}}. \tag{5.2}$$

5.2.3   One-level models

Objects from the database may be described by a vector of p means of variables or a vector of values from one measurement. Then the information about the within-object variability ($\mathbf{U}_g$) is lost. Such a situation precludes the two-level model and reduces to the simplified one-level model (Zadora 2009).

5.2.3.1   Multivariate data

In the case of multivariate data ($p > 1$) the following equations should be used: for the numerator,

$$f(\bar{y}|\mathbf{C}_1, \bar{\mathbf{x}}_1, H_1) = (2\pi)^{-p/2} |\mathbf{C}_1|^{-1/2} \exp\left\{-\frac{1}{2}(\bar{y} - \bar{\mathbf{x}}_1)^T (\mathbf{C}_1)^{-1}(\bar{y} - \bar{\mathbf{x}}_1)\right\};$$

and for the denominator,

$$f(\bar{y}|\mathbf{C}_2, \bar{\mathbf{x}}_2, H_2) = (2\pi)^{-p/2} |\mathbf{C}_2|^{-1/2} \exp\left\{-\frac{1}{2}(\bar{y} - \bar{\mathbf{x}}_2)^T (\mathbf{C}_2)^{-1}(\bar{y} - \bar{\mathbf{x}}_2)\right\}.$$

Then

$$LR = \frac{f(\bar{y}|\mathbf{C}_1, \bar{\mathbf{x}}_1, H_1)}{f(\bar{y}|\mathbf{C}_2, \bar{\mathbf{x}}_2, H_2)} = \frac{|\mathbf{C}_1|^{-1/2} \exp\left\{-\frac{1}{2}(\bar{y} - \bar{\mathbf{x}}_1)^T (\mathbf{C}_1)^{-1}(\bar{y} - \bar{\mathbf{x}}_1)\right\}}{|\mathbf{C}_2|^{-1/2} \exp\left\{-\frac{1}{2}(\bar{y} - \bar{\mathbf{x}}_2)^T (\mathbf{C}_2)^{-1}(\bar{y} - \bar{\mathbf{x}}_2)\right\}}. \tag{5.3}$$

5.2.3.2   Univariate data

In the case of univariate data ($p = 1$), the matrices become scalars (e.g. $\mathbf{C}_g$ becomes c_g^2) and the following equations should be used: for the numerator,

$$f(\bar{y}|c_1^2, \bar{x}_1, H_1) = \left(2\pi c_1^2\right)^{-1/2} \exp\left\{-\left(2c_1^2\right)^{-1}(\bar{y} - \bar{x}_1)^2\right\};$$

and for the denominator,

$$f(\bar{y}|c_2^2, \bar{x}_2, H_2) = \left(2\pi c_2^2\right)^{-1/2} \exp\left\{-\left(2c_2^2\right)^{-1}(\bar{y} - \bar{x}_2)^2\right\}.$$

Then

$$LR = \frac{f(\bar{y}|c_1^2, \bar{x}_1, H_1)}{f(\bar{y}|c_2^2, \bar{x}_2, H_2)} = \frac{\left(c_1^2\right)^{-1/2} \exp\left\{-\left(2c_1^2\right)^{-1}(\bar{y} - \bar{x}_1)^2\right\}}{\left(c_2^2\right)^{-1/2} \exp\left\{-\left(2c_2^2\right)^{-1}(\bar{y} - \bar{x}_2)^2\right\}}. \qquad (5.4)$$

5.3 Between-object distribution modelled by kernel density estimation

The assumption of a normal distribution for θ in the case of univariate data, or of a multivariate normal distribution in the case of multivariate data (i.e. the distribution of the data within a database) may not always be correct. The assumption of normality can be replaced by considering a kernel density estimate (Section 3.3.5) for the between-object distribution. Earlier examples of the use of the KDE in forensic science are available in Aitken (1986), Aitken and Taroni (2004), Berry (1991), Berry et al. (1992), Chan and Aitken (1989), and Evett et al. (1987).

The kernel density function is expressed as follows. For multivariate data θ in group g,

$$K(\theta \mid \bar{x}_{gi}, \mathbf{H}_g) = (2\pi)^{-p/2}|\mathbf{H}_g|^{-1/2} \exp\left\{-\frac{1}{2}(\theta - \bar{x}_{gi})^T \mathbf{H}_g^{-1}(\theta - \bar{x}_{gi})\right\},$$

where $\bar{x}_{gi}$ is the mean of the n measurements for the ith item in group g, $\bar{x}_{gi} = \frac{1}{n}\sum_{j=1}^{n} \mathbf{x}_{gij}$, and $\mathbf{H}_g$ is the kernel bandwidth matrix, $\mathbf{H}_g = h_g^2 \mathbf{C}_g$, where h_g is a smoothing parameter (Section 3.3.5), commonly calculated as

$$h_g = h_{\text{opt}} = \left(\frac{4}{m_g(2p + 1)}\right)^{\frac{1}{p+4}}.$$

For univariate data θ in group g,

$$K(\theta \mid \bar{x}_{gi}, h_g) = (2\pi h_g^2 c_g^2)^{-1/2} \exp\left\{-\frac{1}{2}(\theta - \bar{x}_{gi})^2 \left(h_g^2 c_g^2\right)^{-1}\right\}.$$

5.3.1 Multivariate data

For multivariate data ($p > 1$) the following equations should be used (Zadora and Neocleous 2009; Zadora et al. 2010): for the numerator,

$$f(\bar{y}|\mathbf{U}_1, \mathbf{C}_1, \bar{x}_{11}, \ldots, \bar{x}_{1m_1}, h_1, H_1)$$
$$= (2\pi)^{-p/2} \left|\frac{\mathbf{U}_1}{n} + h_1^2 \mathbf{C}_1\right|^{-1/2} \frac{1}{m_1} \sum_{i=1}^{m_1} \exp\left\{-\frac{1}{2}(\bar{y} - \bar{x}_{1i})^T \left(\frac{\mathbf{U}_1}{n} + h_1^2 \mathbf{C}_1\right)^{-1}(\bar{y} - \bar{x}_{1i})\right\};$$

and for the denominator,

$$f(\bar{\mathbf{y}}|\mathbf{U}_2, \mathbf{C}_2, \bar{\mathbf{x}}_{21}, \dots, \bar{\mathbf{x}}_{2m_2}, h_2, H_2)$$

$$= (2\pi)^{-p/2} \left| \frac{\mathbf{U}_2}{n} + h_2^2 \mathbf{C}_2 \right|^{-1/2} \frac{1}{m_2} \sum_{i=1}^{m_2} \exp \left\{ -\frac{1}{2}(\bar{\mathbf{y}} - \bar{\mathbf{x}}_{2i})^T \left(\frac{\mathbf{U}_2}{n} + h_2^2 \mathbf{C}_2 \right)^{-1} (\bar{\mathbf{y}} - \bar{\mathbf{x}}_{2i}) \right\}.$$

Then

$$LR = \frac{\left| \frac{\mathbf{U}_1}{n} + h_1^2 \mathbf{C}_1 \right|^{-1/2} \frac{1}{m_1} \sum_{i=1}^{m_1} \exp \left\{ -\frac{1}{2}(\bar{\mathbf{y}} - \bar{\mathbf{x}}_{1i})^T \left(\frac{\mathbf{U}_1}{n} + h_1^2 \mathbf{C}_1 \right)^{-1} (\bar{\mathbf{y}} - \bar{\mathbf{x}}_{1i}) \right\}}{\left| \frac{\mathbf{U}_2}{n} + h_2^2 \mathbf{C}_2 \right|^{-1/2} \frac{1}{m_2} \sum_{i=1}^{m_2} \exp \left\{ -\frac{1}{2}(\bar{\mathbf{y}} - \bar{\mathbf{x}}_{2i})^T \left(\frac{\mathbf{U}_2}{n} + h_2^2 \mathbf{C}_2 \right)^{-1} (\bar{\mathbf{y}} - \bar{\mathbf{x}}_{2i}) \right\}}. \tag{5.5}$$

5.3.2   Univariate data

For univariate data ($p = 1$), the matrices become scalars (e.g. $\mathbf{U}_g$ becomes u_g^2) and the following equations should be used: for the numerator,

$$f(\bar{y}|c_1^2, u_1^2, \bar{x}_{11}, \dots, \bar{x}_{1m_1}, h_1, H_1)$$

$$= \left(2\pi \left(\frac{u_1^2}{n} + h_1^2 c_1^2 \right) \right)^{-1/2} \frac{1}{m_1} \sum_{i=1}^{m_1} \exp \left\{ - \left(2 \left(\frac{u_1^2}{n} + h_1^2 c_1^2 \right) \right)^{-1} (\bar{y} - \bar{x}_{1i})^2 \right\};$$

and for the denominator,

$$f(\bar{y}|c_2^2, u_2^2, \bar{x}_{21}, \dots, \bar{x}_{2m_2}, h_2, H_2)$$

$$= \left(2\pi \left(\frac{u_2^2}{n} + h_2^2 c_2^2 \right) \right)^{-1/2} \frac{1}{m_2} \sum_{i=1}^{m_2} \exp \left\{ - \left(2 \left(\frac{u_2^2}{n} + h_2^2 c_2^2 \right) \right)^{-1} (\bar{y} - \bar{x}_{2i})^2 \right\}.$$

Then

$$LR = \frac{\left(\frac{u_1^2}{n} + h_1^2 c_1^2 \right)^{-1/2} \frac{1}{m_1} \sum_{i=1}^{m_1} \exp \left\{ - \left(2 \left(\frac{u_1^2}{n} + h_1^2 c_1^2 \right) \right)^{-1} (\bar{y} - \bar{x}_{1i})^2 \right\}}{\left(\frac{u_2^2}{n} + h_2^2 c_2^2 \right)^{-1/2} \frac{1}{m_2} \sum_{i=1}^{m_2} \exp \left\{ - \left(2 \left(\frac{u_2^2}{n} + h_2^2 c_2^2 \right) \right)^{-1} (\bar{y} - \bar{x}_{2i})^2 \right\}}. \tag{5.6}$$

5.3.3   One-level models

Objects from the database may be described by a vector of p means of variables or a vector of values from one measurement. Then the information about the within-object variability ($\mathbf{U}_g$) is lost. Such a situation precludes the two-level model and reduces to the simplified one-level model (Zadora 2009).

5.3.3.1 Multivariate data

For multivariate data ($p > 1$) the following equations should be used: for the numerator,

$$f(\bar{\mathbf{y}}|\mathbf{C}_1, \bar{\mathbf{x}}_{11}, \ldots, \bar{\mathbf{x}}_{1m_1}, h_1, H_1)$$

$$= (2\pi)^{-p/2} \left|h_1^2 \mathbf{C}_1\right|^{-1/2} \frac{1}{m_1} \sum_{i=1}^{m_1} \exp\left\{-\frac{1}{2}(\bar{\mathbf{y}} - \bar{\mathbf{x}}_{1i})^T (h_1^2 \mathbf{C}_1)^{-1}(\bar{\mathbf{y}} - \bar{\mathbf{x}}_{1i})\right\};$$

and for the denominator,

$$f(\bar{\mathbf{y}}|\mathbf{C}_2, \bar{\mathbf{x}}_{21}, \ldots, \bar{\mathbf{x}}_{2m_2}, h_2, H_2)$$

$$= (2\pi)^{-p/2} \left|h_2^2 \mathbf{C}_2\right|^{-1/2} \frac{1}{m_2} \sum_{i=1}^{m_2} \exp\left\{-\frac{1}{2}(\bar{\mathbf{y}} - \bar{\mathbf{x}}_{2i})^T (h_2^2 \mathbf{C}_2)^{-1}(\bar{\mathbf{y}} - \bar{\mathbf{x}}_{2i})\right\}.$$

Then

$$LR = \frac{\left|h_1^2 \mathbf{C}_1\right|^{-1/2} \frac{1}{m_1} \sum_{i=1}^{m_1} \exp\left\{-\frac{1}{2}(\bar{\mathbf{y}} - \bar{\mathbf{x}}_{1i})^T (h_1^2 \mathbf{C}_1)^{-1}(\bar{\mathbf{y}} - \bar{\mathbf{x}}_{1i})\right\}}{\left|h_2^2 \mathbf{C}_2\right|^{-1/2} \frac{1}{m_2} \sum_{i=1}^{m_2} \exp\left\{-\frac{1}{2}(\bar{\mathbf{y}} - \bar{\mathbf{x}}_{2i})^T (h_2^2 \mathbf{C}_2)^{-1}(\bar{\mathbf{y}} - \bar{\mathbf{x}}_{2i})\right\}}. \tag{5.7}$$

5.3.3.2 Univariate data

For univariate data ($p = 1$), the matrices become scalars (e.g. $\mathbf{C}_g$ becomes c_g^2) and the following equations should be used: for the numerator,

$$f(\bar{y}|c_1^2, \bar{x}_{11}, \ldots, \bar{x}_{1m_1}, h_1, H_1) = \left(2\pi h_1^2 c_1^2\right)^{-1/2} \frac{1}{m_1} \sum_{i=1}^{m_1} \exp\left\{-\left(2h_1^2 c_1^2\right)^{-1}(\bar{y} - \bar{x}_{1i})^2\right\};$$

and for the denominator,

$$f(\bar{y}|c_2^2, \bar{x}_{21}, \ldots, \bar{x}_{2m_2}, h_2, H_2) = \left(2\pi h_2^2 c_2^2\right)^{-1/2} \frac{1}{m_2} \sum_{i=1}^{m_2} \exp\left\{-\left(2h_2^2 c_2^2\right)^{-1}(\bar{y} - \bar{x}_{2i})^2\right\}.$$

Then

$$LR = \frac{\left(h_1^2 c_1^2\right)^{-1/2} \frac{1}{m_1} \sum_{i=1}^{m_1} \exp\left\{-\left(2h_1^2 c_1^2\right)^{-1}(\bar{y} - \bar{x}_{1i})^2\right\}}{\left(h_2^2 c_2^2\right)^{-1/2} \frac{1}{m_2} \sum_{i=1}^{m_2} \exp\left\{-\left(2h_2^2 c_2^2\right)^{-1}(\bar{y} - \bar{x}_{2i})^2\right\}}. \tag{5.8}$$

5.4 Examples

There are numerous software packages available for performing all the calculations in this book. The recommended package is **R** (Appendix D). Suitable **R** routines which allow the user to calculate *LR* according to the equations given can be found in Appendix D and are available from the book's website. They have the advantage that they can be modified as required by the user. For applications involving Bayesian networks, **Hugin Researcher**™ software may be useful (see Appendix E).

5.4.1   Univariate casework data – normal between-object distribution – Bayesian network

In this example a Bayesian network model (Appendix E) is applied for casework purposes to data obtained during the analysis of flammable liquids (Section 1.3). The aim was to solve a classification problem, namely, whether the sample falls within the category of kerosene or diesel fuel. Research on the usefulness of this model is described in Zadora (2010).

5.4.1.1   Database creation

Twenty-two samples (11 samples of kerosene and 11 samples of diesel fuel), which were applied to carpeting to act as an accelerant during the research, were analysed (Borusiewicz *et al.* 2006). After a pre-specified period of time (between 20 and 80 min), the fire was extinguished with water and each fire debris sample was placed in a separate, clean, and airtight glass jar. Samples collected from each experiment were analysed in the laboratory three times, that is, three adsorption tubes with Tenax TA (polyoxide 2,6-diphenyl-*p*-phenylene) were placed in each jar. The jars were kept at a temperature of 60°C for 16 hours. Any analytes adsorbed on the Tenax TA were subjected to thermal desorption (at 325°C for 20 min) with the use of an automated thermal desorber (Turbo Matrix, Perkin Elmer, USA) and then concentrated on a cryogenic trap (-30°C for 20 min) also filled with Tenax TA. The analytes were again desorbed (325°C for 20 min) and analysed by gas chromatography (Auto System XL, Perkin Elmer, USA) coupled with a mass spectrometer (TurboMass Gold, Perkin Elmer, USA). The carrier gas was helium. A capillary column was used (Elite 1, 30 m $\times$ 0.25 mm $\times$ 0.25 μm). The chromatographic analysis was conducted according to the following temperature programme: initial temperature 30°C held for 3 min; increasing by 5°C·min^{-1} to 120°C; increasing by 15°C·min^{-1} to 270°C; and the final temperature of 270°C was held for 2.5 min. Electron ionisation (EI+; 70 eV) was used for the MS detection. Chromatograms and peak areas were collected using the TurboMass 4.0.4.14 software (Perkin Elmer, USA).

Following the research results presented in Zadora (2010), the example in this section is based on the *C14* variable (Zadora *et al.* 2005): the peak areas of *n*-tetradecane ($C_{14}H_{30}$) were divided by the corresponding peak areas of *n*-dodecane ($C_{12}H_{26}$) and the logarithm ($\log_{10}$) of the ratios was calculated.

The database consists of 22 objects (11 from the kerosene category and 11 from the diesel fuel category). The file GCMS_flammable_liquids_database.txt contains the values of the *C14* variable in column 3. Column 1 (Class) gives the category names (kerosene, diesel_fuel), while column 2 (Item) assigns a serial number for each flammable liquid sample. There are 22 rows, each giving the mean of the three measurements performed for each sample.

5.4.1.2   Graphical summary

The descriptive statistics for the *C14* variable are presented in the form of box-plots (Figure 5.1; Section 3.3.3) and Q-Q plots (Figure 5.2; Section 3.3.5). Figure 5.2 indicates that

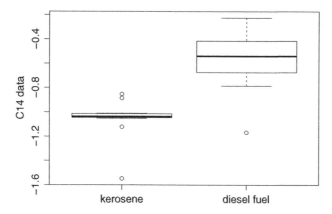

Figure 5.1 Box-plots for the *C14* variable for kerosene and diesel fuel (data available in GCMS_flammable_liquids_database.txt).

the assumption of a normal distribution for the diesel fuel data is reasonable, but not for the kerosene data. However, a normal between-object distribution was also assumed for kerosene due to the small number of samples. Parameters (means and variances) of the normal distributions considered in the node VC (Figure 5.3(a)) were calculated for each category separately under the $H_1 = K$ and $H_2 = DF$ hypotheses (where K stands for kerosene and DF for diesel fuel).

5.4.1.3   LR modelling

LR values were calculated using a Bayesian network model (Figure 5.3; file dfk_c14.oobn). A detailed description of this calculation can be found in Appendix E and Taroni *et al.* (2010).

It was assumed that the prior probabilities $P(H_1)$ and $P(H_2)$ were equal to 0.5 (node *H* in Figure 5.3(a)). Figure 5.3(a) presents a BN model after entering historical data, that is, parameters (means and variances) of the normal distributions considered in the node VC in Figure 5.3(a): $N(-1.057, 0.032)$ for $H_1 = K$ and $N(-0.567, 0.072)$ for $H_2 = DF$.

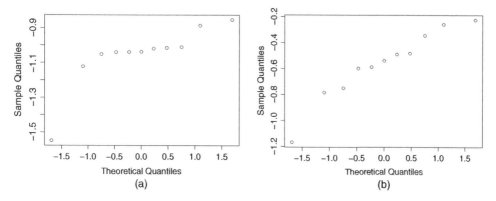

Figure 5.2 Q-Q plots for (a) kerosene and (b) diesel fuel (data available in GCMS_flammable_liquids_database.txt).

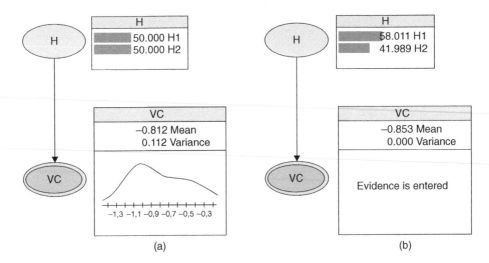

Figure 5.3 Bayesian network model for solving the classification problem for kerosene and diesel fuel samples based on continuous type data: (a) BN model after entering background data obtained for the *C14* variable with the marginal distribution for *C14*; (b) results of entering hard evidence (-0.853 for *C14*) into the VC node to give $P(H_1 \mid VC = -0.853) = 58.011\%$ and $P(H_2 \mid VC = -0.853) = 41.989\%$. (Reproduced by permission of Wiley.)

The node for VC in Figure 5.3(a) gives the marginal distribution of VC. It is a mixture of two normal distributions as shown in the density plot. Since $P(H_1) = P(H_2) = 0.5$, the mean is the simple mean of the two conditional means, $(-1.057 + (-0.567))/2 = -0.812$. The calculation of the variance is rather more technical, requiring the determination of the variance of conditional expectations and the expectation of conditional variances, and its calculation is omitted.

5.4.1.4   Results

The posterior probabilities (equation (2.1)), $P(H_1|E) = 0.58011$ and $P(H_2|E) = 0.41989$, were obtained in node H after entering information into node VC on the evidence value of the value of *C14* determined from the fire debris samples, $E = -0.853$ (Figure 5.3(b)). It was assumed that the prior probabilities ($P(H_1)$ and $P(H_2)$) in node H were equal for each of the two states (Figure 5.3(a)). Therefore, the ratio of posterior probabilities (0.58011/0.41989) can be treated as *LR* (equation (2.1)), to give an *LR* of approximately 1.4. It can be concluded that a sample with a *C14* value of -0.853 is 1.4 times more likely if the sample belongs to the kerosene (H_1) category than to the diesel fuel (H_2) category. This is, however, rather weak evidence.

Nevertheless, the equation for the LR calculation value in this particular case can also be expressed as

$$LR = \frac{f(E|H_1)}{f(E|H_2)} = \frac{f(VC \mid H_1)}{f(VC \mid H_2)}. \tag{5.9}$$

The terms $f(\text{VC} \mid H_1)$ and $f(\text{VC} \mid H_2)$ are the values of the probability density functions $N(-1.057, 0.032)$ and $N(-0.567, 0.072)$, evaluated at $\text{VC} = -0.853$, for H_1 and H_2, respectively. These values are the heights of the density functions at -0.853, which could, for example, be calculated using the **R** (Appendix D) command dnorm, where d denotes density function and norm the normal distribution.

Note that **R** uses the mean and standard deviation as the parameters of the normal distribution.

```
> dnorm(-0.853, -1.057, 0.032^0.5)
[1] 1.163952
```

This means that $f(\text{VC}|H_1) = 1.163952$.

```
> dnorm(-0.853, -0.567, 0.072^0.5)
[1] 0.8424662
```

This means that $f(\text{VC}|H_2) = 0.8424662$. Then, using equation (5.9), *LR* calculated as:

```
> 1.163952/0.8424662
[1] 1.381601
```

The value 1.381601 ($LR \simeq 1.4$) obtained is the same value obtained from the ratio of $P(H_1|E) = 0.58011$ and $P(H_2|E) = 0.41989$ (see node H in Figure 5.3(b)):

```
> 0.58011/0.41989
[1] 1.381576
```

5.4.2   Univariate research data – kernel density estimation procedure – R software

This example illustrates the application of **R** code presented in Section D.9.4 in Appendix D to the classification problem. The data for research purposes were obtained from an analysis of glass fragments by GRIM technique before and after the annealing process (Section 1.2.2) generating the $dRI = RI_a - RI_b$ variable. The aim of the research was to evaluate the usefulness of the univariate *LR* model based on *dRI* variable of glass fragments by estimating the rates of correct classifications.

5.4.2.1   Database creation

The database consists of 111 glass objects (56 from the package glass category (p), 32 from the car window category (c), and 23 from the building window category (w)), for which *RI* before (RI_b) and after (RI_a) the annealing process was determined by performing four replicate measurements for each glass object according to the procedure described in Section 1.2.2. The file dRI_cwp_database.txt gives the *dRI* values for the 111 glass fragments in column 4. Column 1 (Factor) of the matrix gives the category names (c, w, p), column 2 (Item) the glass fragment serial numbers, while column 3 (Piece) indexes the individual measurements on each glass fragment. There are 444 rows (4 measurements performed for each of 111 glass samples).

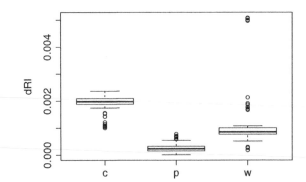

Figure 5.4 Box-plots of the data in `dRI_cwp_database.txt` for each category: car windows (c), building windows (w), and package glass (p).

5.4.2.2   Graphical summary

The box-plots (Section 3.3.3) illustrating the population data for each category of objects are given in Figure 5.4 and are based on the data gathered in `dRI_cwp_database.txt`.

To show the data distribution, Q-Q plots (Section 3.3.5) were plotted (Figure 5.5). These plots indicate that the data distribution is far from normal, especially for categories w and c.

5.4.2.3   LR modelling

Taking the descriptive statistics into account, all calculations were performed using KDE for probability density function estimation according to equation (5.6).

For LR calculations within a classification problem, it is necessary to have a database containing objects belonging to two categories. However, the database `dRI_cwp_database.txt` consists of objects belonging to three categories. In this case, then, we proceed by setting up three separate two-category classification problems, with a data file for each. For each problem we have a univariate LR model based on the *dRI* variable with the following hypotheses:

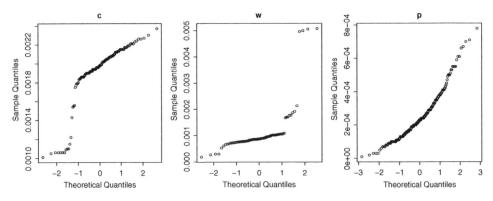

Figure 5.5 Q-Q plots of the data in `dRI_cwp_database.txt` for each category: car windows (c), building windows (w), and package glass (p).

- for the c versus w problem,
  H_1 : the object belongs to category c,
  H_2 : the object belongs to category w;

- for the c versus p problem,
  H_1 : the object belongs to category c,
  H_2 : the object belongs to category p;

- for the w versus p problem,
  H_1 : the object belongs to category w,
  H_2 : the object belongs to category p.

The files containing the data are dRI_c_w_database.txt, dRI_c_p_database.txt, and dRI_w_p_database.txt.

The calculations were performed by running the code dRI_classification_code.R, which uses the functions given in Section D.9.3:

- UC1_one_level_calculations.R, UC2_one_level_calculations.R, UC1_two_level_calculations.R, and UC2_two_level_calculations.R, which contain the UC function to calculate the within-object variability (**U**, calculated according to equation (3.3)) and between-object variability (**C**, calculated according to equation (3.4)) of the variables for objects belonging to each of the categories 1 or 2. There are different functions when one- (equations (5.3), (5.4), (5.7), and (5.8)) or two-level (equations (5.1), (5.2), (5.5), and (5.6)) LR models are considered. However, the code automatically chooses between them by recognising the number of measurements for each of the objects in the database.

- LR_classification_one_level_KDE.R and LR_classification_two_level_KDE.R, which contain the LR.KDE.function to compute the LR when KDE is used to estimate the between-object distribution. *LR* values are calculated according to equations (5.5) and (5.6) for the two-level model and equations (5.7) and (5.8) for the one-level model.

5.4.2.4   Results

The levels of correct and false classifications of objects (Section 2.4) into the relevant categories in each of the three classification problems were estimated.

In order to obtain the results for all three classification problems, the user needs to run the code dRI_classification_code.R three times, changing the database file name each time (dRI_c_w_database.txt, dRI_w_p_database.txt, or dRI_c_p_database.txt). This can be done in the line

```
data = read.table("dRI_c_p_database.txt", header = TRUE)
##dRI_c_w_database.txt and dRI_w_p_database.txt
```

in the file dRI_classification_code.R.

The code can be run in the **R** Console after setting the proper working directory (Appendix D) and by inserting the command:

```
source(file = "dRI_classification_code.R")
```

The code provides the user with the following files as an outcome:

- dRI_c_w_classification_research_KDE.txt, which contains *LR* values (cal-culated when KDE is used for between-object variability estimation; equation (5.6)). For easier interpretation of results, the second column of the matrix gives the real category the objects belong to (not the category that the objects belong to according to the calculated *LR* values).

- dRI_c_w_research_descriptive_statistics.eps, which is a graphical file presenting the descriptive statistics (box-plots, Q-Q plots, and KDE functions).

- dRI_c_w_research_LR_distribution.eps, which is a graphical file presenting the $\log_{10} LR$ distribution for both categories.

- dRI_c_w_classification_research_correct_rate.txt, which gives the per-centage of correctly classified objects for each of two categories and generally in the classification problem.

For the remaining two classification problems (c versus p, w versus p) the same set of files is obtained but with the appropriate file name changes, for example dRI_c_w_classification_research_KDE.txt is replaced by dRI_c_p_classification_research_KDE.txt.

The results suggest that it is easiest to discriminate between the car window and package glass categories, as discrimination is perfect for the datasets used. This compares to 92% correct answers for the w versus p problem and 89% for the c versus w problem. The outstanding performance of the LR model for classification between the c and p categories is due to the production process as well as mechanical features of these glass categories. Compared to package glass, car windows are toughened, which implies introducing some tension to the glass objects. Such tension can be removed in the course of the annealing process, which influences the *RI* values. Due to the *dRI* values being significantly higher than for non-toughened glass, it becomes easier to distinguish between the c and w categories.

It is important to note that the size of the database used was small, which favoured the very high classification rates. Care should be taken when extracting conclusions from these results, and a bigger database should be used in order to confirm these rates in a more robust manner.

However, it is not only the rate of incorrect answers that is important. Far more crucial are the *LR* values and the strength of the support for either hypothesis. The LR models used may be evaluated using empirical cross entropy (Chapter 6).

5.4.3   Multivariate research data – kernel density estimation – R software

In this example **R** code (given in Section D.9.4) is applied for research purposes to the classification problem for multivariate data obtained from the SEM-EDX analysis of glass fragments (Section 1.2.1). The aim of the research was to evaluate the usefulness of a graphical LR model (Section 3.7.2) by estimating the rates of correct classifications.

5.4.3.1   Database creation

The database consists of data obtained from the SEM-EDX analysis (Sections 1.2.1 and 4.4.6) of glass fragments. The raw data on the elemental content of Na, Mg, Al, Si, K, Ca, and Fe obtained in the course of the analysis were transformed by dividing by the oxygen content and taking $\log_{10}$ ratios (Section 4.4.6). The database consists of 278 glass objects (79 from the package glass category (p), and 199 from the combined building and car windows category (cw)), each measured 4 times. The file `SEMEDX_cw_p_database.txt` gives the seven variables for the elemental content of 278 glass fragments (`logNaO`, `logMgO`, `logAlO`, `logSiO`, `logKO`, `logCaO`, `logFeO`: columns 4–10). Column 1 (`Factor`) of the matrix gives the category names (`cw, p`), column 2 (`Item`) the glass fragment serial numbers, while column 3 (`Piece`) indexes the individual measurements on each glass fragment. There are 1112 rows (four measurements performed on each of 278 glass samples).

5.4.3.2   Graphical summary

The box-plots (Section 3.3.3) showing the population data for each category are given in Figure 5.6. To show the data distribution, Q-Q plots and KDE curves (Section 3.3.5) were

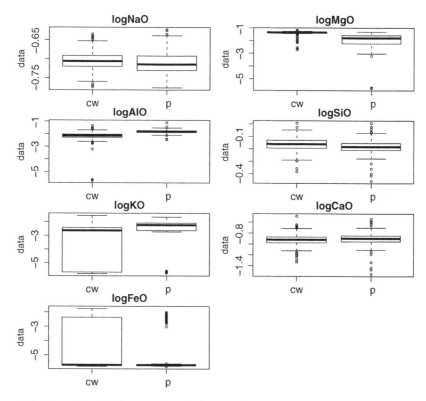

Figure 5.6   Box-plots for $\log_{10}$ ratios of elemental concentrations to oxygen obtained from SEM-EDX for each category (car windows and building windows (cw) and package glass (p)). Data available in `SEMEDX_cw_p_database.txt`.

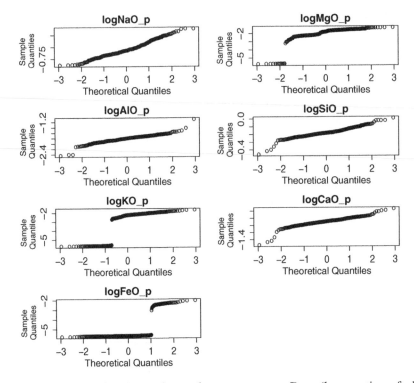

Figure 5.7   Q-Q plots for the package glass category p. Data ($\log_{10}$ ratios of elemental concentrations to oxygen obtained from SEM-EDX) are available in SEMEDX_cw_p_database.txt.

plotted (Figures 5.7–5.10). These plots indicate that the between-object distribution for a majority of the variables is far from normal.

5.4.3.3   LR modelling

Calculations were carried out using KDE for probability density function estimation as the between-object variability cannot be assumed normal according to Figures 5.7–5.10. The graphical LR model described in Section 3.7.2 was applied. The LR calculations involved one trivariate model involving *logNaO*, *logSiO*, and *logCaO*, one bivariate model for *logAlO* and *logKO*, and two univariate models for *logMgO* and *logFeO*. The results for each model were multiplied together and the final *LR* values were obtained based on equation (3.16). The computations were done using the code in SEMEDX_glass_classification_code.R to test the hypotheses

- H_1 : the object belongs to category p,

- H_2 : the object belongs to category cw.

The code uses the functions given in Section D.9.3:

- UC1_one_level_calculations.R, UC2_one_level_calculations.R, UC1_two_level_calculations.R, and UC2_two_level_calculations.R, which

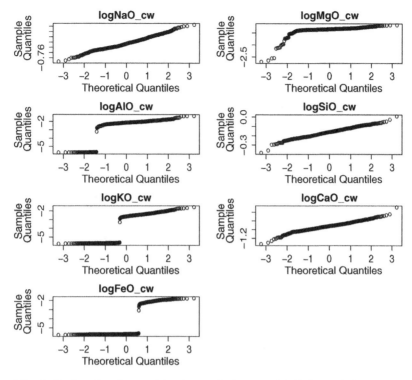

Figure 5.8 Q-Q plots for the combined car windows and building windows category cw. Data ($\log_{10}$ ratios of elemental concentrations to oxygen obtained from SEM-EDX) are available in SEMEDX_cw_p_database.txt.

contain the UC function to calculate the within-object variability (**U**, calculated according to equation (3.3)) and between-object variability (**C**, calculated according to equation (3.4)) of the variables for objects belonging to each of the categories 1 and 2. There are different functions when one- (equations (5.3), (5.4), (5.7), and (5.8)) or two-level (equations (5.1), (5.2), (5.5), and (5.6)) LR models are considered. However, the code automatically chooses between them by recognising the number of measurements for each of the objects in the database.

- LR_classification_one_level_KDE.R and LR_classification_two_level _KDE.R, which contain the LR.KDE.function to compute the *LR* values when KDE is used to estimate the between-object distribution. *LR* values are calculated according to equations (5.5) and (5.6) for the two-level model and equations (5.7) and (5.8) for the one-level model.

5.4.3.4 Results

The rates of correct and false classifications of objects (Section 2.4) into the categories p and cw were estimated. The code SEMEDX_glass_classification_code.R can be run in the **R** Console after setting the proper working directory (Appendix D) and by inserting the command:

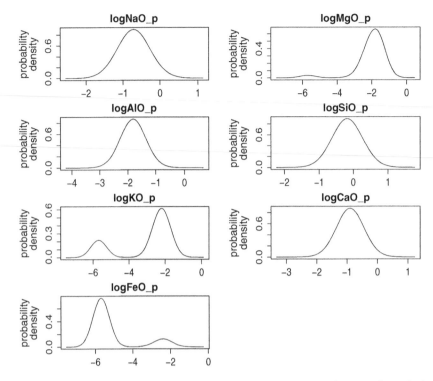

Figure 5.9 KDE plots for the package glass category p. Data ($\log_{10}$ ratios of elemental concentrations to oxygen obtained from SEM-EDX) are available in SEMEDX_cw_p_ database.txt.

```
source(file = "SEMEDX_glass_classification_code.R")
```

The code provides the user with the following files as an outcome:

- model_p_cw_classification_research_KDE.txt, which contains *LR* values (calculated when KDE is used for between-object variability estimation; equations (5.5) and (5.6)) for the graphical model. For easier interpretation of results, the second column of the matrix gives the real category the objects belong to (not the category that the objects belong to according to the calculated *LR* values).

- model_p_cw_classification_research_correct_rate.txt, which gives the percentage of correctly classified objects for each category and generally in the classification problem.

- logNaO_p_cw_research_descriptive_statistics.eps, which is a graphical file presenting the descriptive statistics (box-plots, Q-Q plots, and KDE functions) for the variables (similar graphical files are obtained for the remaining variables).

- model_p_cw_research_LR_distribution.eps, which is a graphical file presenting the distribution of the $\log_{10} LR$ for the model for both categories.

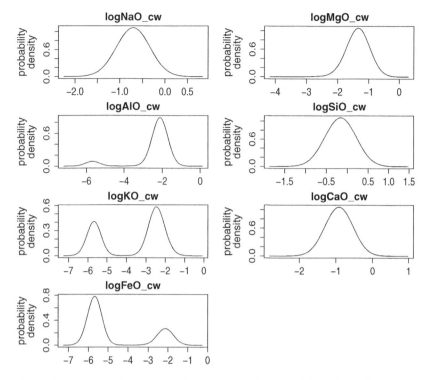

Figure 5.10 KDE plots for the combined car windows and building windows category cw. Data ($\log_{10}$ ratios of elemental concentrations to oxygen obtained from SEM-EDX) are available in SEMEDX_cw_p_database.txt.

The results of the calculations show that the model correctly classifies objects into their categories in 95% of cases. The efficiency of the trivariate model for *logNaO*, *logSiO*, and *logCaO* as well as the efficiency of the graphical model are evaluated by empirical cross entropy in Section 6.6.3.

5.4.4 Multivariate casework data – kernel density estimation – R software

In this example the **R** code presented in Section D.9.3 is applied for casework purposes to the classification problem for multivariate data obtained during the SEM-EDX analysis of glass fragments (Section 1.2.1).

5.4.4.1 Database creation

The database used in this section is the same as that used in Section 5.4.3 (SEMEDX_cw_p_database.txt). In this example there are now also four evidence glass fragments that were collected from a crime scene, each measured five times. The data describing their Na, Mg, Al, Si, K, Ca, and Fe content are available in SEMEDX_evidence.txt. This has the same structure as the database, but of course only 20 rows, one for each of the five

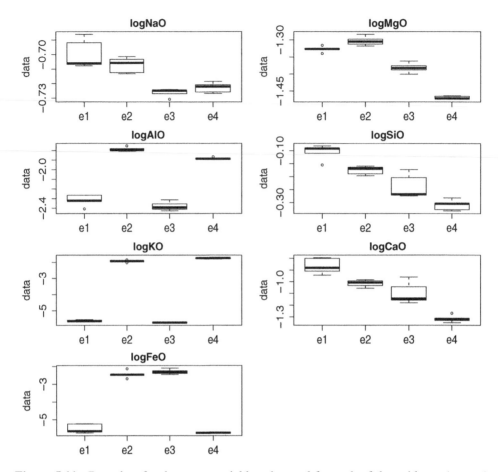

Figure 5.11 Box-plots for the seven variables observed for each of the evidence (e1-e4) glass fragments. Data are available in the file SEMEDX_evidence.txt.

measurements on the four fragments. Also, in the SEMEDX_evidence.txt the first column is labelled Name, whereas in SEMEDX_cw_p_database.txt it is denoted as Factor.

5.4.4.2   Graphical summary

The box-plots (Section 3.3.3) showing the population data for each category are given in Figure 5.6. To show the data distribution, Q-Q plots and KDE curves (Section 3.3.5) were plotted (Figures 5.7–5.10). These plots indicate that the probability distribution for a majority of the variables is far from normal. For the evidence fragments the box-plots are shown in Figure 5.11.

5.4.4.3   LR modelling

Since the between-object variability cannot be assumed normal according to Figures 5.7–5.10, the calculations were carried out using KDE for probability density function estimation. The graphical LR model described in Section 3.7.2 was applied. The LR calculations involved one trivariate model for *logNaO*, *logSiO*, and *logCaO*, one bivariate model for *logAlO* and *logKO*, and two univariate models for *logMgO* and *logFeO*. The results for each model were multiplied together and the final *LR* values were obtained based on equation (3.16). The computations were done using the code in SEMEDX_glass_classification_casework_ code.R to test the hypotheses

- H_1 : the object belongs to category cw,

- H_2 : the object belongs to category p.

The code uses the functions given in Section D.9.3:

- UC1_one_level_calculations.R, UC2_one_level_calculations.R, UC1_ two_level_calculations.R, and UC2_two_level_calculations.R, which contain the UC function to calculate the within-object variability (**U**, calculated according to equation (3.3)) and between-object variability (**C**, calculated according to equation (3.4)) of the variables for objects belonging to each of the categories 1 or 2. There are different functions when one- (equations (5.3), (5.4), (5.7), and (5.8)) or two-level (equations (5.1), (5.2), (5.5), and (5.6)) LR models are considered. However, the code automatically chooses between them by recognising the number of measurements for each of the objects in the database.

- LR_classification_one_level_KDE.R and LR_classification_two_level _KDE.R, which contain the LR.KDE.function to compute the *LR* values when KDE is used to estimate the between-object distribution. *LR* values are calculated according to equations (5.5) and (5.6) for the two-level model and equations (5.7) and (5.8) for the one-level model.

5.4.4.4   Results

The code SEMEDX_glass_classification_casework_code.R can be run in the **R** Console after setting the proper working directory (Appendix D) and by inserting the command:

```
source(file = "SEMEDX_glass_classification_casework_code.R")
```

The code provides the user with the following files as an outcome:

- model_cw_p_classification_casework_KDE.txt, which contains *LR* values (calculated when KDE is used for between-object variability estimation; equations (5.5) and (5.6)) for the graphical model. For easier interpretation of results, the second column of the matrix gives the category the objects belong to based on the *LR* values obtained.

- logNaO_cw_p_casework_descriptive_statistics.eps, which is a graphical file presenting the descriptive statistics (box-plots, Q-Q plots, and KDE functions) for the variables (similar graphical files are obtained for the remaining variables).

The results of the calculations show that the chemical composition detected for glass fragments e1, e2, and e3 more likely supports the hypothesis H_1 (that the object comes from category cw) than H_2 (that it comes from category p) since the *LR* values obtained are greater than 1.

Name	Item	Piece	logNaO	logMgO	logAlO	logSiO	logKO	logCaO	logFeO
e1	1	1	-0.686077911	-1.315786607	-2.26651392	-0.081264285	-5.642246341	-0.80037674	-5.697546977
e1	1	2	-0.691784751	-1.32786715	-2.322313288	-0.089618837	-5.652149605	-0.794214341	-5.754697745
e1	1	3	-0.706655111	-1.325218778	-2.408522617	-0.091067613	-5.553598382	-0.880295215	-5.234547453
e1	1	4	-0.705978625	-1.339800722	-2.26651392	-0.109532999	-5.580244837	-0.94203627	-5.656432567
e1	1	5	-0.707687479	-1.326480229	-2.322313288	-0.154876185	-5.682506086	-0.908719641	-5.242567312
e2	2	1	-0.703326054	-1.317820683	-1.796747739	-0.196183207	-2.018596488	-1.056385049	-2.455281518
e2	2	2	-0.712528686	-1.304604456	-1.753921371	-0.168854321	-1.932544966	-1.006274746	-2.489291118
e2	2	3	-0.701408599	-1.283553485	-1.796736481	-0.15922076	-1.92735235	-0.991262103	-2.135842256
e2	2	4	-0.713285194	-1.297634959	-1.781934798	-0.163233135	-1.852515872	-0.982594249	-2.465322456
e2	2	5	-0.70566786	-1.312191349	-1.80951599	-0.188690675	-1.91067688	-1.033248884	-2.690545321

Figure 5.12 Data on evidence glass objects available in `evidence.txt`.

The chemical composition detected for glass fragment e4 more likely supports the hypothesis H_2 (*LR* value less than 1).

5.5   R software

5.5.1   Routines for casework applications

Casework calculations in a classification problem involve computing the likelihood ratio with the aim of classifying an object into one of two categories.

The routines described in this section calculate the LR for one (one-level model) or two (two-level model) sources of data variability: within-object (**U**) and between-object (**C**) variability assuming a normal between-object data distribution or using KDE for its estimation. For the one-level model only between-object variability is taken into account. For the full **R** routines see Section D.9.3 in Appendix D.

The calculations concern a two-level univariate problem for aluminium content obtained from the SEM-EDX analysis of glass fragments (column 6 in the data files) measured in two fragments of evidence glass. The data describing each of these objects is available in `evidence.txt` and presented in Figure 5.12. In order properly to estimate the parameters, a database of 278 glass objects (`cwp_database.txt`) containing 199 fragments of float glass (denoted as cw) and 79 fragments of package glass (denoted as p) was used (part of which is shown in Figure 5.13).

Factor	Item	Piece	logNaO	logMgO	logAlO	logSiO	logKO	logCaO	logFeO
cw	1	1	-0.67538764	-1.298750859	-2.327178567	-0.161035884	-2.443684136	-0.931543872	-5.682566225
cw	1	2	-0.673112056	-1.285925834	-2.357807842	-0.107347177	-2.36510708	-0.854244122	-5.658837837
cw	1	3	-0.656437632	-1.267804835	-2.375014805	-0.066392192	-2.30919752	-0.800878266	-5.63825624
cw	1	4	-0.656437632	-1.267804835	-2.375014805	-0.066392192	-2.30919752	-0.800878266	-5.63825624
cw	2	1	-0.692372693	-1.227743407	-2.573787385	-0.143228615	-2.421177222	-0.925684277	-5.676449727
.									
.									
.									
p	277	3	-0.703833984	-2.256958153	-1.654898161	-0.155812773	-2.063138127	-0.845759479	-5.682926885
p	277	4	-0.706054312	-2.266309118	-1.664249127	-0.164066517	-2.092054202	-0.855585261	-5.686814955
p	278	1	-0.689712259	-1.647873241	-1.734731944	-0.183714313	-2.203913408	-0.877889396	-5.69057998
p	278	2	-0.69021692	-1.650778155	-1.737484493	-0.17467945	-2.200207927	-0.864488456	-5.6868745
p	278	3	-0.688494671	-1.649162636	-1.732067189	-0.177861979	-2.173810404	-0.86946938	-5.687915225
p	278	4	-0.691140292	-1.630599397	-1.731781733	-0.147631158	-2.143137309	-0.82172978	-5.674616226

Figure 5.13 Extract from database of float glass (cw) and package glass fragments (p) available in `cwp_database.txt`.

Data files (`cwp_database.txt`, `evidence.txt`) and **R** code files must be stored in the same file folder. Data file names should have the `*.txt` file extension and **R** code file names the `*.R` file extension. The rows in the data files correspond to *n* measurements carried out on each of *m* objects (so that there are *nm* rows in total). The **R** routines presented in this book require that the number of measurements (*n*) carried out for each object is the same throughout the database. The columns of the `cwp_database.txt` matrix are as follows:

- column 1 (`Factor`), category name;

- column 2 (`Item`), object serial number;

- column 3 (`Piece`), measurement index from 1 to *n*,

- from column 4, variables (e.g. columns 4–10: `logNaO`, `logMgO`, `logAlO`, `logSiO`, `logKO`, `logCaO`, `logFeO` variables).

Instead of the column labelled `Factor` in the database file (`cwp_database.txt`), the evidence data file (`evidence.txt`) must contain a column labelled `Name` with unique evidence object names.

The basic code for the casework application (`classification_casework_code.R`, Section D.9.3) uses the following functions given in Section D.9.3:

- `UC1_one_level_calculations.R`, `UC2_one_level_calculations.R`, `UC1_two_level_calculations.R`, and `UC2_two_level_calculations.R`, which contain the UC function to calculate the within-object variability (**U**, calculated according to equation (3.3)) and between-object variability (**C**, calculated according to equation (3.4)) of the variables for objects belonging to each of the categories 1 or 2. There are different functions when one- or two-level LR models are considered. However, the code automatically chooses between them by recognising the number of measurements for each of the objects in the database.

- `LR_classification_one_level_Nor.R` and `LR_classification_two_level_Nor.R`, which contain the LR.Nor.function to compute the LR when the between-object distribution is assumed normal. *LR* values are calculated according to equations (5.1) and (5.2) for the two-level model and equations (5.3) and (5.4) for the one-level model.

- `LR_classification_one_level_KDE.R` and `LR_classification_two_level_KDE.R`, which contain the LR.KDE.function to compute the LR when KDE is used for between-object distribution estimation. *LR* values are calculated according to equations (5.5) and (5.6) for the two-level model, and equations (5.7) and (5.8) for the one-level model.

In order to run the code `classification_casework_code.R` to obtain LR results for the models described, the user must include file names for the object data (evidence data) analysed and the database data:

```
data = read.table("ENTER THE FILE NAME.txt", header = TRUE)
population = read.table("ENTER THE FILE NAME.txt", header = TRUE)
```

	logAlO_Nor	category_Nor		logAlO_KDE	category_KDE
e1	1700	cw	e1	9.846	cw
e2	0.09884	p	e2	0.1789	p
	(a)			(b)	

Figure 5.14 Results of LR calculations (for *logAlO*) for the classification problem of glass fragments: (a) when between-object distribution is assumed normal; (b) when KDE is used to estimate the between-object distribution (available in `logAlO_cw_p_classification_casework_Nor.txt` and `logAlO_cw_p_classification_casework_KDE.txt`).

Also required are the index (indices) of the column(s) containing the data for the relevant variable(s):

```
variables = c(ENTER THE INDEX (INDICES) OF THE COLUMN(S), COMMA
SEPARATED)
```

For *logNaO* and *logMgO*, for example, the correct input would be:

```
variables = c(4,5)
```

The code can be run in the **R** Console after setting the proper working directory (Appendix D) and by inserting the command:

```
source(file = "classification_casework_code.R")
```

Running the code for the *logAlO* variable results in the following files as an outcome:

- `logAlO_cw_p_classification_casework_Nor.txt` containing *LR* values calculated when the between-object distribution is assumed normal (equations (5.1)–(5.4)). The second column of the matrix gives information about the hypothesis supported by the *LR* value (Figure 5.14(a)). In general, if other models (univariate or multivariate) are considered, the beginning of the file name will change according to the new variable(s) name(s) and the categories specified in the column and row names of the data files.

- `logAlO_cw_p_classification_casework_KDE.txt` containing *LR* values calculated when KDE is used (equations (5.5)–(5.8)). The second column of the matrix gives information about the hypothesis supported by the *LR* value (Figure 5.14(b)). In general, if other models (univariate or multivariate) are considered, the beginning of the file name will change according to the new variable(s) name(s) and the categories specified in the column and row names of the data files.

- `logAlO_cw_p_casework_descriptive_statistics.eps`, which provides descriptive statistics and data distributions in graph form (Figure 5.15). In general, if other models (univariate or multivariate) are considered, the beginning of the file name

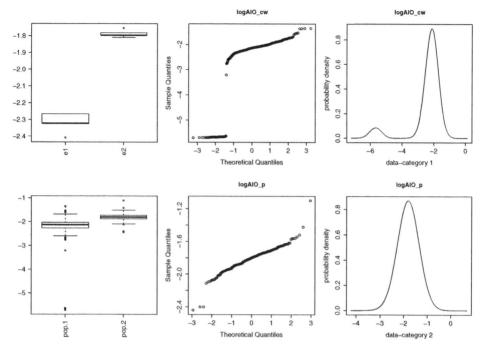

Figure   5.15 Graphs   (logAlO_cw_p_casework_descriptive_statistics.eps) obtained by running the code classification_casework_code.R: (left) box-plots; (middle) Q-Q plots; (right) KDE plots.

will change according to the new variable(s) name(s) and the categories specified in the column and row names of the data files.

5.5.2   Routines for research applications

The aim of LR calculations in a classification problem for research applications is to gain information about the performance of LR models by estimating the levels of correctly classified objects.

The code fragments described in this section calculate the likelihood ratio for one (one-level model) or two (two-level model) sources of data variability: within-object (**U**) and between-object (**C**) variability assuming a normal between-object data distribution or using KDE for its estimation. For one-level model only between-object variability is taken into account. For the full **R** routines see Section D.9.4 in Appendix D.

In order to work out the routines, an example is given of a univariate classification problem concerned with aluminium content (column 6 in the data file), measured on 278 glass fragments (cwp_database.txt, Figure 5.13).

Data files (cwp_database.txt) and **R** code files must be stored in the same file folder. Data file names should have the *.txt file extension and R code file names the *.R file extension. The rows in the data files correspond to n measurements carried out on each of m objects (so that there are nm rows in total). The **R** routines presented in this book require that the number of measurements (n) carried out for each object is the same throughout the database. The columns of the `cwp_database.txt` matrix are as follows:

- column 1 (`Factor`), category name;

- column 2 (`Item`), object serial number;

- column 3 (`Piece`), measurement index from 1 to n,

- from column 4, variables (e.g. columns 4–10: `logNaO`, `logMgO`, `logAlO`, `logSiO`, `logKO`, `logCaO`, `logFeO` variables).

The basic code for the research application (`classification_research_code.R`, Section D.9.4) uses the following functions given in Section D.9.3:

- `UC1_one_level_calculations.R`, `UC2_one_level_calculations.R`, `UC1_two_level_calculations.R`, and `UC2_two_level_calculations.R`, which contain the `UC` function to calculate the within-object variability (**U**, calculated according to equation (3.3)) and between-object variability (**C**, calculated according to equation (3.4)) of the variables for objects belonging to each of the categories 1 or 2. There are different functions when one- or two-level *LR* models are considered. However, the code automatically chooses between them by recognising the number of measurements for each of the objects in the database.

- `LR_classification_one_level_Nor.R` and `LR_classification_two_level_Nor.R`, which contain the `LR.Nor.function` to compute the LR when the between-object distribution is assumed normal. *LR* values are calculated according to equations (5.1) and (5.2) for the two-level model and equations (5.3) and (5.4) for the one-level model.

- `LR_classification_one_level_KDE.R` and `LR_classification_two_level_KDE.R`, which contain the `LR.KDE.function` to compute the LR when KDE is used for between-object distribution estimation. *LR* values are calculated according to equations (5.5) and (5.6) for the two-level model and equations (5.7) and (5.8) for the one-level model.

The code classifies each object into one of the categories. Therefore, there are m LR computations for a database of m objects. The database used as a background information for proper parameter estimation is then diminished by 1 object, which is actually classified, and thus now comprises $m - 1$ objects. This is known as a jackknife procedure.

In order to run the code `classification_research_code.R` to obtain LR results for the models described, the user must include file name for the database data:

```
data = read.table("ENTER THE FILE NAME.txt", header = TRUE)
```

Also required are the index (indices) of the column(s) containing the data for the relevant variable(s):

```
variables = c(ENTER THE INDEX (INDICES) OF THE COLUMN(S), COMMA
SEPARATED)
```

For *logNaO* and *logMgO*, for example, the correct input would be:

```
variables = c(4,5)
```

In order to become familiar with the routines, the database of 278 glass objects in
cwp_database.txt can be downloaded in order to perform the calculations. Our example
concerns a univariate comparison problem based on aluminium content (column 6 in the data
file) measured in 278 glass samples.

The code can be run in the **R** Console after setting the proper working directory
(Appendix D) and by inserting the command:

```
source(file = "classification_research_code.R")
```

Running the code for the variable *logAlO* results in the following files as an outcome
(Figures 5.16–5.18):

- logAlO_cw_p_classification_research_Nor.txt containing *LR* values calculated when the between-object distribution is assumed normal (Figure 5.16(a); equations (5.1)–(5.4)). In general, if other models (univariate or multivariate) are considered,
  the beginning of the file name will change according to the new variable name(s) and
  the categories specified in the column and row names of the data files.

- logAlO_cw_p_classification_research_KDE.txt containing *LR* values calculated when KDE is used for distributional parameters estimation (Figure 5.16(b);

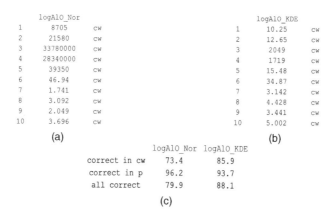

	logAlO_Nor				logAlO_KDE	
1	8705	cw		1	10.25	cw
2	21580	cw		2	12.65	cw
3	33780000	cw		3	2049	cw
4	28340000	cw		4	1719	cw
5	39350	cw		5	15.48	cw
6	46.94	cw		6	34.87	cw
7	1.741	cw		7	3.142	cw
8	3.092	cw		8	4.428	cw
9	2.049	cw		9	3.441	cw
10	3.696	cw		10	5.002	cw

(a)   (b)

	logAlO_Nor	logAlO_KDE
correct in cw	73.4	85.9
correct in p	96.2	93.7
all correct	79.9	88.1

(c)

Figure 5.16 Extracts from the results of the LR calculations (for the *logAlO* variable)
for the classification problem of glass fragments: (a) when the between-object distribution is assumed normal; (b) when KDE is used for estimating the between-object distribution (available in logAlO_cw_p_classification_research_Nor.txt, logAlO_cw_
p_classification_research_KDE.txt, and cw_p_classification_research_
correct_rate.txt).

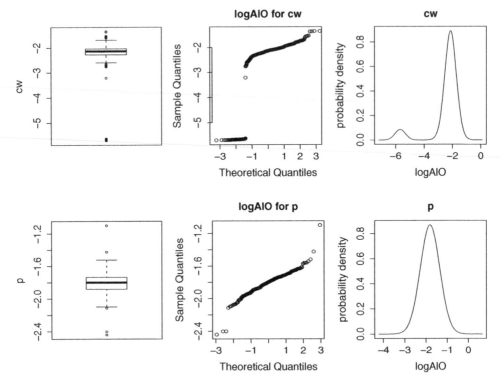

Figure 5.17 Graphs (logAlO_cw_p_research_descriptive_statistics.eps) obtained by running the code classification_research_code.R: (left) box-plots; (middle) Q-Q plots; (right) KDE plots.

equations (5.5)–(5.8)). In general, if other models (univariate or multivariate) are considered, the beginning of the file name will change according to the new variable name(s) and the categories specified in the column and row names of the data files.

- cw_p_classification_research_correct_rate.txt, which gives the percentage of correctly classified objects for each of the two categories and generally in the classification problem (both when normal distribution is assumed and when KDE is used; Figure 5.16(c)).

- logAlO_cw_p_research_descriptive_statistics.eps, which provides some idea of the descriptive statistics and data distribution by plotting graphs and saving them in the relevant files denoted by the variable name (Figure 5.17). In general, if other models (univariate or multivariate) are considered, the beginning of the file name will change according to the new variable name(s) and the categories specified in the column and row names of the data files.

- logAlO_cw_p_research_LR_distribution_Nor.eps, which is a graphical file showing the distribution of $\log_{10} LR$ (Figure 5.18(a)) obtained when a normal distribution is assumed. In general, if other models (univariate or multivariate) are considered,

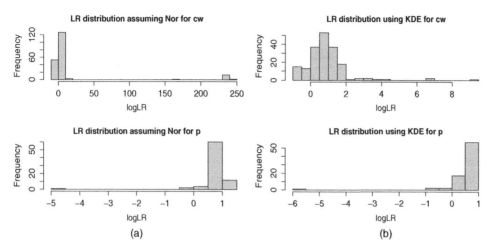

Figure 5.18 Graphs obtained by running the code `classification_research_code.R` showing the $\log_{10} LR$ distribution of the *logAlO* variable when: (a) a normal distribution is assumed (available in `logAlO_cw_p_research_LR_distribution_Nor.eps`); (b) KDE is used (available in `logAlO_cw_p_research_LR_distribution_KDE.eps`).

the beginning of the file name will change according to the new variable name(s) and the categories specified in the column and row names of the data files.

- `logAlO_cw_p_research_LR_distribution_KDE.eps`, which is a graphical file showing the distribution of $\log_{10} LR$ (Figure 5.18(b)) obtained when KDE is used. In general, if other models (univariate or multivariate) are considered, the beginning of the file name will change according to the new variable name(s) and the categories specified in the column and row names of the data files.

References

Aitken CGG 1986 Statistical discriminant analysis in forensic science. *Journal of Forensic Sciences* **26**, 237–247.

Aitken CGG, Taroni F 2004 *Statistics and the Evaluation of Evidence for Forensic Scientists*, 2nd edn. John Wiley & Sons, Ltd, Chichester.

Aitken CGG, Lucy D 2004 Evaluation of trace evidence in the form of multivariate data. *Applied Statistics* **53**, 109–122.

Berry DA 1991 Inferences in forensic identification and paternity cases using hypervariable DNA sequences (with discussion). *Statistical Science* **6**, 175–205.

Berry DA, Evett IW, Pinchin R 1992 Statistical inference in crime investigations using deoxyribonucleic acid profiling (with discussion). *Applied Statistics* **41**, 499–531.

Borusiewicz R, Zieba-Palus J, Zadora G 2006 The influence of the type of accelerant, type of burned material, time of burning and availability of air on the possibility of detection of accelerants traces. *Forensic Science International* **160(2-3)**, 115–126.

Chan KPS, Aitken CGG 1989 Estimation of the Bayes' factor in a forensic science problem. *Journal of Statistical Computation and Simulation* **33**, 249–264.

Evett IW, Cage PE, Aitken CGG 1987 Evaluation of the likelihood ratio for fibre transfer evidence in criminal cases. *Applied Statistics* **36**, 174–180.

Taroni F, Bozza S, Biedermann A, Garbolino P and Aitken CGG 2010 *Data Analysis in Forensic Science: A Bayesian Decision Perspective*. John Wiley & Sons, Ltd, Chichester.

Zadora G 2009 Classification of glass fragments based on elemental composition and refractive index. *Journal of Forensic Sciences* **54**, 49–59.

Zadora G 2010 Evaluation of evidential value of physicochemical data by a Bayesian Network approach. *Journal of Chemometrics* **24**, 346–366.

Zadora G, Neocleous T 2009 Likelihood ratio model for classification of forensic evidences. *Analytica Chimica Acta* **64**, 266–278.

Zadora G, Borusiewicz R, Zieba-Palus J 2005 Differentiation between weathered kerosene and diesel fuel using automatic thermal desorption-GC-MS analysis and the likelihood ratio approach. *Journal of Separation Sciences* **28**(13), 1467–1475.

Zadora G, Neocleous T, Aitken CGG 2010 A two-level model for evidence evaluation in the presence of zeros. *Journal of Forensic Sciences* **55**, 371–384.

6

Performance of likelihood ratio methods

6.1 Introduction

In previous chapters, methods for the computation of likelihood ratios in forensic science for the statistical evaluation of evidence have been given. However, as has been seen in previous examples, the computation of *LR* values still remains a challenge. There are many reasons for this challenge, among them:

- the scarcity of the databases used as populations;

- the mismatch in the conditions of the materials in the population databases and in the evidence;

- the degraded quality or quantity of the evidential materials.

Moreover, if the conditions are extremely degrading, it may be that the magnitudes of *LR* values supporting the wrong proposition are large, leading to what is known as *strong misleading evidence* (Royall 1997), an effect that is highly undesirable.

It is essential for the interpretation of evidence in casework that the LR model performs well. Misleading *LR* values in court may lead fact finders to the wrong decisions. This idea is the main motivation behind the establishment of validation procedures for evidence evaluation methods. Validation procedures enable the control of procedures and hence ease the use of LR methods in casework. If the *LR* generated by a given method performs badly, then it should not be used in casework. If it performs well it should be used. Therefore, the validation of evidence evaluation methods should be based on a careful process of performance measurement.

Statistical Analysis in Forensic Science: Evidential Value of Multivariate Physicochemical Data, First Edition.
Grzegorz Zadora, Agnieszka Martyna, Daniel Ramos and Colin Aitken.
© 2014 John Wiley & Sons, Ltd. Published 2014 by John Wiley & Sons, Ltd.
Companion website: www.wiley.com/go/physicochemical

The performance of analytical methods in chemistry (Chapter 1) has been widely studied over several decades, and there exist commonly accepted metrics for measuring the performance. As a consequence, validation procedures for analytical methods have become quite standardised, with common procedures that can be easily checked and accredited by the appropriate institutions. However, currently standards like these do not exist for evidence interpretation and evaluation methods in forensic science.

In this chapter, several procedures are presented that may be used to assess the performance of *LR* values for any forensic discipline. Their use is illustrated with examples in the context of the evaluation of physicochemical data. The examples include **R** routines (Appendix D) that allow any forensic examiner working with likelihood ratios to measure the performance of their methods in a straightforward way.

6.2   Empirical measurement of the performance of likelihood ratios

The first step in measuring the performance of any evidence evaluation method is the simulation of a number of hypothetical forensic cases. In this way, measuring the performance of the method in those simulated cases gives an idea of the performance in future casework when the evidence evaluation method is used to compute *LR* values. This approach is called *empirical* because it is based on the observation of the outcomes of suitable experimental set-ups.

As the simulated cases will represent the performance of future cases in operational casework, the conditions considered in the design of the simulations have to be as similar as possible to a typical case where the evidence evaluation method is used. For instance, if the method is to be used for computing *LR* values for comparisons of pieces of glass from building windows, then the simulated cases should consider building windows as evidence.

A *validation* database is defined as the data needed to simulate the cases. The term 'validation' is used because the simulated cases are used to assess empirically the performance of a method, and this will typically be considered for its validation prior to its use in casework. The description of this validation process is outwith the scope of this book, but it constitutes an important stage in the use of *LR*-based methods in real cases.

For each simulated forensic case, the proposition that is true in that case – the *prosecution* (H_1) or *defence* (H_2) proposition – must be known. As a consequence, the true origin of the data in the validation database must be known. For example, if the validation database consists of physicochemical data obtained from glass objects, each observation should be associated with a *label* that determines the origins of the object. These labels are called the *ground-truth* labels.

In summary, for each simulated case, the following elements are considered:

- recovered materials, drawn from the validation database. For instance, physicochemical data of a glass object of the database simulating some pieces of glass recovered from a suspect.

- control materials, drawn from the validation database. For instance, physicochemical data of a glass object simulating a piece of glass found at the scene of the crime.

- the population of the simulated case. For instance, the same set of physicochemical data as used for *LR* computation in casework in the laboratory.

- ground-truth label. If the propositions are at the source level (Section 2.1), the ground-truth label should indicate that the prosecution proposition is true if both recovered and control materials from the validation database originate from the same object, and that the defence hypothesis is true if they come from different objects.

Once a validation database has been selected, a number of *LR* values may be generated for all the simulated cases defined. This set of *LR* values and their corresponding ground-truth labels is called a *validation* set of *LR* values. Those *LR* values are then said to be representative of the cases in casework, in the sense that the conditions of the evidence and the population in the simulated cases and in casework have been selected to be comparable.

According to the nomenclature used in previous chapters, H_1 denotes the prosecution proposition and H_2 denotes the defence proposition. Then, depending on the ground-truth label in each simulated case, the corresponding comparison in each case is respectively referred to as a true-H_1 and a true-H_2 comparison. Similarly, the *LR* values in the validation set are respectively referred to as true-H_1 and true-H_2 *LR* values depending on which proposition is true for each *LR* value. There are N_1 true-H_1 and N_2 true-H_2 likelihood ratio values in the validation set of likelihood ratio values, a total of $N = N_1 + N_2$.

It is remarkable that the validation of an LR method is typically conducted in the forensic laboratory, and before that method is used in a given real case. In that sense, the conditions of the validation database are really critical in relation to the conditions of the case. For instance, imagine that a validation process has been conducted for a glass evidence evaluation method, and a validation database of building windows has been used. Using the method in cases where the evidence consists of glass from containers can be risky, because the conditions of the validation database are not representative of the conditions in casework. Therefore, the performance measured in the validation process may not correspond to the performance in casework.

Summarising, in an empirical approach to performance measurement, first a validation set of *LR* values from a validation database is obtained, consisting of the *LR* values from several simulated cases and their corresponding ground-truth labels. Then, a measure of performance is computed from the validation set of *LR* values. Thus, if the conditions in casework are comparable to the conditions of the simulated cases, this measure represents the performance of the evidence evaluation method in casework.

In the following sections some typical measures for the assessment of the performance of likelihood ratios are introduced.

6.3 Histograms and Tippett plots

One of the most common tools for visualising data are histograms (Section 3.3.3), which have been used through this book. In this chapter they are used in order to obtain information about the performance of a validation set of *LR* values. With this objective, it is useful to draw histograms separately for true-H_1 and true-H_2 likelihood ratio values. Figure 6.1 shows histograms of a validation set of *LR* values generated from simulated data for the sake of illustration. Notice that the logarithm of the *LR* is plotted on the horizontal axis of the histogram rather than the *LR*, because it helps visualisation. Although the base of the

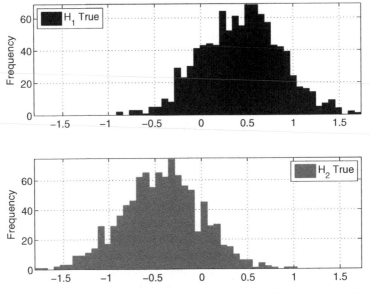

Figure 6.1   Histogram of $\log_{10}(LR)$ values in an artificial validation set of LR values. Two separate histograms are shown: one for true-H_1 values (top) and one for true-H_2 values (bottom). True-H_1 and true-H_2 denote the observations for which the respective hypotheses are true.

logarithm is irrelevant for the purpose used here, base 10 is almost always used throughout this chapter as it is often easier to interpret.

Some observations about the performance of the validation set of LR values can be made from examining histograms:

- The range of $\log_{10}(LR)$ values in the validation set is shown on the horizontal axis of the histogram. Then it can be seen whether there exist true-H_1 likelihood ratio values that are less than 1 ($\log_{10}(LR) < 0$) and to what extent, and whether there are true-H_2 likelihood ratio values that are higher than 1 ($\log_{10}(LR) > 0$). In these two situations it is said that the method used to obtain the validation set of LR values yields *misleading evidence*.

- Histograms show the *strength* or *weight* of the evidence, defined as $\left|\log_{10}(LR)\right|$. If $\left|\log_{10}(LR)\right|$ is large, for an LR value supporting the wrong proposition, it can be said that the LR value yields *strongly misleading evidence*. There are theoretical criteria for considering a value of $\left|\log_{10}(LR)\right|$ to be large; these are outwith the scope of this book, but they can be found in Royall (2000). Histograms show whether the LR values strongly support the correct or incorrect proposition in a case and help detect strongly misleading evidence.

- In the ideal scenario, true-H_1 likelihood ratio values should be as high as possible and greater than 1 ($\log_{10}(LR) > 0$) and true-H_2 values should be as low as possible and less than 1 ($\log_{10}(LR) < 0$). In this scenario there would be a complete separation of the histograms (i.e. no overlap), in the sense that true-H_1 values would be always greater than true-H_2 values. In a real scenario, however, there will most probably be some misleading evidence in the validation set of LR values, and therefore some overlap. The degree of overlap is a measure of the *discriminating power* of the set of LR values, which is an important property of the validation set. However, there are more convenient ways of measuring and analysing the discriminating power (see below).

An extension of the visualisation of the LR values with histograms are *Tippett* plots, which have been classically used for empirical performance assessment since Evett and Buckleton (1996) introduced them. Tippett plots are the cumulative version of histograms, and they represent the cumulative proportion of LR values in the validation set. As with histograms, two curves are plotted for a validation set of LR values: one for true-H_1 values and one for true-H_2 values. Both curves are plotted with respect to the $\log_{10}(LR)$ values on the horizontal axis. Figure 6.2 shows the Tippett plots of the example validation set of LR values.

Tippett plots have the following interpretation:

- the true-H_1 curve is the proportion of true-H_1 $\log_{10}(LR)$ values in the validation set that are greater than the value on the horizontal axis;

- the true-H_2 curve is the proportion of true-H_2 $\log_{10}(LR)$ values in the validation set that are greater than the value on the horizontal axis.

In this sense, Tippett plots follow the definition of the complement of cumulative distributions in probability theory.

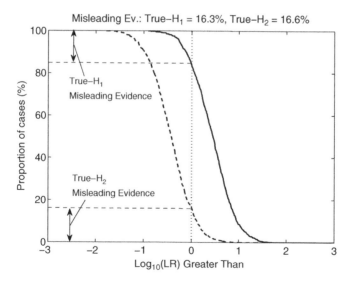

Figure 6.2 Tippett plots of the validation set of LR values artificially generated. The curve for true-H_1 values is the solid line on the right, and the curve for true-H_2 values is the dashed line on the left. The rates of misleading evidence are indicated in the title of the plot.

With Tippett plots, all the insights about performance for histograms also apply, but with a slightly different and more practical visualisation:

- The range of $\log_{10} LR$ values for the validation set is shown on the horizontal axis of the Tippett plots. However, the so-called rates of misleading evidence (i.e. the proportions of cases where the LR yields misleading evidence) can be visualised much more easily than for histograms. For the true-H_2 values, the rate of misleading evidence is given by the value of the true-H_2 curve at the value on the horizontal axis of 0, which means $\log_{10}(LR) = 0$. Conversely, for the true-H_1 values, the rate of misleading evidence is given by 1 minus the value of the true-H_1 curve at $\log_{10}(LR) = 0$. This difference in the computation of rates of misleading evidence for true-H_1 and true-H_2 curves is due to the following fact: both curves are monotonically decreasing, because if the $\log_{10}(LR)$ value τ on the horizontal axis increases, the proportion of $\log_{10}(LR)$ values greater than τ decreases in both cases. However, $\log_{10}(LR)$ values should be greater than 0 when H_1 is true, but less than 0 when H_2 is true. Therefore, when H_2 is true, the crossing of the curve with 0 is in fact the rate of misleading evidence, but the crossing of the true-H_1 curve with 0 is exactly the opposite (1 minus its value) of the corresponding rate of misleading evidence of true-H_1 values. The rates of misleading evidence are highlighted in Figure 6.2.

- Strongly misleading evidence can be seen in Tippett plots. For the true-H_1 curve, if it is not equal to 1 (or 100% as a percentage) for very low values on the horizontal axis (meaning very low $\log_{10}(LR)$ values), it would mean that there is a proportion of true-H_1 LR values that strongly support the H_2 proposition, which means strongly misleading evidence in favour of H_2. Conversely, for the true-H_2 curve, if it is not equal to 0 for very high values on the horizontal axis (meaning very low $\log_{10}(LR)$ values), it would mean that there is a proportion of true-H_2 LR values that strongly support H_1, also meaning strongly misleading evidence, but this time in favour of H_1.

- A measure of discriminating power in Tippett plots is the vertical separation of the true-H_1 and the true-H_2 curves at a given value on the $\log_{10}(LR)$ axis. The more separated the curves at a given $\log_{10}(LR)$ value, the higher the discriminating power at that value.

Additional discussion about Tippett plots can be found in Ramos (2007) and Ramos *et al.* (2013).

6.4   Measuring discriminating power

In forensic science the question of how a method performs in order to use measurements to distinguish between sources of materials has been a matter of discussion since the middle of the twentieth century (Aitken and Taroni 2004). In order to address this problem, *discriminating power* has been defined for discrete data in the context of DNA population genetics as the probability that two samples extracted from a population will match if they come from different sources. Aitken and Taroni (2004) give a detailed review of the use of discriminating power for discrete data.

In the context of evidence evaluation using LR values, discriminating power is associated with distinguishing cases where H_1 and H_2 are respectively true. Discriminating power is defined here for the set of validation LR values, following the definitions in Brümmer and du Preez (2006) and Brümmer (2010). It is much easier to explain the definition if a popular measure of performance is first introduced: false positive and false negative rates (Section 2.4).

6.4.1   False positive and false negative rates

False positive and false negative error rates are a common measure for evaluating decisions in forensic science (Taroni *et al.* 2010). Although decisions as to which of H_1 or H_2 is believed to be true should not be made by the forensic examiner (Evett 1998), or without knowledge of the prior probabilities (Taroni *et al.* 2010), false positive and false negative LR values have been reported as a measure of discriminating power to assess the performance of the likelihood ratio.

In an LR context, it has been argued that a false negative error occurs when $LR < 1$ for a true-H_1 comparison, and a false positive error occurs when $LR > 1$ for a true-H_2 comparison. The term *false negative* refers to the fact that true-H_1 likelihood ratio values should be greater than 1; if they are less than 1 they are falsely supporting the wrong proposition, H_2. Conversely, true-H_2 likelihood ratio values should be less than 1; if they are greater than 1 they are falsely supporting H_1; this is a *false positive*.

However, aside from this interpretation in the LR context, false positive and false negative values have mainly been interpreted in the context of decisions (Duda *et al.* 2001). Imagine that a threshold τ (not necessarily 0) were set for the LR in order to decide that H_1 is true when $\log_{10}(LR) > \tau$, and that H_2 is true when $\log_{10}(LR) \leq \tau$. Then a false positive (also known as a *false acceptance* in some areas of decision theory such as access control) is defined as the case where H_2 is true but $\log_{10}(LR) > \tau$ and H_1 is decided. Conversely, a false negative (also known as a *false rejection*) is defined as the case where H_1 is true but $\log_{10}(LR) \leq \tau$ and H_2 is decided. Then the threshold τ determines the decisions that are actually taken. In the typical reporting of false positives and false negatives to describe the discriminating power of LR values, the value of $\tau = 0$ (or equivalently $LR = 1$) is therefore the most common. Nevertheless, reporting false positives and false negatives in an experiment to assess the performance of likelihood ratios should not be interpreted as if the forensic examiner were making decisions about the propositions in casework.

False positives and false negatives are related to Tippett plots and histograms. If the threshold τ is set, the false positive rate in a histogram is the number of $\log_{10}(LR)$ values greater than τ in the true-H_2 histogram divided by the total number of LRs in the true-H_2 histogram. Conversely, the false negative rate in a histogram is the number of $\log_{10}(LR)$ values less than or equal to τ in the true-H_1 histogram divided by the total number of LRs in the true-H_1 histogram. As an example, for histograms in the example in Figure 6.1 and for $\tau = 0$ ($LR = 1$), the false positive rate is 0.166 (or 16.6%), and the false negative rate is 0.163 (or 16.3%).

As Tippett plots and histograms are strongly related, it seems reasonable that false positive and false negative rates can be seen from Tippett plots. In the example in Figure 6.2, and for $\tau = 0$, the false negative rate corresponds to 1 minus the rate of misleading evidence when H_1 is true (the proportion of cases when H_1 is true and $LR < 1$). Analogously, the false positive rate is equal to the rate of misleading evidence when H_2 is true (the proportion of cases when H_2 is true and $LR > 1$).

Different values of τ yield different false positive and false negative rates that can be easily seen in Tippett plots. For instance, for a given threshold τ, the false positive rate is the proportion of true-H_2 cases in the Tippett plot greater than $\log_{10}(LR) = \tau$ (i.e. the value of the true-H_2 Tippett plot when the value on the horizontal axis is τ). Conversely, the false negative rate is 1 minus the proportion of true-H_1 cases in the Tippett plot greater than $\log_{10}(LR) = \tau$ (i.e. 1 minus the value of the true-H_1 Tippett plot when the value on the horizontal axis is τ).

From Tippett plots it can be seen that false positive rates decrease as the threshold τ increases, and false negative rates increase as τ increases. This is in accordance with the idea that if τ increases, then the number of true-H_2 $\log_{10}(LR)$ values that are greater than τ can only decrease, and therefore the false positive rate decreases with τ. Conversely, if τ increases, then the number of true-H_1 values that will be less than τ can only increase, and therefore the false negative rate increases with τ. This can be observed with the monotonic nature of the curves in a Tippett plot. Therefore, varying the threshold τ, when the false positive rate increases the false negative rate decreases, and vice versa.

The use of false positives and false negatives is not appropriate in an LR context of evidence evaluation. The reason is that an explicit use of a threshold τ is needed in order to report false positive and false negative rates. Under this approach a decision would be assumed to be made based only on the LR value, which is not acceptable in forensic science (Evett 1998). However, if they are computed in an experiment aimed at measuring performance, false positive and false negative rates are useful in order to illustrate and measure the concept of discriminating power, as shown in the following section.

6.4.2   Discriminating power: A definition

Discriminating power is defined here as a measure of performance of a validation set of LR values, following Brümmer (2010). It is defined in relation to the false positive (FP) and false negative (FN) rates of the set of LR values, with respect to the threshold τ used to compute those rates. In the previous section it was shown that FP and FN rates change as a function of the threshold τ, and that this function is basically defined by the Tippett plots. The discriminating power or discrimination performance of a validation set of LR values is then defined by the following two propositions:

1. Two different validation sets of LR values S_1 and S_2 have the same discriminating power if, for every possible threshold τ in S_1, a threshold τ' can be found in S_2 such as the false acceptance and false rejection rates are the same for both sets. Conversely, for every possible threshold in S_2, say τ', there exists a threshold in S_1, say τ, leading to the same false acceptance and false rejection rates.

2. A set of LR values S_1 is said to have better discriminating power than a set S_2 *at a single threshold*, if, for a given threshold τ in S_1, a threshold τ' can be found for S_2 such that the false acceptance and false rejection rates in S_2 at threshold τ' are both higher than in S_1 at threshold τ.

An example of our definition of discriminating power is given in Figure 6.3. Two validation sets of LR values are shown as Tippett plots. The set S_1 (Figure 6.3(a)) clearly has better discriminating power than the set S_2 (Figure 6.3(b)) according to our definition. This is because for any threshold selected in S_1, a threshold can be found in S_2 with both higher FP

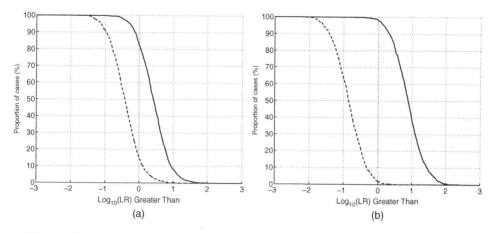

Figure 6.3   Tippett plots of two validation sets of *LR* values, (a) set S_1 and (b) set S_2.

and FN rates. (The reader should try to check this as an exercise. Hint: use Tippett plots as explained before, and set the threshold τ as the value on horizontal axis.)

Another example is shown in Figure 6.4, where both sets of *LR* values, S_1 again and S_3, have the same discriminating power. It can be checked that for every threshold τ_1 in S_1, it is possible to find a threshold τ_2 in S_3 showing the same FP and FN rates (again, the reader is recommended to check this).

Discriminating power as defined here is a critical performance characteristic of any validation set of *LR* values. Imagine that one wants to separate cases where H_1 is true from cases where H_2 is true, and the *LR* is used as a way to do this. Also, imagine that a threshold τ is set, and for any case, if $\log_{10}(LR) > \tau$, then it is decided that H_1 is true, and if $\log_{10}(LR) \leq \tau$, then it is decided that H_2 is true. Then the better the discriminating power of the *LR* values, the lower the FP and FN rates that will be obtained when making decisions. In

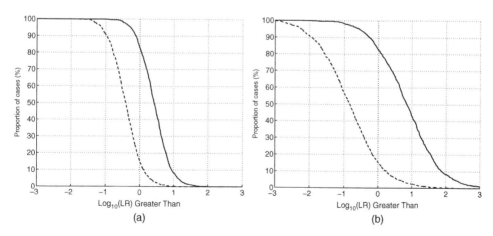

Figure 6.4   Tippett plots of two validation sets of *LR* values, (a) set S_1 and (b) set S_3.

other words, better discriminating power means that a decision maker who uses the LR for making decisions will make fewer errors.

In order to visualise discriminating power from histograms, the relative overlap RO is defined as

$$RO = \frac{A_O}{A_T},$$

where A_O is the overlapping area of the true-H_1 and true-H_2 histograms, and A_T the total area of the histograms. This is an approximate and intuitive view of the discriminating power: the lower the relative overlap between the true-H_1 and true-H_2 histograms of a validation set of LR values, the better the discriminating power will tend to be.

6.4.3   Measuring discriminating power with DET curves

A commonly used graphical representation of discriminating power, especially useful when comparing multiple systems, is the *receiver* (or *relative*) *operating characteristic* (ROC) plot, in which the FP rate is plotted against the FN rate for all possible values of the threshold τ. As our definition of discriminating power is related to the trade-off between the FP and FN rates, this representation clearly measures the discriminating power of a validation set of LR values. A variant of the ROC curve, the so-called *detection error trade-off* (DET) plot, will also be used in this book (Martin *et al.* 1997). DET curves rely on the use of a non-linear, Gaussian scale that facilitates the comparison of different validation sets. Figure 6.5 shows an example of a DET plot.

As a rule of thumb, the closer the DET curve is to the origin of the coordinates (bottom left-hand corner in the plot), the higher the discriminating power of the set of LR values. This corresponds to the idea that the discriminating power is better if the FP and FN rates are both lower.

A summary measure of discriminating power is the equal error rate (EER), which is the FP or FN rate at the threshold τ_{EER} where both are equal. The EER is easily seen as the intersection point between the DET curve and the line where the FP rate is equal to the FN rate.

The relationship between DET curves and Tippett plots is shown in Figure 6.6. It can be checked there that every pair of FP and FN values at a given threshold τ in Tippett plots is just a point in the DET plot. Then the DET plot can be drawn by plotting all the FP–FN points obtained from the Tippett plots at different values of τ (the reader is encouraged to check this as an exercise).

Notice that the EER need not correspond to a threshold of $\tau = 0$, but the value of the threshold where the error rates are equal will depend on the distribution of the $\log_{10}(LR)$ values.

From the definition of discriminating power given above, it can be seen that two validation sets of LR values with the same discriminating power have the same DET curve (Martin *et al.* 1997). However, two sets of LR values with the same discriminating power may have very different Tippett plots. This is exemplified in Figures 6.4 and 6.5. It can be checked that the validation sets of LR values S_1 and S_3 in Figure 6.4 have the same DET curve, shown in Figure 6.5. The explanation of this is that the important factor for discriminating power is the relationship between the FP and FN rates, and not the range in which the values of the

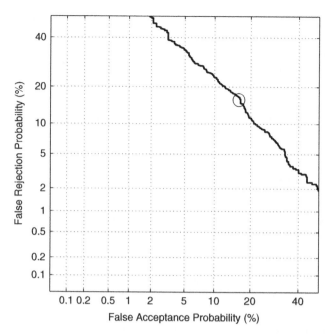

Figure 6.5 An example of a DET plot for an artificial validation set of *LR* values. For any value of the threshold τ, the FP and FN values are shown. The FP and FN rates for the threshold $\tau = 0$ are highlighted with a circle. This plot corresponds to a validation set where the FP and FN are both roughly 15% when $\tau = 0$. The point where the FP and FN are equal (the equal error rate) need not correspond to a threshold of $\tau = 0$.

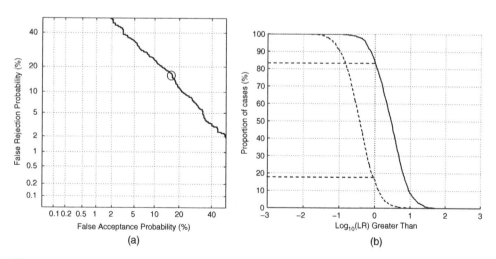

Figure 6.6 Comparison of (a) DET plots and (b) Tippett plots for the same artificial validation set of *LR* values. The FP and FN rates for the threshold $\tau = 0$ are highlighted in the DET plot in (a) with a circle. Notice that these rates correspond to the rates of misleading evidence in the Tippett plot in (b), marked with horizontal dashed lines.

LR sets lie. In other words, two sets of *LR* values with very different numerical ranges may have the same discriminating power if the relationship between their FP and FN rates is the same. It can be checked that this is what happens in the example in Figure 6.4. If a threshold $\tau = 0$ is established in both *LR* sets represented as Tippett plots, the FP and FN rates obtained are the same in both of them. Moreover, for any threshold τ_1 established in the *LR* values in Figure 6.4(a), another threshold τ_2 can be found in the *LR* values in Figure 6.4(b) for which the same FP and FN values in both sets of *LR* values were obtained. (The reader is encouraged to check this by inspection of the Tippett plots as an exercise.)

6.4.4   Is discriminating power enough?

Discriminating power is an important measure of performance of every validation set of *LR* values, because it measures the ability of the *LR* values to distinguish between cases where H_1 is true and cases where H_1 is false. However, in the context of evidence evaluation using *LR* values, it may not be enough for the measurement of the performance of *LR* values.

Consider the example in Figure 6.7, which shows the Tippett and DET plots of two validation sets of *LR* values, namely S_c and S_u. If only discriminating power is considered, then S_u is much better than S_c, because its DET curve is closer to the origin of coordinates than the DET curve of S_c.

However, from the Tippett plots, it can be seen that the range of *LR* values for S_u is biased: many of the true-H_1 values quite strongly support the wrong proposition H_2, leading to strongly misleading evidence in most of the cases where H_1 is true. This misleading evidence may lead fact finders to wrong decisions, because the *LR* values often support the wrong proposition when H_1 is true. Therefore, it seems that S_u is not appropriate for evidence evaluation. On the other hand, both rates of misleading evidence for S_c seem much more controlled.

This example illustrates that, although a set of *LR* values (S_u in our case) may present good discriminating power, the model that generated those *LR* values may not be appropriate for evidence evaluation. Therefore, although discriminating power is a desirable characteristic of a set of *LR* values, it may not be enough to determine whether a model is appropriate for evidence evaluation or not. In our example, S_u presents a problem not related to discriminating power. It will become evident below that the problem is related to a critical performance characteristic that should be considered in addition to the discriminating power. This is called *calibration*.

6.5   Accuracy equals discriminating power plus calibration: Empirical cross-entropy plots

For many years, Bayesian statisticians have been seriously concerned about the elicitation of probabilistic assessments (Garthwaite *et al.* 2005; Lindley *et al.* 1979; Savage 1971), which can be understood given the Bayesian interpretation of probability as a degree of belief (Lindley 2006; O'Hagan 2004). In this topic of research, one of the main questions under study has always been the performance of probabilistic assessments, which can be summarised as follows: if someone is eliciting probability assessments (according to a given model and data, or based on personal experience), how may their performance be evaluated?

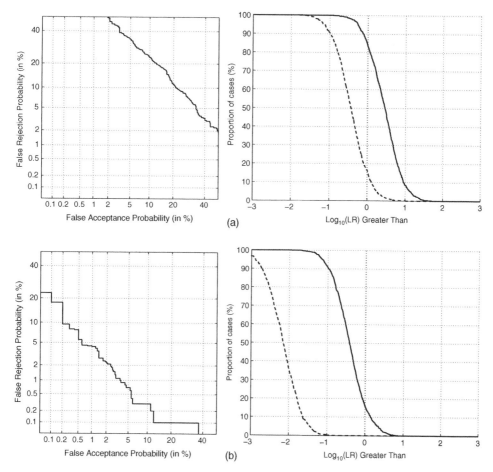

Figure 6.7 Example illustrating that discriminating power is not enough to determine whether an evidence evaluation is good or bad. DET plots (left) and Tippett plots (right) of two validation sets of LR values: (a) $\mathcal{S}_c$ and (b) $\mathcal{S}_u$.

6.5.1   Accuracy in a classical example: Weather forecasting

Let us take a classical example of the problem: weather forecasting. Imagine a forecaster who every day provides a probability of rain the next day. The evaluation of the performance of a single forecast can be intuitively seen. Imagine that the forecaster assigns a probability of rain for the next day of 0.9 (or 90% as a percentage). Two days later, it is known that it did actually rain on the day of the forecast. Then, for that particular forecast, the forecaster performed well. If an external evaluator of performance assigned a cost (or penalty) to that particular forecast, that penalty should be low. However, if the forecaster had assigned a probability of rain the next day of 0.1, then that forecast would not have been a good one, since the probability associated with what actually happened (it rained) would have been low. These examples suggest that, in order to evaluate the performance of a single forecast, two elements are needed: the probability assigned by the forecaster (the probability of rain the next day) and the actual outcome (it rained or it did not rain), which was unknown by the forecaster, but is known when the performance is evaluated after the event.

According to this general idea, the performance of probabilistic assessments is addressed by the use of so-called *strictly proper scoring rules* (SPSRs) (deGroot and Fienberg 1983; Gneiting and Raftery 2007; Savage 1971). An SPSR is a function of both the probability distribution assigned to a given unknown variable, and the actual value of the variable. The value of the SPSR is interpreted as a *loss* or *cost* assigned to the probability distribution depending on the actual value of the variable. In this example, the *logarithmic* SPSR is used. This is defined as follows:

$$
\begin{array}{ll}
-\log_2 (P) & \text{if it rained;} \\
-\log_2 (1 - P) & \text{if it did not rain,}
\end{array}
\tag{6.1}
$$

where, for the weather forecasting example, P represents the probability (Appendix A) of rain on a given day. The base of the logarithm is irrelevant for the exposition; logarithms to base 2 are used for information-theoretic reasons explained in Ramos (2007). The intuition behind SPSRs is illustrated by the logarithmic SPSR in Figure 6.8. The figure shows the two possible values of the logarithmic SPSR, depending on whether it rained (H_1) or not (H_2), as a function of the probability P of rain from a single forecast on a given day. According to equation (6.1), if it actually rained on the day of the forecast, the SPSR assigns a high penalty to low values of P, and vice versa. This corresponds to the fact that, if the weather forecaster assigned a high probability of rain P, and it actually rained, then the penalty should be low, and vice versa. In the limit, if the forecaster assigned a categorical probability of $P = 0$ (i.e. it is impossible that it will rain tomorrow), and it actually rained, the penalty will be infinite for the logarithmic SPSR. This is a desirable property of the use of logarithmic scoring rule, if it is assumed that someone who categorically makes a wrong judgement should be the worst possible forecaster. From Figure 6.8, analogous reasoning follows for the case where it did not rain. (This is left as an exercise for the reader.)

Strictly proper scoring rules measure the quality of a single forecast. However, the overall performance of a given set of forecasts from a given forecaster is desirable. This is done by

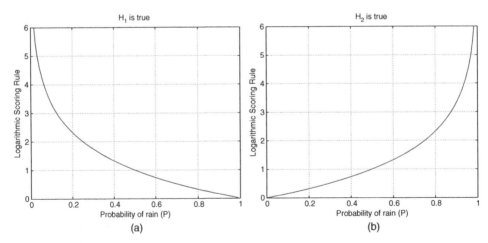

Figure 6.8   Logarithmic strictly proper scoring rule: (a) $-\log_2(P)$; (b) $-\log_2(1 - P)$.

averaging the values of the SPSR over a *validation* set, in this case of probabilities P, as is done in deGroot and Fienberg (1983). In weather forecasting this can be done from a record of the probabilities of rain assigned by the forecaster, plus the ground-truth labels obtained from a weather database where the days on which it actually rained are already registered. Then an average of the SPSR can be computed for all those forecasts P, and a global measure of quality L can be obtained:

$$L = -\frac{1}{N_1} \sum_{i:H_1 \text{ is true}} \log_2 (P_i) - \frac{1}{N_2} \sum_{j:H_2 \text{ is true}} \log_2 (1 - P_j), \qquad (6.2)$$

where N_1 is the number of days on which it rained and N_2 is the number of days on which it did not rain, and the values of the SPSR of the forecasts for which H_1 is true (indexed as P_i in equation (6.2)) and for which H_2 is true (indexed as P_j in equation (6.2)) are separately averaged and then summed.

The global measure of qualty L can be interpreted as the *accuracy* of the probabilistic assessments of the weather forecaster. Accuracy is defined as the closeness of a given magnitude to its true value. The interpretation of L as accuracy is as follows. A *perfect* or *oracle* forecaster is defined as one who knows the ground-truth labels. It can be said that the knowledge of the oracle forecaster allows her or him to elicit perfect forecasts: $P = 1$ for days on which it rains, and $P = 0$ for days on which it does not rain. From equation (6.2), $L = 0$ for the oracle forecaster. A forecaster who is not an oracle does not know whether or not it will rain on a given day and therefore will not make perfect forecasts, leading to an SPSR greater than 0. Then the deviation of the non-perfect forecaster from the behaviour of the oracle forecaster is the penalty assigned by the SPSR, and thus is a measure of the accuracy of the given forecaster.

6.5.2 Calibration

The so-called property of *calibration* of a set of probabilistic assessments has been extensively studied by Bayesian statisticians (Dawid 1982; deGroot and Fienberg 1983). An intuitive definition of calibration is given by Lindley *et al.* (1979, p. 147) in the weather forecast context: 'If the meteorologist is using the scale properly, however, we would expect that rain would occur on two-thirds of the days to which he assigns a rain probability of 2/3. This criterion is called calibration.'

In the notation used above, calibration can be defined as follows. If a forecaster elicits probabilistic assessments about rain on some days, namely P, then, for all probabilistic assessments for which $P = p$ the proportion of days on which it actually rained should be p. This definition is useful if the values of the probabilities that the forecaster can assign are discrete, as in deGroot and Fienberg (1983). If the forecaster can elicit any continuous value in the [0, 1] range, then the definition should include some kind of partitioning (or *binning*) of that range. Then it can be said that a set of probabilistic assessments are well calibrated if, for all assessments P with values within a region (or *bin*) defined by $p \pm \Delta$ the proportion of cases on which it actually rained is close to p.

Some of the good properties of calibration can be guessed intuitively. If a forecaster elicits well-calibrated probabilistic assessments, they are constraining their opinions to the actual proportion of occurrence of events, a behaviour that seems reasonable at first glance. However, there are many other desirable properties of well-calibrated probabilistic assessments. As highlighted in Ramos and Gonzalez-Rodriguez (2013), if a set of probabilistic assessments has low discriminating power, meaning that a forecaster is not proficient at separating days on which it rains from days on which it does not rain, then it can be said that the forecaster is giving forecasts that are not too useful. In that case, if the calibration of the set of probabilities is good, then the value of those probabilities should not be too close to 0 or 1. In other words, a forecaster who is not proficient should not express opinions too strongly. Conversely, if the discriminating power of a set of probabilities is good, if the forecaster is well calibrated then they would be able to provide probabilities close to 1 or 0. Thus, only forecasters who are really proficient are allowed by calibration to express strong opinions. This property is one of the reasons for calibration being dubbed *reliability* in the statistics literature (Dawid 2004; deGroot and Fienberg 1983).

6.5.3   Adaptation to forensic inference using likelihood ratios

If the weather forecasting example can be seen as a probabilistic inference problem in a Bayesian context, it can be seen to be equivalent to the problem of inference in a forensic case using likelihood ratios, just by taking the following analogies into account:

- In weather forecasting, the unknown variable of interest is the fact that it rained a given day or not. In forensic inference, the aim is to know which of the prosecution proposition H_1 or the defence proposition H_2 is actually true.

- In weather forecasting, all the knowledge available to the forecaster is considered in order to elicit the forecast. In forensic inference, all the available knowledge is split into E (the evidence) and I (the rest of the information).

- Given the analogies above, the problem of forensic inference can be stated in the same terms as in weather forecasting. On the one hand, the aim of the weather forecaster is to assign a posterior probability P that represents his or her opinion about the value of the unknown variable (it rains, it does not rain), from their knowledge. On the other hand, in forensic inference the aim of the fact finder is to obtain a posterior probabilities $P(H_1|E, I)$ and $P(H_2|E, I)$ that represent his or her opinion about the value of the proposition (H_1, H_2) from the information obtained from the evidence E and the rest of information in the case I.

Despite these analogies, the use of the methodology of performance assessment based on SPSRs in forensic science is not straightforward. In weather forecasting, the aim is to measure the performance of the probabilistic assessments of the forecaster, namely P. As the person responsible for such assessments is the forecaster, performance can be measured directly from a validation set of posterior probabilities assigned by the forecaster (and their corresponding ground-truth labels). However, in forensic evidence evaluation only the measure of the performance of the probabilistic assessments of the forensic examiner is of interest. Therefore, it is in order to focus just on measuring the performance of the *LR*, not the performance of the posterior probability $P(H_1|E, I)$. This is because the posterior probability depends both on the *LR* and on the prior probability $P(H_1|I)$, the latter not being the province of the forensic examiner. Unfortunately, the SPSR framework is based on measuring the performance of the posterior probabilities, and therefore if it is directly applied to forensic science, the performance of the prior probabilities would be also measured, not just the LR given by the forensic examiner.

In order to avoid those problems, a method is proposed that allows the use of SPSRs to measure the performance of a set of LR values (Ramos *et al.* 2013; Ramos and Gonzalez-Rodriguez 2013). This method follows Brümmer and du Preez (2006), and basically involves setting up an experiment as follows:

1. Obtain a validation set of LR values, following the procedure described in Section 6.2. Notice that LR values are computed for the validation set, meaning that no prior probability is considered in the process.

2. The forensic examiner cannot fix a value of the prior probabilities, even in the experiment, because that value is not in his or her competence (Chapter 2). Instead, accuracy in the experiment can be computed for many values in a wide range of prior probabilities, following the same procedure as in Brümmer and du Preez (2006). That way, in the experiment, the SPSR is computed when the validation set of LR values is used for each of the prior probabilities in that range.

3. Plot a measure of accuracy (the average of an SPSR) for all the values in the range of prior probabilities considered in the experiment. Thus, the forensic examiner never fixes a value for the prior probabilities, but can know the performance (accuracy) if their LR values were used in a correct Bayesian framework within that range of prior probabilities.

Notice the difference between the use of an LR in casework and the procedure described to measure the performance of a set of LR values in a controlled experiment, typically prior to casework. In casework, it is totally unrealistic to think that the fact finder is assessing a prior probability, at least in current practice. However, the role of the SPSR assessment methodology is to consider the likelihood ratio in a formal way, as part of a Bayesian inference process. Therefore, in the experiment, the forensic examiner acts as if the fact finder were assessing a prior, but its particular value will never be known. This is the reason for considering a wide range of prior probabilities, and not a particular value of the prior probability.

Following this procedure, accuracy can be represented as a function of the prior probability. First, our measure of accuracy is defined. A variant of L in equation (6.2) is chosen, namely *empirical cross-entropy* (ECE), as the average value of the logarithmic scoring rule, weighted as follows:

$$ECE = -\frac{P(H_1|I)}{N_1} \sum_{i:H_1 \text{ is true}} \log_2 P(H_1|E_i, I)$$

$$-\frac{P(H_2|I)}{N_2} \sum_{j:H_2 \text{ is true}} \log_2 P(H_2|E_j, I), \tag{6.3}$$

where E_i and E_j denote the evidence in each of the comparisons (cases) in the validation set where either H_1 or H_2 is true, and N_1 and N_2 are the numbers of cases. It is informative to express the ECE explicitly in terms of the prior odds, which can be shown to be:

$$ECE = \frac{P(H_1|I)}{N_1} \sum_{i:H_1 \text{ is true}} \log_2 \left(1 + \frac{1}{LR_i \times O(H_1|I)} \right)$$

$$+ \frac{P(H_2|I)}{N_2} \sum_{j:H_2 \text{ is true}} \log_2 \left(1 + LR_j \times O(H_1|I) \right), \tag{6.4}$$

where $O(H_1|I) = \frac{P(H_1|I)}{P(H_2|I)}$ are the prior odds in favour of H_1.

As can be seen in equations (6.3) and (6.4), the averages in ECE are weighted by the value of the prior probabilities. This weighting allows the ECE to be interpreted in an information-theoretic way, but this topic is outwith the scope of this work (see Ramos 2007; Ramos et al. 2013). However, it can be shown that the interpretation of the ECE as a measure of the accuracy of the method and its properties related to calibration remain the same as for the average of the SPSR, namely L (Brümmer 2010).

Equation (6.4) shows that the ECE depends on the validation set of LR values in the experiment (i.e. the LR values and their corresponding ground-truth labels). However, the ECE also depends on the value of the prior odds $O(H_1|I)$, since an SPSR depends on the posterior probabilities. Thus, following the procedure described above, the ECE can be represented in a prior-dependent way. An example of such a representation can be seen in Figure 6.9. Logarithms to base 10 are used for the prior odds because they are typically used

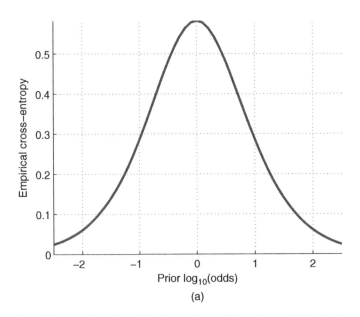

(a)

Figure 6.9   ECE as a function of the logarithm of the prior odds (prior log-odds).

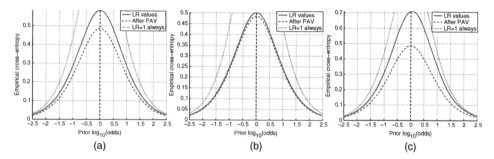

Figure 6.10   Examples of ECE plots: (a) typical ECE plot of a fairly well-calibrated validation set of *LR* values, although calibration could still be improved; (b) the ECE plot of a well-calibrated validation set of *LR* values; (c) badly calibrated validation set of *LR* values.

for evidence evaluation. However, base-2 logarithms are used for computation of the ECE because of its information-theoretic interpretation (Cover and Thomas 2006; Ramos 2007; Ramos *et al*. 2013).

The ECE in Figure 6.9 represents the accuracy for all the possible values of the prior probability, but calibration is not explicitly measured in such a representation. Therefore, an explicit measurement of discriminating power and calibration is given in a so-called ECE plot (Ramos *et al*. 2013), which shows three comparative performance curves together (Figure 6.10):

- solid curve: accuracy. This curve is the ECE of the *LR* values in the validation set, as a function of the prior log-odds. The lower this curve, the more accurate the method. This is the same representation as shown in Figure 6.9.

- dashed curve: calibrated accuracy. This curve is the ECE of the validation set of *LR* values after being perfectly calibrated, as a function of the prior log-odds. Therefore, this shows the performance of a validation set of optimally calibrated *LR* values obtained by a transformation applied to the original validation set of *LR* values. In order to obtain this curve the value of the ground-truth labels should be known. Therefore, this curve is not possible to obtain in practice, and represents a *ceiling of performance* useful for measuring calibration. Details on the procedure of obtaining these calibrated *LR* values are outwith the scope of this book, and can be found in Brümmer (2010) and Ramos *et al*. (2013). The transformation is essentially carried out by an algorithm called *pool adjacent violators* (PAV), and therefore this curve in the ECE plots is sometimes referred to as *accuracy after PAV* (Appendix C).

- dotted curve: neutral reference. This represents the comparative performance of a so-called *neutral LR method*, defined as one that always delivers $LR = 1$ for each case. This neutral method is taken as a *floor of performance*: the accuracy should always be better than the neutral reference. Therefore, the solid curve in an ECE plot should be always lower than the dotted curve, for all values of the prior log-odds (the names *floor* and *ceiling* may be seen as the opposite of the usual physical connotations but are chosen to represent the lowest and highest levels of performance).

Thus, and according to Ramos *et al.* (2013) and Ramos and Gonzalez-Rodriguez (2013), the following are represented (in ECE plots):

- accuracy: solid curve. The lower the curve, the better the accuracy.

- discriminating power: dashed curve. The lower the curve, the better the discriminating power. That the ECE after PAV represents the discriminating power is justified in Brümmer (2010) (with a theoretical development) and Ramos and Gonzalez-Rodriguez (2013) (more adapted to forensic science).

- calibration: difference between the solid and dashed curves. The closer the dashed and the solid curves, the better the calibration.

In Figure 6.10, three ECE plots are shown. Figure 6.10(a) shows the performance of a fairly well-calibrated set of *LR* values, with solid and dashed curves quite close to each other, although closeness could still be improved. Figure 6.10(b) shows a well-calibrated *LR* set, with solid and dashed curves very close to each other. In the ECE plot of Figure 6.10(c), the solid curve is far from the dashed curve, indicating bad calibration. Interestingly enough, the three sets of *LR* values presents the same discriminating power, because the dashed curve is the same for all three.

Notice that the accuracy of a set of *LR* values can be explicitly decomposed between discriminating power and calibration. The proof of this is set out in Brümmer (2010). Accuracy remains as the global performance measure: the better the accuracy, the better the validation set of *LR* values. There are two possible reasons (which may both occur) why accuracy may be poor:

- the discriminating power is poor. This means that the validation set of *LR* values is poor at separating *LR* values for which H_1 is true from *LR* values for which H_2 is true.

- the calibration is poor. This means that the *LR* values provide poor probabilistic measures of the value of the evidence. Even if the *LR* values have high discriminating power, poor calibration can degrade the accuracy considerably.

6.6   Comparison of the performance of different methods for LR computation

In this section, the techniques for performance assessment described above are exemplified with several experiments using examples presented in previous chapters of the book.

6.6.1   MSP-DAD data from comparison of inks

The experiment presented here is the same as that presented in Section 4.4.1, which compared several models for LR computation in a comparison problem, from chromaticity coordinates analysis of inks using MSP-DAD (Section 1.5). Thus, in this problem the propositions are defined as H_1 (the inks compared come from the same source) and H_2 (the inks come from different sources). Three univariate models have been used for each of the chromaticity coordinates, x, y, and z ($x + y + z = 1$). Here, the performances of the three methods are analysed and compared with the performance assessment tools presented in this chapter.

The histograms, Tippett plots, DET plots, and ECE plots can be obtained by using the **R** code `x_y_z_MSP_inks_histogram_Tippett_DET_ECE_code.R`, which is available in Section D.10 and on the website. The code uses the data files containing the *LR* values obtained from the pairwise comparisons carried out to estimate the rates of false positive and false negative answers as per the example in Section 4.4.1:

- `x_comparison_research_Nor_different.txt` and `x_comparison_research_Nor_same.txt`,

- `y_comparison_research_Nor_different.txt` and `y_comparison_research_Nor_same.txt`,

- `z_comparison_research_Nor_different.txt` and `z_comparison_research_Nor_same.txt`.

Running the code requires the following functions to be included in the same file folder (available in Section D.10 in Appendix D):

- `histogram_function.R`,

- `Tippett_function.R`,

- `DET_function.R`,

- `ECE_function.R`.

The code provides the user with the following files as an outcome:

- `x_histogram.eps`, `x_Tippett.eps`, `x_DET.eps`, `x_ECE.eps`,

- `y_histogram.eps`, `y_Tippett.eps`, `y_DET.eps`, `y_ECE.eps`,

- `z_histogram.eps`, `z_Tippett.eps`, `z_DET.eps`, `z_ECE.eps`.

6.6.1.1   Histograms and Tippett plots

Histograms of the MSP results are shown in Figure 6.11. From the histograms, several effects can be observed:

- The three validation sets of *LR* values present a similar distribution both for true-H_1 and true-H_2 values. This indicates that the data for the three chromaticity coordinates yield similar strength of support over the cases simulated.

- For the three chromaticity coordinates, there is a large amount of very low true-H_2 *LR* values. This is because a cut-off limit of $\log_{10}(LR) = -20$ has been applied to all the *LR* values in all the validation sets. The reason for this is that many of the true-H_2 *LR* values presented extremely small values, which may result in support for the H_2 proposition which is too strong. This fact can be observed in the histograms, allowing

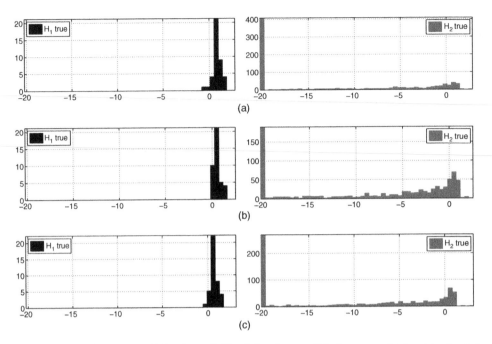

Figure 6.11 Histograms of the $\log_{10}(LR)$ of the three validation sets obtained in the ink analysis experiment using MSP-DAD data (Section 4.4.1). Three different univariate models have been used, for chromaticity coordinates (a) x, (b) y, and (c) z.

the forensic examiner a more careful analysis of those cases in order to improve the models in the future.

- The proportion of overlap between the true-H_1 and true-H_2 histograms is quite high for all the three chromaticity coordinates. This is because many of the true-H_1 LR values lie in the range of true-H_2 values. This will cause a degradation of the discriminating power of the three sets of LR values.

Similar comments about the range and discriminating power of the LR values can be made based on the Tippett plots in Figure 6.12, since they are just cumulative versions of the histograms in Figure 6.11. However, the rates of misleading evidence can be more easily observed from the Tippett plots. It can be seen that those values of misleading evidence are fairly similar for the three chromaticity coordinates, though slightly lower for the x component under the defence proposition, indicating that it may give a slightly better discriminating power than the rest of the components.

6.6.1.2   Discriminating power: DET plots

The DET plots comparing the three chromaticity coordinates are shown in Figure 6.13. Recall that DET plots measure discriminating power, and the closer the curve to the origin of coordinates (bottom left-hand corner), the better the discriminating power. It is seen that

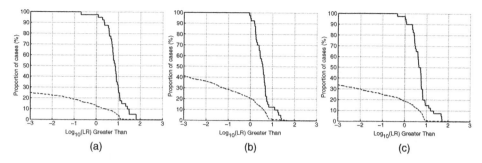

Figure 6.12 Tippett plots for the three validation sets of *LR* values obtained in the ink analysis experiment using MSP-DAD data (Section 4.4.1). Three different univariate models have been used, for chromaticity coordinates (a) *x*, (b) *y*, and (c) *z*.

following that simple visual criterion, the best discriminating power is given by the validation set of *LR* values corresponding to the *x* chromaticity component.

Notice that this conclusion generalises the one drawn from the rates of misleading evidence in Tippett plots. In particular, each point in a DET curve can be seen as a measure of two rates: the proportion of true-H_1 $\log_{10}(LR)$ values that are below a given threshold τ (the proportion of false rejections) and the proportion of true-H_2 $\log_{10}(LR)$ values that are greater than the same threshold τ (the proportion of false acceptances). The particular values of these rates

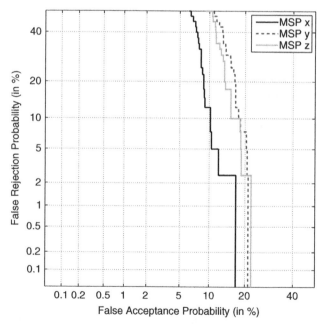

Figure 6.13 DET plots comparing the three validation sets of *LR* values obtained in the ink analysis experiment using MSP-DAD data (Section 4.4.1).

where the threshold τ for the $\log_{10}(LR)$ is 0 (i.e. the threshold for the LR is 1) are the rates of misleading evidence shown in Tippett plots.

6.6.1.3   Accuracy: ECE plots

Histograms and Tippett plots provide an initial measure of performance for the LR values in terms of their range of values and their discriminating power. In DET plots, the discriminating power is characterised more informatively, permitting an easy comparison among different sets of LR values.

Here ECE plots are presented in order to show the overall performance of the set of LR values in terms of accuracy, which takes into account the discriminating power of the set of LR values, as well as the calibration. Figure 6.14 shows the ECE plots of the three validation sets of LR values for the different chromaticity coordinates in the MSP-DAD data. Several conclusions can be drawn from the ECE plots:

- The accuracy (solid curve) is best for the x component, because its solid curve is the lowest of all for the entire range of prior log-odds. This could also have been guessed from previous measures, since DET plots indicated better discriminating power for the x component, and Tippett plots showed a similar range of LR values for all components. However, in ECE plots this fact is explicitly measured by the accuracy curve.

- The discriminating power (dashed curve) of the x component is also best. This can be seen in the fact that the dashed curve is lowest for the x component. This is exactly the same information as given by the DET curve, and it can be seen that the DET curve and the dashed curve in the ECE plot are closely related (Brümmer 2010). Therefore, in ECE plots the information about the discriminating power that is given by a DET plot is already included.

- The calibration of all the chromaticity coordinates is good. This is because calibration is better as the solid and dashed curves come closer. It can be seen in Figure 6.14 that this difference between the curves is similar for the three chromaticity coordinates over the entire range of prior log-odds. This good calibration is a consequence of the acceptable range of LR values for all the validation sets: there is no strongly misleading evidence, and the maximum values of the misleading LR values are severely limited, especially

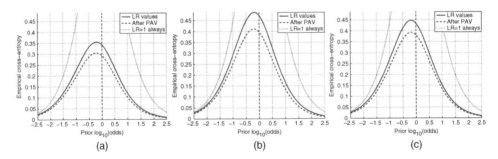

Figure 6.14   ECE plots for each of the three validation sets of LR values obtained in the ink analysis experiment using MSP-DAD data (Section 4.4.1). Three different univariate models have been used, for chromaticity coordinates (a) x, (b) y, and (c) z.

for the true-H_1 cases. Therefore, the calibration of each of the LR sets does not degrade heavily. However, there is still room for improvement, perhaps from making the LR model more conservative, in order to limit the LR values further.

- For the entire range of the prior odds, the accuracy of the LR values (solid curve) is better than the neutral reference (dotted curve). This is the aim of any evidence evaluation method. It indicates that the method is both discriminating and sufficiently well calibrated to be more accurate than not evaluating the evidence. The neutral reference is therefore a floor of performance that may indicate when a method is good enough to be used in casework (if the experimental set-up has been properly designed).

Our conclusion for this example after examining the ECE plots is that the LR values given by the model from each of the chromaticity coordinates separately are well calibrated, which means that the accuracy is better than the neutral reference in all cases. Therefore, they can be used in evidence evaluation. However, the discriminating power of the LR values is poor in all cases, and therefore the LR values are not too useful as a source of information to the fact finder, even though they are well calibrated. As a consequence, future efforts should be directed towards improving the models or the data in order to extract more information from the evidence and to arrive at better discriminating power without loss of calibration.

6.6.2   Py-GC/MS data from comparison of car paints

This is the same experiment as that presented in Section 4.4.3. Here, Py-GC/MS (pyrolytic gas chromatography mass spectrometry; Section 1.4) is used in order to obtain data from car paints in a comparison problem. Thus, in this problem the propositions are defined as H_1 (the paints compared come from the same source) and H_2 (the paints come from different sources). Seven variables are obtained, and a univariate model based on a kernel density function for the between-source variation is applied. The LR values obtained by each model are then multiplied in order to obtain a global LR value for all the variables.

As explained in Section 4.4.3, the multiplication of the seven LR values assumes that the seven variables are independent. This is not a valid assumption, since there is correlation between the variables. The effect of this incorrect assumption will affect the performance of the LR values, as shown below.

The histograms, Tippett plots, DET plots, and ECE plots can be obtained by using the **R** code `model_car_paints_histogram_Tippett_DET_ECE_code.R`, which is available in Section D.10 and on the website. The code uses the data files containing the LR values obtained from the pairwise comparisons carried out to estimate the rates of false positive and false negative answers as per the example in Section 4.4.3 (`model_comparison_research_KDE_different.txt` and `model_comparison_res earch_KDE_same.txt`).

Running the code requires the following functions to be included in the same file folder (available in Section D.10 in Appendix D):

- `histogram_function.R`,

- `Tippett_function.R`,

- `DET_function.R`,

- `ECE_function.R`.

The code provides the user with the `model_histogram.eps`, `model_Tippett.eps`, `model_DET.eps`, `model_ECE.eps` files as an outcome.

6.6.2.1   Histograms and Tippett plots

Histograms of the combined model for the data are shown in Figure 6.15. From the histograms, it can be observed that a very high proportion of true-H_2 LR values are extremely low. This is because the model yields too strong evidence in support of the H_2 proposition. In fact, a cut-off limit of $\log_{10}(LR) = -20$ has been applied to all the LR values in the set. Otherwise, those small values would have been much smaller. However, the main problem is that there is a quite significant proportion of true-H_1 LR values that also have extremely low values, which means strongly misleading evidence. The latter is one of the worst effects that may appear in a validation set of LR values, and therefore should be avoided. Moreover, there are some true-H_2 LR values with values greater than 10^5, indicating that there is some proportion of strongly misleading evidence also in true-H_2 values.

The main reason for the strongly misleading evidence is the independence assumption in the combination of the seven variables. If the variables are not independent but are considered so, the multiplication of their LR values may give a very large (and misleading) value. Imagine, for instance, that all the variables are the same (which implies total correlation), and that a high value for the LR is obtained for all of them. Then the independence assumption provides an overall value that is the individual LR to the power 7. This effect in general means that if independence is incorrectly assumed, the multiplication of the LR values yields too much

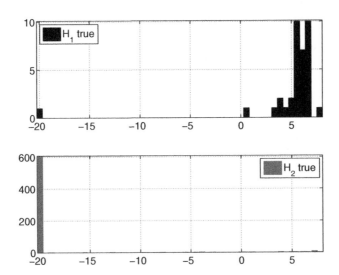

Figure 6.15   Histograms of the $\log_{10}(LR)$ of the validation set obtained in the car paint experiment (Section 4.4.3). Results for H_1 true and for H_2 true are shown in the top and bottom histograms, respectively. Seven LR values from different variables are multiplied, assuming independence among them. The high concentration of $\log_{10}(LR)$ values around -20 is because $\left|\log_{10}(LR)\right|$ values have been limited to 20 for illustration.

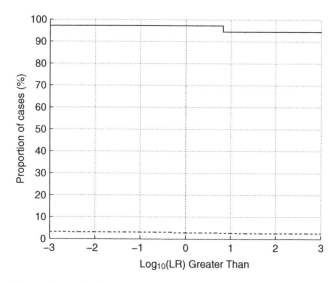

Figure 6.16   Tippett plots of the validation set obtained in the car paint experiment (Section 4.4.3).

support, or in other words, the *LR* values obtained are over-confident. This is exactly what is happening in this example, which can be seen in the histograms: if misleading evidence is over-confident, it may become strong, and therefore it will degrade the performance of the *LR* set.

Although the histograms reveal strongly misleading evidence, the overlap between the histograms is not large. Although the proportion of strongly misleading evidence is not negligible, some discriminating power can be expected in the analysis below.

From the Tippett plots in Figure 6.16 the same effects as in the histograms can be observed, as the cumulative proportions reveal very strong misleading evidence for true-H_2 and especially for true-H_1 values. However, if the rates of misleading evidence are observed, they are both below 3%. This clearly indicates that the discriminating power of the set of *LR* values is good around the threshold of $LR = 1$. However, the presence of strongly misleading evidence indicates poor performance globally for the set of *LR* values, even though the discriminating power is good. This is an example where the discriminating power is not enough to measure the overall performance of *LR* values.

6.6.2.2 Discriminating power: DET plots

The DET plot shown in Figure 6.17 reveals a discrimination performance that appears to be good, because the curve somehow approximates the origin of coordinates, which in fact is a desirable effect. However, the curve clearly diverges from the coordinate axes, and is never able to yield low false acceptance or false rejection rates. The latter is an indication that there are true-H_1 *LR* values higher than most of the true-H_2 values, and also that there are true-H_2 values that are lower than the majority of true-H_1 values. In this case, such a bizarre effect is a consequence of the presence of strongly misleading evidence, because too small true-H_1

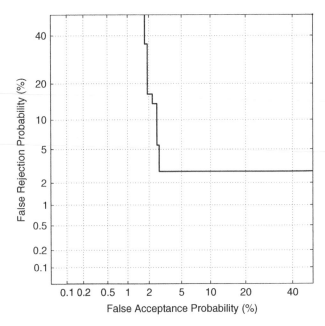

Figure 6.17 DET plot of the validation set obtained in the car paint experiment (Section 4.4.3).

values are smaller than most of the true-H_2 *LR* values, and conversely, too big true-H_2 *LR* values are bigger than most of the true-H_1 *LR* values. Therefore, strongly misleading evidence may also degrade the discriminating power of a set of *LR* values, resulting in a DET curve that does not converge to the coordinate axes.

6.6.2.3   Accuracy: ECE plots

The ECE plot of the validation set obtained in the car paint experiment is shown in Figure 6.18. The following effects are observed:

- The accuracy (solid curve) is extremely bad, with the curve showing a very high value. This is in fact due to the effect of the strongly misleading evidence, which is one of the main factors degrading accuracy as measured by the ECE.

- The discriminating power (dashed curve) is reasonably good, as was seen before in the rates of misleading evidence and DET plot. However, as strongly misleading evidence is present, even with reasonable discriminating power, the total accuracy is not acceptable.

- The calibration is bad, as evidenced by an enormous separation between the solid curve (accuracy) and the dashed curve (discriminating power). Thus the problem with the accuracy of the model does not come from a lack of discriminating power, but from bad calibration. This should prompt investigations to locate the problem. In our case, it is clear that one of the main problems comes from the incorrect assumption of independence, which yields over-confident *LR* values. Here it is shown that one of the

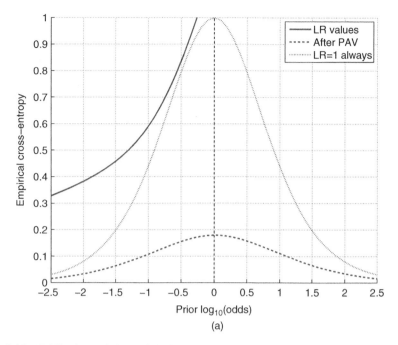

Figure 6.18 ECE plots of the validation set obtained in the car paint experiment (Section 4.4.3).

main degrading factors of calibration is the presence of strongly misleading evidence, and therefore calibration takes into account this highly undesirable assumption of independence, resulting in the poor accuracy of the set of *LR* values.

- For the entire range of the prior odds, the accuracy of the *LR* values (solid curve) is worse than the neutral reference (dotted curve). This means that the evidence evaluation method is even worse than not evaluating the evidence. Although this may sound strange, it is a consequence of the presence of strongly misleading evidence: if a fact finder were to use the *LR* value provided by this method in court, there is a significant risk that they would reach the wrong conclusion because of this method, and therefore it should not be used.

In conclusion, the model combining the seven variables under the assumption of independence yields over-confident *LR* values, and this means strongly misleading evidence. This is the main reason for bad calibration, which degrades the accuracy to the point of being even worse than the neutral reference. In conclusion, the model could lead the fact finder to wrong decisions in casework, and therefore it must not be used.

6.6.3 SEM-EDX data for classification of glass objects

In this experiment, the performance of the *LR* values obtained in Section 5.4.3 for the classification of glass objects is analysed. The experiment uses multivariate SEM-EDX (Section 1.2.1) data to give an *LR* value in support of whether a given glass comes from a container, or a car or building window. Thus, in this problem the propositions are defined as H_1 (the glass comes from a container) and H_2 (the glass comes from a car or building window).

Two *LR* models for classification are compared in this section. The first is a multivariate model with assumptions of normality for the within-source variation and using kernel density functions for the between-source variation, which only takes into account the variables *logNaO*, *logSiO*, and *logCaO* (we refer to it as the Na–Si–Ca model). The second is the graphical model (GM) used in Section 5.4.3, which takes into account all seven variables that can be obtained from a glass object using the SEM-EDX method.

The histograms, Tippett plots, DET plots, and ECE plots can be obtained by using the **R** code `NaSiCa_model_SEMEDX_histogram_Tippett_DET_ECE_code.R`, which is available in Section D.10 and on the website. The code uses the data files containing the *LR* values calculated for objects from categories p and cw as per the example in Section 5.4.3:

- `model_p_cw_classification_research_KDE.txt`,

- `NaSiCa_p_cw_classification_research_KDE.txt`.

Running the code requires the following functions to be included in the same file folder (available in Section D.10 in Appendix D):

- `histogram_function.R`,

- `Tippett_function.R`,

- `DET_function.R`,

- `ECE_function.R`.

The code provides the user with the following files as an outcome:

- `model_histogram.eps`, `model_Tippett.eps`, `model_DET.eps`, `model_ECE.eps`,

- `NaSiCa_histogram.eps`, `NaSiCa_Tippett.eps`, `NaSiCa_DET.eps`, `NaSiCa_ECE.eps`.

6.6.3.1   Histograms and Tippett plots

Histograms for the glass classification models are shown in Figure 6.19. Several observations can be made:

- The proportion of overlap between the histograms is higher for the Na–Si–Ca *LR* values than for the GM *LR* values. This signals higher discriminating power for the *LR* values obtained with the GM model. This consequence makes sense, since the GM model uses all seven variables from the data, whereas the Na–Si–Ca model only uses three. Thus, the more variables used in the evidence evaluation process, the higher the discriminating power should be, because there is more information about the origins of the glass.

- The range of *LR* values for the Na–Si–Ca model seems to be much more limited than for the GM model. For the Na–Si–Ca model, the true-H_1 values range from 10^{-4} to 10^5, but in the GM model they range from 10^{-5} to 10^{20}. Moreover, for true-H_2 values, the

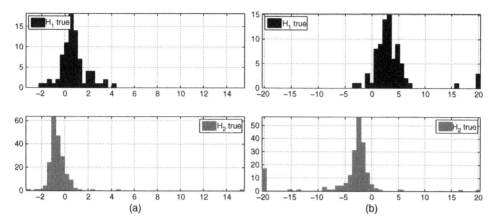

Figure 6.19 Histograms of the $\log_{10}(LR)$ of the validation sets obtained in the glass classification experiment (Section 5.4.3): (a) multivariate model with $logNaO$, $logSiO$, and $logCaO$ variables; and (b) graphical model accounting for all seven SEM-EDX variables (Section 3.7.2; equation (3.16)).

Na–Si–Ca model presents LR values in the range from 10^{-3} to 10^5 (except for one true-H_2 value giving strongly misleading evidence), but in the GM model they range from 10^{-20} to 10^{20}. Notice also that the $[10^{-20}, 10^{20}]$ bounds are artificial cut-offs imposed on all LR values, and therefore the GM model may present even stronger LR values. Therefore, there is evidence that the LR values are over-confident for the case of the GM model. There are several reasons for this. On the one hand, the GM model presents higher discriminating power than the Na–Si–Ca model, and therefore, and according to calibration, the LR values should be stronger for GM than for Na–Si–Ca. However, there is a second reason: the GM model assumes independence between some groups of variables according to the graphical factorisation in Section 5.4.3. Those assumptions are sometimes made even when there is evident correlation between the variables. Even though the independence assumption is made in cases where the correlations are weak, the resulting LR values tend to be over-confident because those correlation effects are ignored.

- There is some misleading evidence in both sets of LR values, which can be seen as normal in any evidence evaluation technique. However, there are also some LR values yielding strongly misleading evidence. In the case of the Na–Si–Ca model, there is one true-H_2 likelihood ratio value of around 10^{14}, which is strongly misleading evidence. However, in the case of the GM model, there are more true-H_2 likelihood ratio values yielding misleading evidence, even 10^{20}. The worse performance of the GM model due to the increase in the strength of the strongly misleading evidence can be attributed to the over-confidence explained before.

The Tippett plots in Figure 6.20 show the same effects as seen in the histograms. First, the rates of misleading evidence are much better for the GM model than for the Na–Si–Ca model, indicating the higher discriminating power of the GM model. However, a higher impact of the strongly misleading evidence for the GM model is also seen in the Tippett plots:

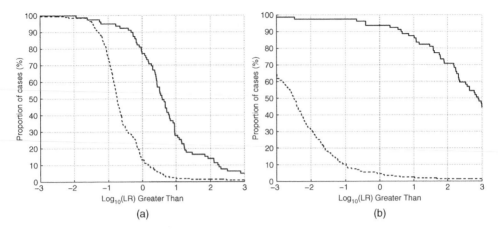

Figure 6.20 Tippett plots of the validation sets obtained in the glass classification exper-iment (Section 5.4.3): (a) multivariate model with $logNaO$, $logSiO$, and $logCaO$ variables; and (b) graphical model accounting for all seven variables (Section 3.7.2; equation (3.16)).

the true-H_1 curve reaches values close to 100% for much lower LR values for GM than for Na–Si–Ca. Also, the true-H_2 curves do not reach values close to 0% in the range represented. This indicates that there is strongly misleading evidence in the LR values, with very large true-H_2 values for both models and very low true-H_1 values for the GM model. Therefore, the performance is highly degraded.

6.6.3.2   Discriminating power: DET plots

Figure 6.21 shows the comparative DET curves of the two models. The GM model presents a curve that is closer to the origin of coordinates, reflecting better discrimination performance than the Na–Si–Ca model. This is in accordance with the observations made from the his-tograms, and its explanation is the same: as more information about the origin of a given glass object is taken into account, the discriminating power of the model should be better. As GM uses all seven variables and Na–Si–Ca only three, it seems reasonable that the GM model has better discriminating power than the Na–Si–Ca model.

Notice also that in Figure 6.21 the same effect as in the example in Section 6.6.2 can be observed: there is some divergence of the DET curves with respect to the coordinate axes. Again, this is due to the presence of strongly misleading evidence in the LR sets of both models, which causes a degradation in the discriminating power shown in the DET plots as the curves approach the coordinate axes.

6.6.3.3   Accuracy: ECE plots

Finally, the ECE plots of both models for classification of glass using SEM-EDX data are shown in Figure 6.22. The following effects can be seen:

- The accuracy (solid curve) is better than the neutral reference for the region of the prior odds close to 0. This indicates that, if the prior log-odds in a given case are close to

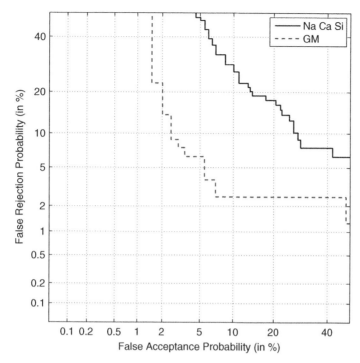

Figure 6.21 DET plots of the validation sets obtained in the glass classification experiment (Section 5.4.3): (a) multivariate model with *logNaO*, *logSiO*, and *logCaO* variables; and (b) graphical model accounting for all seven SEM-EDX variables (Section 3.7.2; equation (3.16)).

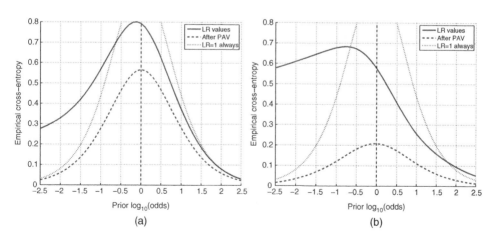

Figure 6.22 ECE plots of the validation sets obtained in the glass classification experiment (Section 5.4.3): (a) multivariate model with *logNaO*, *logSiO*, and *logCaO* variables; and (b) graphical model accounting for all seven SEM-EDX variables (Section 3.7.2; equation (3.16)).

0, then both models yield *LR* values that tend to help the fact finder arrive at the right decision.

- As the prior log-odds become stronger (further from the 0 value), the accuracy degrades. In the case of the Na–Si–Ca model, the accuracy is worse than the neutral reference for prior log-odds lower than about -0.7. This means that if the prior odds are lower than $1/10^{0.7} \approx 0.2$, then the model will lead the fact finder to wrong decisions on average. This is due to the presence of strongly misleading evidence. If the prior odds are assumed to represent a good amount of information from sources other than the evidence, then low prior odds mean that the fact finder has good reasons to think that the glass under analysis comes from a window (H_2). However, strongly misleading evidence may change the fact finder's mind towards a decision that the glass came from a container (H_1) in a case where the glass actually comes from a window. This is an effect that is highly undesirable. In the case of the GM model this strongly misleading evidence also occurs in cases where H_1 (the glass comes from a container) is true, and therefore the accuracy degrades in both directions of the prior log-odds axis. Also, the accuracy becomes worse than the neutral reference when the prior log-odds are very high or very low.

- The discriminating power (dashed curve) of the GM model is better than for the Na–Si–Ca model. This is in agreement with what was observed, and also with the fact that DET curves are highly related to the dashed curve in ECE plots.

- The calibration of the *LR* values is bad for both models, because the solid curves are far from the dashed curves in most regions of the prior log-odds. In the case of the Na–Si–Ca model, this bad calibration mainly occurs for low values of the prior log-odds, due to the presence of strongly misleading evidence in true-H_2 values. For the GM model, the miscalibration effect appears in the whole range of the prior log-odds, due to the presence of strongly misleading evidence both for true-H_1 and true-H_2 *LR* values.

In conclusion, it is seen from the ECE plots that the lack of calibration due to strongly misleading evidence leads both models to be potentially misleading in some cases, depending on the prior odds. As the forensic examiner must not deal with the prior odds, it is extremely risky to use these models in casework, because the examiner will never know in which region of the prior odds his or her models will be working. Therefore, the recommendation is that the models should not be used in casework until the problem of strongly misleading evidence that causes bad calibration has been resolved for the entire range of the prior odds.

Ideally, a forensic examiner should only use in casework methods that offer better accuracy than the neutral reference for all values of the prior odds. Otherwise, there will be some cases in which the method will yield misleading evidence, and the risk of this occurring is too high to permit the method to be used.

6.7   Conclusions: What to measure, and how

In this chapter the concept of the empirical measurement of performance of likelihood ratios has been reviewed. Several methods to measure performance of a so-called validation set of *LR* values have been also described. Several important tools that represent and quantitatively assess different performance characteristics of such a validation set values have been presented. They may be summarised in the following scheme:

- Accuracy – the main performance characteristic, which determines whether a model will be good or bad (on average) with regard to its purpose of assisting the fact finder in the inference process in a case. It is the sum of two characteristics: discriminating power and calibration. Additionally, it:

  - can be quantitatively measured by ECE plots, and by comparison with a neutral reference a minimum desirable floor of performance can be established;

  - cannot be measured by DET plots;

  - cannot be measured by histograms or Tippett plots, although some intuition about the accuracy of the *LR* values can be obtained from them.

- Discriminating power – measures the ability of the *LR* model to extract valuable information from the evidence in order to support the propositions in a case. Additionally, it:

  - can be quantitatively measured by ECE plots;

  - can be quantitatively measured by DET plots;

  - can be observed from histograms and Tippett plots, with a quantitative measure in the form of rates of misleading evidence.

- Calibration – measures the ability of the LR to behave as a probabilistic weight of the evidence. Additionally, it:

  - can be quantitatively measured by ECE plots;

  - cannot be measured by DET plots;

  - cannot be measured by histograms or Tippett plots, although some intuition about the calibration of the *LR* values can be obtained from them.

As a final conclusion, it can be seen that ECE plots allow quantitative measurement of all the performance characteristics that have been considered: accuracy, discriminating power, and calibration. Moreover, the use of a neutral reference yields a practical floor of performance in order to decide whether a method performs sufficiently well to be used in casework. Although histograms, Tippett plots, and DET plots also give complementary views on the overall performance of a set of *LR* values, the use of ECE plots as a main tool for measuring performance is recommended.

6.8 Software

The **R** routines for generating histograms, Tippett plots, DET plots, and ECE plots are provided in Appendix D. Plotting graphs requires the files containing the *LR* values calculated within experiments for estimating the rates of false positive and false negative answers in the case of comparison problems (Chapter 4) and correct classifications in the case of classification problems (Chapter 5).

References

Aitken CGG and Taroni F 2004 *Statistics and the Evaluation of Evidence for Forensic Scientists*, 2nd edn. John Wiley & Sons, Ltd, Chichester.

Brümmer N 2010 *Measuring, refining and calibrating speaker and language information extracted from speech*. PhD thesis, School of Electrical Engineering, University of Stellenbosch, South Africa. Available at http://sites.google.com/site/nikobrummer/ (last accessed 2 October 2012).

Brümmer N and du Preez J 2006 Application independent evaluation of speaker detection. *Computer Speech and Language* **20**(2–3), 230–275.

Cover TM and Thomas JA 2006 *Elements of Information Theory*, 2nd edn. Wiley-Interscience, Hoboken, NJ.

Dawid AP 1982 The well-calibrated Bayesian. *Journal of the American Statistical Association* **77**(379), 605–610.

Dawid AP 2004 Statistics and the law. Technical report, University College London. Available at http://www.ucl.ac.uk/

deGroot MH and Fienberg SE 1983 The comparison and evaluation of forecasters. *The Statistician* **32**, 12–22.

Duda RO, Hart PE and Stork DG 2001 *Pattern Classification*. John Wiley & Sons, Inc., New York.

Evett IW 1998 Towards a uniform framework for reporting opinions in forensic science casework. *Science and Justice* **38**(3), 198–202.

Evett IW and Buckleton JS 1996 Statistical analysis of STR data. In B. Brinkmann, A Carracedo, and W. Bär (eds), *Advances in Forensic Haemogenetics* 6, Springer-Verlag, Berlin. **6**, 79–86.

Garthwaite PH, Kadane JB and O'Hagan A 2005 Statistical methods for eliciting probability distributions. *Journal of the American Statistical Association* **100**(470), 680–701.

Gneiting T and Raftery A 2007 Strictly proper scoring rules, prediction and estimation. *Journal of the American Statistical Association* **102**, 359–378.

Lindley DV 2006 *Understanding Uncertainty*. Wiley-Interscience, Hoboken, NJ.

Lindley DV, Tversky A and Brown RV 1979 On the reconciliation of probability assessments. *Journal of the Royal Statistical Society, Series A* **142**(2), 146–180.

Martin A, Doddington G, Kamm T, Ordowski M and Przybocki M 1997 The DET curve in assessment of decision task performance. *Proc. of Eurospeech*, pp. 1895–1898.

O'Hagan T 2004 Dicing with the unknown. *Significance* **1**(3), 132–133.

Ramos D 2007 *Forensic evaluation of the evidence using automatic speaker recognition systems* PhD thesis Depto. de Ingenieria Informatica, Escuela Politécnica Superior, Universidad Autónoma de Madrid, Spain. Available at http://atvs.ii.uam.es.

Ramos D and Gonzalez-Rodriguez J 2013 Reliable support: measuring calibration of likelihood ratios. *Forensic Science International* **230**(1–3), 156–169.

Ramos D, Gonzalez-Rodriguez J, Zadora G and Aitken CGG 2013 Information-theoretical assessment of the performance of likelihood ratio computation methods. *Journal of Forensic Sciences*. In press.

Royall R 1997 *Statistical Evidence, A Likelihood Paradigm*. Chapman & Hall, London.

Royall R 2000 On the probability of observing statistical misleading evidence. *Journal of the American Statistical Association* **95**(451), 760–768.

Savage L 1971 The elicitation of personal probabilities and expectations. *Journal of the American Statistical Association* **66**(336), 783–801.

Taroni F, Bozza S, Biedermann A, Garbolino P and Aitken CGG 2010 *Data Analysis in Forensic Science: A Bayesian Decision Perspective*. John Wiley & Sons, Ltd, Chichester.

Appendix A

Probability

A.1 Laws of probability

Probability is a measure of uncertainty (Lindley 1991). It can also be a subjective measure of belief. For example, in a sporting context, there are many references to the probability that a certain team will win a particular football match or a certain horse will win a particular race. Probability can also refer to events that have happened in the past if their outcome is unknown by an enquirer. Questions are asked about a personal belief about the outcome of a sporting event in the past, such as *what is the probability a certain team won a particular match?* Of course, the question can be answered by checking records, but until then the knowledge is uncertain and this uncertainty can be represented by a probability.

The subjectivity can be extended beyond a sporting context to a legal context (Aitken and Taroni 2004; Lucy 2005). There is uncertainty in a criminal trial. In practice, it cannot be said that it is certain that the defendant is guilty or non-guilty. The uncertainty can be measured by probability. It may be cloaked in verbal phrases, such as *it is almost certain the defendant is guilty* or *it is clear that the defendant did not commit the crime*, but these are only verbal expressions of probability. If the issue of guilt or otherwise is considered the ultimate issue, there are also intermediate issues to which probability can be applied. A match is declared between a defendant's DNA profile and the DNA profile extracted from a blood stain at a crime scene. Then a proposition that the blood stain came from the defendant is uncertain and that uncertainty can be represented by a probability. This is different from the probability associated with the event of a match if the blood stain came from the defendant, and the difference is discussed later in the context of conditional probabilities.

Probability can also be objective. It is objective in the sense that the probability of an event may be determined in such a way that anyone given the same information will arrive at the same value for the probability as anyone else given the same information. This is best discussed in the context of an example.

Statistical Analysis in Forensic Science: Evidential Value of Multivariate Physicochemical Data, First Edition.
Grzegorz Zadora, Agnieszka Martyna, Daniel Ramos and Colin Aitken.
© 2014 John Wiley & Sons, Ltd. Published 2014 by John Wiley & Sons, Ltd.
Companion website: www.wiley.com/go/physicochemical

Example

Imagine there are five otherwise indistinguishable balls numbered from 1 to 5 in an urn. A ball is chosen at random from the urn. The phrase *at random* means in this context that all balls are equally likely to be chosen. Of course, a discussion of *probability* with the use of the phrase *equally likely* has, of necessity, to be circular: probability is defined in terms of likelihood which, in turn, may be defined in terms of probability. As all balls are equally likely to be chosen, the probability the ball numbered 1 is chosen is $1/5$ (it is one ball out of five). Anyone given this information (five balls all equally likely to be chosen) will arrive at the same conclusion. The example can be extended. A ball is chosen at random from the urn (the previously chosen ball having been returned so there are still five balls, all equally likely to be chosen). The probability the ball that is chosen has an odd number is determined by the ratio of the number of balls with an odd number divided by the total number of balls from which the selection has been made. There are three balls which are odd-numbered, those balls labelled 1, 3, and 5. There are five balls in total. Thus the probability the chosen ball is odd is $3/5$. Consider the event that the ball chosen is even-numbered. The probability the ball is chosen has an even number is determined by the ratio of the number of balls with an even number divided by the total number of balls from which the selection has been made. There are two balls which are even-numbered, those balls labelled 2 and 4. There are five balls in total. Thus the probability the chosen ball is even is $2/5$.

Note that the last two probabilities in the example add up to 1. This provides an illustration of the *first law of probability*. The two events, *draw an even-numbered ball* and *draw an odd-numbered ball*, are said to be *complementary*. A ball drawn from the urn is either even-numbered or odd-numbered. No balls are blank and no balls have more than one number on them or a number greater than 5 or less than 1. An event which is certain has probability 1. Conversely, it is impossible to draw a ball which has no number on it, more than one number, or a number greater than 5 or less than 1. Such an event, an event which is impossible, has probability 0.

Some notation is needed. Let R be an event (e.g. drawing an even-numbered ball from an urn, a football team winning a match). Let $P(R)$ denote the probability that R occurs. The first law of probability states that probability can take any value between 0 and 1, inclusive of those values. Thus, $0 \leq P(R) \leq 1$. If R is impossible, $P(R) = 0$. If R is certain, $P(R) = 1$. The law has been illustrated with the drawing of balls from an urn and using ratios as probabilities. It is also applicable to subjective probabilities.

Let us remain with objective probabilities. In general, assume all possible events in a certain scenario are equally likely. Denote the set of all possible events by Ω and the number of events in Ω by $|\Omega|$. Conditions are imposed that are only satisfied by a subset of Ω. Denote this subset by A and the number of events in A by $|A|$. An event is chosen at random from Ω. The probability it belongs to A is

$$P(A) = \frac{\text{events meeting the requirements of event } A}{\text{all possible outcomes of an experiment } \Omega} = \frac{|A|}{|\Omega|}.$$

There is now a change of terminology. Original events in the example above will be referred to as *fundamental events*. The word *event* will now be used to refer to subsets of Ω

formed from groups of fundamental events. Assume that A and B are two random events. If event A is not associated with event B, A and B are considered *independent*. If A is known to have happened, the probability B will happen will not alter from the value it took when there was no information about A. The relationship is symmetric. If B is known to have happened, the probability A will happen will not alter from the value it took when there was no information about B.

Consider two coins. Let A be the event that a toss of the first coin gives a head. Let B be the event that a toss of the second coin gives a head. These two events are independent: the outcome of the toss of the first coin has no effect on the outcome of the toss of a second coin.

If there is an association between A and B it may not be causal. To take an informal example, consider a dataset recording the number of doctors in towns and another dataset recording the number of deaths per year in towns. There is an association between these two sets of data: as the number of doctors increases so the number of deaths increases. However, this is not a direct causal relationship. An increase in the number of doctors does not cause an increase in the number of deaths per year. The relationship is indirect and related to the size of the town. Large towns have larger numbers of doctors and larger numbers of deaths per year than small towns.

If the events cannot occur at the same time in a single experiment, they are *mutually exclusive*. For instance, drawing an odd-numbered ball from the urn above with five balls in it means that an even-numbered ball has not been drawn. It is not possible to draw one ball from the urn and for it to be both odd- and even-numbered. Such events are mutually exclusive. A set of events are *mutually exclusive and exhaustive* when they are exclusive and exhaust all the possible experimental outcomes, so that one of them is certain to happen.

The *second law of probability* is concerned with the so-called *disjunction* or *union* of two events. For two events A and B, the disjunction is denoted by $A \cup B$, read as *either A or B happens (or possibly both)*. If the events are mutually exclusive,

$$P(A \cup B) = P(A) + P(B),$$

that is, $P(A \cup B)$ is the sum of the probabilities of A and B.

Example

Consider again the five otherwise indistinguishable balls numbered from 1 to 5. Here, the question of interest is the probability of drawing a ball with an extreme number, which in this context means 1 (event A) or 5 (event B).

The possible outcomes of a draw of a ball from the urn within A and B are:

$A = \{1\}$, one possible outcome;

$B = \{5\}$, one possible outcome;

$\Omega = \{1, 2, 3, 4, 5\}$, five possible outcomes.

Events A and B are mutually exclusive, so the corresponding probability of their disjunction is equal to: $P(A \cup B) = P(A) + P(B) = \frac{1}{5} + \frac{1}{5} = \frac{2}{5}$.

When events are not mutually exclusive the probability of the disjunction of events is reduced by the probability of the occurrence of their common outcome or *intersection*, denoted $P(A \cap B)$ or $P(AB)$, read as *the probability of the occurrence of A and B*, to give the second law of probability:

$$P(A \cup B) = P(A) + P(B) - P(A \cap B).$$

As AB is counted within A and within B it has been counted twice so has to be discounted once.

Example

What is the probability of choosing a ball with a number greater than or equal to 3 (event C) or with an even number (event D)?

Events C and D are not mutually exclusive, because within the set of fundamental events forming C there is an even number, 4, which is part of D:

$C = \{3, 4, 5\}$;
$D = \{2, 4\}$;
$\Omega = \{1, 2, 3, 4, 5\}$;

so

$C \cup D = \{2, 3, 4, 5\}$;
$C \cap D = \{4\}$;

and

$$P(C \cup D) = P(C) + P(D) - P(C \cap D) = \tfrac{3}{5} + \tfrac{2}{5} - \tfrac{1}{5} = \tfrac{4}{5}.$$

Consider an event T in Ω. The event that consists of all the fundamental events not in T is said to be the *complement* of T and is denoted $\bar{T}$, read as *T-bar*. For example, the complement of $C = \{3, 4, 5\}$ in the scenario with an urn and five numbered balls is $\bar{C} = \{1, 2\}$. An event and its complement are mutually exclusive and exhaustive. It is not possible for a fundamental event to be in an event and also in its complement, by definition. Also, the union of an event and its complement is exhaustive, again by definition. From the first law of probability, $P(\Omega) = 1$. Also $T \cup \bar{T} = \Omega$ for any event T in Ω. Thus, $P(T \cup \bar{T}) = P(\Omega) = 1$. Also $P(T \cap \bar{T}) = 0$ since they are mutually exclusive and hence the event of their intersection is impossible, an event with probability 0 (from the first law). Thus $P(T) + P(\bar{T}) = 1$. The ratio of $P(T)$ to $P(\bar{T})$ is known as the *odds* in favour of T or the odds against $\bar{T}$, a phrase familiar to those interested in sports.

The so-called *conjunction* of events is the event formed from the intersection of two events. It is the set of fundamental events that are in both of the two events that form the conjunction. Consider two events R and S. The conjunction is written as $R \cap S$ or RS and read as 'R and S'. The *third law of probability* is concerned with the probability of the conjunction of two events. A new idea is required, that of a *conditional probability*. This enables determination of the probability that an event happens if another event is known to have happened. For two events R and S, the probability of R occurring given that S has occurred is denoted by $P(R \mid S)$. The vertical bar | denotes conditioning. Events to the left of it are uncertain and are those about which probabilities are desired. Events to the right of it are known to have occurred. Their occurrence may or may not be associated with the occurrence of events to the left. If there is an association then the events are said to be *dependent* and if there is not an association then the events are said to be *independent*. For independent events,

$$P(R \cap S) = P(R) \times P(S), \text{ sometimes written as } P(R) \cdot P(S),$$

that is, the probability that R and S both happen is the product of the probability that R happens and the probability that S happens.

Example

Imagine you draw one ball in each of two separate rounds, returning the ball to the urn after each draw. What is the probability of choosing a ball with an even number in the first draw (event R) and an odd number in the second draw (event S)?

$R = \{2, 4\}$;
$S = \{1, 3, 5\}$;
$\Omega = \{1, 2, 3, 4, 5\}$;
$P(R \cap S) = P(R) \cdot P(S) = \frac{2}{5} \cdot \frac{3}{5} = \frac{6}{25}$.

When events are not independent, the probability $P(R \cap S)$ involves a conditional probability. The third law of probability states that

$$P(R \cap S) = P(S) \cdot P(R \mid S) = P(R) \cdot P(S \mid R).$$

A.2   Bayes' theorem and the likelihood ratio

When R and S are independent, $P(R \mid S) = P(R)$, $P(S \mid R) = P(S)$, and $P(R \cap S) = P(R) \cdot P(S)$. A corollary of the third law is that

$$P(R \mid S) = \frac{P(S \mid R)P(R)}{P(S)}.$$

This result is known as *Bayes's theorem*. Note also that

$$P(\bar{R} \mid S) = \frac{P(S \mid \bar{R})P(\bar{R})}{P(S)},$$

and division of the first of these expressions by the second gives

$$\frac{P(R \mid S)}{P(\bar{R} \mid S)} = \frac{P(S \mid R)P(R)}{P(S \mid \bar{R})P(\bar{R})} = \frac{P(S \mid R)}{P(S \mid \bar{R})} \times \frac{P(R)}{P(\bar{R})}. \tag{A.1}$$

The ratio $P(R)/P(\bar{R})$ is the odds in favour of R, sometimes known as the *prior* odds in the sense of prior to knowledge of S. The ratio $P(R \mid S)/P(\bar{R} \mid S)$ is the odds in favour of R given that S has occurred, is known, or is accepted as true, sometimes known as the *posterior* odds in the sense of posterior to the knowledge of S. The ratio $P(S \mid R)/P(S \mid \bar{R})$ is known as the likelihood ratio. It is a measure of how much more likely S is if R has happened than if $\bar{R}$ has happened. It is a ratio of probabilities and can take values from 0 ($P(S \mid R) = 0$) to infinity ($P(S \mid \bar{R}) = 0$). Note that $P(S \mid R) + P(S \mid \bar{R})$ need not equal 1. In the first term it is the probability S is true if R is true that is of interest. In the second term it is the probability S is true if $\bar{R}$ is true that is of interest. The background information R and $\bar{R}$ is different in the two situations. Consider a sporting analogy. Suppose S is the event that a horse wins a race, R is that the going is soft and $\bar{R}$ that the going is not soft. The probability the horse wins may be high in both situations, whether the going is soft ($P(S \mid R)$) or not ($P(S \mid \bar{R})$). The sum of the two probabilities may be greater than 1. This is acceptable as it is the conditioning values R and $\bar{R}$ that are complementary, not the events about which probabilities are required.

Example

Imagine now that, having previously been indistinguishable, the balls numbered 1 and 2 in the urn are white and those numbered 3, 4, and 5 are black. What is the probability that a white ball has been chosen, if it is known that it has a number less than or equal to 3? Let V be the event {1-white, 2-white} and W the event {1, 2, 3}. Then $V \cap W = \{1, 2\}$, and $\Omega = \{1, 2, 3, 4, 5\}$ as before, so

$$P(V \mid W) = \frac{P(V \cap W)}{P(W)} = \frac{\mid V \cap W \mid}{\mid \Omega \mid} / \frac{\mid W \mid}{\mid \Omega \mid} = \frac{2}{5} / \frac{3}{5} = \frac{2}{3}.$$

Note that if V occurs, then W occurs since V is contained in W. Thus $P(W \mid V) = 1$ (because of the first law of probability). However, $P(W \mid V) \neq P(V \mid W)$.

As another example of the result $P(W \mid V) \neq P(V \mid W)$, the probability of an event that if someone's pet is a Labrador (W) it is a dog (V) is written as $P(V \mid W) = P$(it is a dog |it is a Labrador), which equals 1. Compare this with the probability that if someone's pet is a dog, it is a Labrador: $P(W \mid V) = P$(it is a Labrador|it is a dog) $\ll 1$, because many dogs are not Labradors.

Example

This example illustrates the different aspects of conditional probabilities. Data are available on the eye and hair colour of 100 individuals (Table A.1) in a random sample from some population (every person in the population was equally likely to have been chosen for the sample, an impractical ideal). Eye colour is categorised as *dark* (D) or *light* (L); denote the events that a person has dark or light eyes by Ey_D and Ey_L. Hair colour is also categorised as *dark* (D) or *light* (L); denote the events that a person has dark or light hair colour as Hr_D and Hr_L. An individual can be classified into one and only one of four categories: $(Ey_D, Hr_D), (Ey_D, Hr_L), (Ey_L, Hr_D),$ and (Ey_L, Hr_L).

All appropriate ratios of frequencies in Table A.1 are used as estimates of the associated probability (Table A.2) in the population from which the sample was taken:

- An estimate of the probability that a person from the population from which this sample was drawn has dark hair is 30/100 (30 people out of the 100 have dark hair).

Table A.1   Frequencies of eye and hair colour.

	Hair colour (dark)	Hair colour (light)	Totals
Eye colour (dark)	20	20	40
Eye colour (light)	10	50	60
Totals	30	70	100

- An estimate of the probability that a person from the population from which this sample was drawn has light hair is 70/100 (70 people out of the 100 have light hair).

- An estimate of the probability that a person from the population from which this sample was drawn has dark eyes is 40/100 (40 people out of the 100 have dark eyes).

- An estimate of the probability that a person from the population from which this sample was drawn has light eyes is 60/100 (60 people out of the 100 have light eyes).

- A person has dark eyes. An estimate of the probability that a person from the population from which this sample was drawn who has dark eyes also has dark hair is 20/40 (20 people out of the 40 with dark eyes have dark hair).

- A person has dark eyes. An estimate of the probability that a person from the population from which this sample was drawn who has dark eyes also has light hair is 20/40 (20 people out of the 40 with dark eyes have light hair).

- A person has light eyes. An estimate of the probability that a person from the population from which this sample was drawn who has light eyes also has dark hair is 10/60 (10 people out of the 60 with light eyes have dark hair).

- A person has light eyes. An estimate of the probability that a person from the population from which this sample was drawn who has light eyes also has light hair is 50/60 (50 people out of the 60 with light eyes have light hair).

Some other probabilities can also be estimated based on the information in Tables A.1 and A.2:

- A person has dark hair. An estimate of the probability that a person from the population from which this sample was drawn who has dark hair also has dark eyes is 20/30 (20 people out of the 30 with dark hair have dark eyes).

Table A.2   Probabilities of eye and hair colour.

	Hair colour (dark)	Hair colour (light)	Totals
Eye colour (dark)	0.2	0.2	0.4
Eye colour (light)	0.1	0.5	0.6
Totals	0.3	0.7	1

- A person has dark hair. An estimate of the probability that a person from the population from which this sample was drawn who has dark hair also has light eyes is 10/30 (10 people out of the 30 with dark hair have light eyes).

- A person has light hair. An estimate of the probability that a person from the population from which this sample was drawn who has light hair also has dark eyes is 20/70 (20 people out of the 70 with light hair have dark eyes).

- A person has light hair. An estimate of the probability that a person from the population from which this sample was drawn who has light hair also has light eyes is 50/70 (50 people out of the 70 with light hair have light eyes).

- A person has dark hair and dark eyes. An estimate of the probability that a person from the population from which this sample was drawn has dark eyes and dark hair is $20/100 = 0.20$ (20 people out of the 100 in the sample have dark eyes and dark hair).

- A person has dark hair and light eyes. An estimate of the probability that a person from the population from which this sample was drawn has light eyes and dark hair is $10/100 = 0.10$ (10 people out of the 100 in the sample have light eyes and dark hair).

- A person has light hair and light eyes. An estimate of the probability that a person from the population from which this sample was drawn has light eyes and light hair is $50/100 = 0.50$ (50 people out of the 100 in the sample have light eyes and light hair).

- A person has light hair and dark eyes. An estimate of the probability that a person from the population from which this sample was drawn has dark eyes and light hair is $20/100 = 0.20$ (20 people out of the 100 in the sample have dark eyes and light hair).

The product of the estimates of the probability that a person has dark eyes and the probability that a person has dark hair is $\frac{40}{100} \times \frac{30}{100} = 0.12$ which is not equal to 0.20, the estimate of the probability a person from the population from which this sample was drawn has dark eyes and dark hair. This demonstrates that for this sample, eye colour and hair colour are not independent.

A.3   Probability distributions for discrete data

It may be thought very time-consuming and often impractical to have to determine probabilities separately for every possible outcome. Fortunately this is not necessary. Some random phenomena may be modelled mathematically and formulae developed from which it is relatively straightforward to determine probabilities in certain standard situations.

The examples in the previous two sections relate to data that are said to be *discrete* as the possible outcomes are discrete. In the example of the five balls in the urn there are only certain outcomes that can happen and they are distinct. Similarly with the example of eye colour and hair colour: the data are in the form of counts, the number of people in each of the four categories. The possible outcomes for any one person are distinct.

It is not necessary to derive probabilities directly on every occasion. For certain situations formulae have been derived. For example, with repeated independent experimental trials with only two possible outcomes (conventionally denoted *success* and *failure*) and a constant probability for each of the two outcomes, the total number of observations that are successes, for example in a fixed number of trials, is said to have a *binomial distribution* (Evans *et al.* 2000). The distribution gives the formula for the probabilities for the number of successes in a fixed number of trials with a pre-specified success probability. As another example, consider random emissions from a radioactive source, independent emissions at a constant rate; the total number of emissions in a fixed time interval is said to have a *Poisson distribution* (Evans *et al.* 2000). The distribution gives the formula for the probability for the number of emissions from a radioactive source over a fixed period of time, with a pre-specified (and constant) rate of emission.

Example

As an illustration of the binomial distribution, consider the outcomes of three tosses of a fair coin. There are only two possible outcomes to a toss, heads (H: call this a 'success') or tails (T). The tosses are independent, the number of trials is fixed at 3, and there is a constant probability of a success (1/2). The binomial distribution is appropriate for the probabilities for the total number of heads in three tosses of a fair coin. The eight possible different outcomes of such an experiment are illustrated in Figure A.1. Note that some of the outcomes, for example that of two heads, may occur in three different ways (i.e. sequences of results): HHT, HTH, THH. The probabilities of these outcomes if the coin is fair ($P(H) = P(T) = 1/2$) and tosses are independent are shown in Table A.3 and Figure A.2. Formulae are not given here as they are beyond the scope of the book.

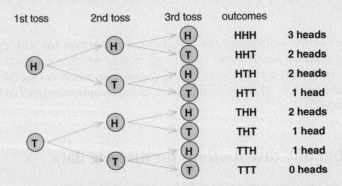

Figure A.1   The possible outcomes of the experiment of tossing a fair coin three times.

Table A.3   Probabilities of obtaining a specified number of heads in three tosses of a fair coin. Symbol Ω denotes the total number of fundamental events which is 8, see Figure A.1.

A	0 heads	1 head	2 heads	3 heads
Number of realisations	1	3	3	1
$P = \frac{|A|}{|\Omega|}$	$\frac{1}{8}$	$\frac{3}{8}$	$\frac{3}{8}$	$\frac{1}{8}$

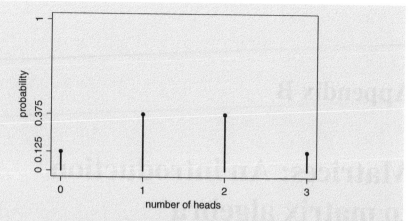

Figure A.2 The discrete distribution of the number of heads likely to occur in the experiment of tossing a fair coin three times.

A.4   Probability distributions for continuous data

When data are in the form of measurements (Chapter 1), the data are said to be continuous (Chapter 3). Probabilities refer to events in the form of intervals for what are known as univariate data. These are data for which only one variable is measured. For example, with glass fragments, the variable could be the refractive index of the glass (Section 1.2.2). Given a fragment of glass, it may be of interest to know if its refractive index (RI) lies in a particular interval. It makes no sense to consider the probability that the RI takes a particular exact value as such a value cannot be measured. Measuring instruments have a measurement accuracy and the probability would refer to the interval specified by the accuracy.

As with discrete data, there are formulae which express the distribution of the probability over its sample space (the space over which measurements may arise). These formulae are known as *probability distribution functions* (Section 3.3.5).

Probabilities can also refer to events in the form of areas for bivariate data, volumes for trivariate data and hypervolumes for data of four or more variables (*multivariate data* is a general term for data consisting of two or more variables). For bivariate data, an example would be the measurements of chromaticity coordinates, for example x and y (Section 1.5). For multivariate data, log ratios of chemical compositions of glass provide an example (Section 4.4.6).

References

Aitken CGG, Taroni F 2004 *Statistics and the Evaluation of Evidence for Forensic Scientists*, 2nd edn. John Wiley & Sons, Ltd, Chichester.

Evans M, Hastings N, Peacock B 2000 *Statistical distributions*. 3rd edn. Wiley, New York, USA.

Lindley DV 1991 *Probability*. In *The Use of Statistics in Forensic Science*. Aitken CGG, Stoney DA (eds). Ellis Horwood, Chichester, 27–50.

Lucy D 2005 *Introduction to statistics for forensic scientists*. Wiley, Chichester.

Appendix B

Matrices: An introduction to matrix algebra

B.1 Multiplication by a constant

Let us define matrices $\mathbf{X}$ and $\mathbf{Y}$ with i rows and j columns ($i \times j$):

$$\mathbf{X}_{[i,j]} = \begin{bmatrix} x_{11} & \cdots & x_{1j} \\ \vdots & \ddots & \vdots \\ x_{i1} & \cdots & x_{ij} \end{bmatrix},$$

$$\mathbf{Y}_{[i,j]} = \begin{bmatrix} y_{11} & \cdots & y_{1j} \\ \vdots & \ddots & \vdots \\ y_{i1} & \cdots & y_{ij} \end{bmatrix}.$$

When multiplying a matrix by a constant each matrix element has to be multiplied:

$$\mathbf{Z} = a \cdot \mathbf{X} = a \cdot \begin{bmatrix} x_{11} & \cdots & x_{1j} \\ \vdots & \ddots & \vdots \\ x_{i1} & \cdots & x_{ij} \end{bmatrix} = \begin{bmatrix} ax_{11} & \cdots & ax_{1j} \\ \vdots & \ddots & \vdots \\ ax_{i1} & \cdots & ax_{ij} \end{bmatrix}$$

Statistical Analysis in Forensic Science: Evidential Value of Multivariate Physicochemical Data, First Edition.
Grzegorz Zadora, Agnieszka Martyna, Daniel Ramos and Colin Aitken.
© 2014 John Wiley & Sons, Ltd. Published 2014 by John Wiley & Sons, Ltd.
Companion website: www.wiley.com/go/physicochemical

Example

A matrix $\mathbf{X} = \begin{bmatrix} 2 & 3 & 1 \\ -4 & 2 & -1 \end{bmatrix}$ multiplied by a constant $a = 2$, gives the matrix $\mathbf{Z}$:

$$\mathbf{Z} = a \cdot \mathbf{X} = 2 \cdot \begin{bmatrix} 2 & 3 & 1 \\ -4 & 2 & -1 \end{bmatrix} = \begin{bmatrix} 4 & 6 & 2 \\ -8 & 4 & -2 \end{bmatrix}.$$

In order to load a matrix $\mathbf{X}$ and carry out multiplication by a constant, the following commands should be typed into **R** (Appendix D):

```
> x = matrix(c(2,3,1,-4,2,-1), nrow = 2, byrow = TRUE)
> a = 2
> z = a * x
```

The output is as follows:

```
> z
     [,1] [,2] [,3]
[1,]    4    6    2
[2,]   -8    4   -2
```

B.2   Adding matrices

Adding (or subtracting) matrices can be done only for matrices of the same order. When $\mathbf{X}$ and $\mathbf{Y}$ are $i \times j$ matrices, the corresponding x_{ij} and y_{ij} elements are summed (or subtracted) to give a resulting matrix $\mathbf{Z}$ of the same size $i \times j$:

$$\mathbf{Z} = \mathbf{X} + \mathbf{Y} = \begin{bmatrix} x_{11} & \cdots & x_{1j} \\ \vdots & \ddots & \vdots \\ x_{i1} & \cdots & x_{ij} \end{bmatrix} + \begin{bmatrix} y_{11} & \cdots & y_{1j} \\ \vdots & \ddots & \vdots \\ y_{i1} & \cdots & y_{ij} \end{bmatrix}$$

$$= \begin{bmatrix} x_{11} + y_{11} & \cdots & x_{1j} + y_{1j} \\ \vdots & \ddots & \vdots \\ x_{i1} + y_{i1} & \cdots & x_{ij} + y_{ij} \end{bmatrix} = \begin{bmatrix} z_{11} & \cdots & z_{1j} \\ \vdots & \ddots & \vdots \\ z_{i1} & \cdots & z_{ij} \end{bmatrix}.$$

Example

$$\mathbf{Z} = \mathbf{X} + \mathbf{Y} = \begin{bmatrix} 2 & 3 & 1 \\ -4 & 2 & -1 \end{bmatrix} + \begin{bmatrix} 3 & 5 & 4 \\ 2 & 1 & 1 \end{bmatrix} = \begin{bmatrix} 2+3 & 3+5 & 1+4 \\ -4+2 & 2+1 & -1+1 \end{bmatrix}$$

$$= \begin{bmatrix} 5 & 8 & 5 \\ -2 & 3 & 0 \end{bmatrix}.$$

In **R**:

```
> x = matrix(c(2,3,1,-4,2,-1), nrow = 2, byrow = TRUE)
> y = matrix(c(3,5,4,2,1,1), nrow = 2, byrow = TRUE)
> z = x + y
```

The output is as follows:

```
> z

     [,1] [,2] [,3]
[1,]    5    8    5
[2,]   -2    3    0
```

The properties of adding and subtracting matrices are consistent with those characteristic of simple addition and subtraction operations involving numbers. This means that addition is commutative $(\mathbf{X} + \mathbf{Y} = \mathbf{Y} + \mathbf{X})$ and associative $(\mathbf{A} + (\mathbf{X} + \mathbf{Y}) = (\mathbf{A} + \mathbf{X}) + \mathbf{Y})$.

B.3   Multiplying matrices

Matrices can be multiplied only if the number of columns of the first matrix, $\mathbf{X}_{[i,j]}$, is equal to the number of rows of the second matrix, $\mathbf{Y}_{[j,k]}$. The resulting matrix $\mathbf{Z}$ then has i rows and k columns:

$$
\mathbf{Z} = \mathbf{X} \cdot \mathbf{Y} =
\begin{bmatrix}
x_{11} & \cdots & x_{1j} \\
\vdots & \ddots & \vdots \\
x_{i1} & \cdots & x_{ij}
\end{bmatrix}
\cdot
\begin{bmatrix}
y_{11} & \cdots & y_{1k} \\
\vdots & \ddots & \vdots \\
y_{j1} & \cdots & y_{jk}
\end{bmatrix}
$$

$$
=
\begin{bmatrix}
x_{11}y_{11} + x_{12}y_{21} + \cdots + x_{1j}y_{j1} & \cdots & x_{11}y_{1k} + x_{12}y_{2k} + \cdots + x_{1j}y_{jk} \\
\vdots & \ddots & \vdots \\
x_{i1}y_{11} + x_{i2}y_{21} + \cdots + x_{ij}y_{j1} & \cdots & x_{i1}y_{1k} + x_{i2}y_{2k} + \cdots + x_{ij}y_{jk}
\end{bmatrix}
$$

$$
=
\begin{bmatrix}
z_{11} & \cdots & z_{1k} \\
\vdots & \ddots & \vdots \\
z_{i1} & \cdots & z_{ik}
\end{bmatrix}.
$$

Matrix multiplication is not commutative $(\mathbf{XY} \neq \mathbf{YX})$, but it is associative $(\mathbf{A} \cdot (\mathbf{X} \cdot \mathbf{Y}) = (\mathbf{A} \cdot \mathbf{X}) \cdot \mathbf{Y})$, and distributive over addition $(\mathbf{A} \cdot (\mathbf{X} + \mathbf{Y}) = \mathbf{A} \cdot \mathbf{X} + \mathbf{A} \cdot \mathbf{Y})$. The general idea of matrix multiplication is illustrated in Figure B.1.

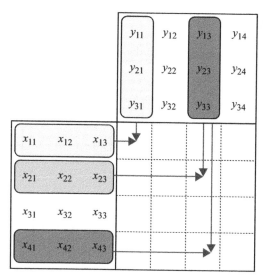

Figure B.1 Graphical illustration of the multiplication of two matrices. For example, in cell (2, 3) (row 2, column 3) the entry is $x_{21}y_{13} + x_{22}y_{23} + x_{23}y_{33}$.

Example

$$\mathbf{Z} = \mathbf{X} \cdot \mathbf{Y} = \begin{bmatrix} 2 & 3 & 1 \\ -4 & 2 & -1 \end{bmatrix} \cdot \begin{bmatrix} 1 & 3 \\ 4 & -1 \\ 2 & 4 \end{bmatrix}$$

$$= \begin{bmatrix} 1 \cdot 2 + 3 \cdot 4 + 1 \cdot 2 & 2 \cdot 3 + 3 \cdot (-1) + 1 \cdot 4 \\ -4 \cdot 1 + 2 \cdot 4 + (-1) \cdot 2 & -4 \cdot 3 + 2 \cdot (-1) + (-1) \cdot 4 \end{bmatrix}$$

$$= \begin{bmatrix} 16 & 7 \\ 2 & -18 \end{bmatrix}.$$

In order to multiply two matrices in **R**, the '∗' sign for simple numbers multiplication must be replaced by % ∗ % for matrix multiplication:

```
> x = matrix(c(2,3,1,-4,2,-1), nrow = 2, byrow = TRUE)
> y = matrix(c(1,3,4,-1,2,4), nrow = 3, byrow = TRUE)
> z = x %*% y
```

The output is as follows:

```
> z

     [,1] [,2]
[1,]   16    7
[2,]    2  -18
```

B.4   Matrix transposition

The operation of transposing a matrix $\mathbf{X}$ of order i,j exchanges columns with rows to give a matrix of order j,i, denoted $\mathbf{X}^T$:

$$\mathbf{X}^T = \begin{bmatrix} x_{11} & \cdots & x_{1j} \\ \vdots & \ddots & \vdots \\ x_{i1} & \cdots & x_{ij} \end{bmatrix}^T = \begin{bmatrix} x_{11} & \cdots & x_{i1} \\ \vdots & \ddots & \vdots \\ x_{1j} & \cdots & x_{ij} \end{bmatrix}.$$

Example

$$\mathbf{Z} = \mathbf{X}^T = \begin{bmatrix} 2 & 3 & 1 \\ -4 & 2 & -1 \end{bmatrix}^T = \begin{bmatrix} 2 & -4 \\ 3 & 2 \\ 1 & -1 \end{bmatrix}.$$

In **R**:

```
> x = matrix(c(2,3,1,-4,2,-1), nrow = 2, byrow = TRUE)
> z = t(x)
```

The output is as follows:

```
> z

     [,1] [,2]
[1,]    2   -4
[2,]    3    2
[3,]    1   -1
```

B.5   Determinant of a matrix

The determinant of a matrix ($\det(\mathbf{X})$ or $|\mathbf{X}|$) exists only for square matrices (matrices in which the number of rows equals the number of columns).

The determinant of a 2×2 matrix $\mathbf{X}$ is calculated as the difference between the products of diagonal elements:

$$\det(\mathbf{X}) = \det \begin{bmatrix} x_{11} & x_{12} \\ x_{21} & x_{22} \end{bmatrix} = \begin{vmatrix} x_{11} & x_{12} \\ x_{21} & x_{22} \end{vmatrix} = x_{11} \cdot x_{22} - x_{12} \cdot x_{21}.$$

Example

$$\det(\mathbf{X}) = \det \begin{bmatrix} 1 & 2 \\ 3 & 4 \end{bmatrix} = 1 \cdot 4 - 2 \cdot 3 = -2.$$

For a matrix $\mathbf{X}$ of higher dimensions it can be calculated as follows. Choose one row or column of $\mathbf{X}$. Suppose we choose the second column in the 3×3 matrix $\mathbf{X}$ given below as:

$$\mathbf{X} = \begin{bmatrix} 1 & 4 & 2 \\ 1 & 4 & 1 \\ 1 & 2 & 2 \end{bmatrix}.$$

The elements of the selected row or column are multiplied by their so-called cofactors, $\mathbf{C}_{ij}$, which are defined by introducing the concept of minors, $\mathbf{M}_{ij}$. Minors are related to cofactors by the expression $\mathbf{C}_{ij} = (-1)^{i+j}\mathbf{M}_{ij}$ for a particular entry, x_{ij}. The $\mathbf{M}_{ij}$ are the determinants of the submatrices of matrix $\mathbf{X}$ obtained by removing the ith row and jth column of the matrix $\mathbf{X}$.

Example

$$\det(\mathbf{X}) = 4 \cdot (-1)^{1+2} \cdot \begin{vmatrix} 1 & 1 \\ 1 & 2 \end{vmatrix} + 4 \cdot (-1)^{2+2} \cdot \begin{vmatrix} 1 & 2 \\ 1 & 2 \end{vmatrix} + 2 \cdot (-1)^{3+2} \cdot \begin{vmatrix} 1 & 2 \\ 1 & 1 \end{vmatrix}$$

$$= -4 \cdot 1 + 4 \cdot 0 + (-2) \cdot (-1) = -2.$$

In **R**:

```
> x = matrix(c(1,4,2,1,4,1,1,2,2), nrow = 3, byrow = TRUE)
> z = det(x)
```

The output is as follows:

```
> z
[1] -2
```

B.6   Matrix inversion

The *inverse* of a matrix, denoted by $\mathbf{X}^{-1}$, is a matrix such that $\mathbf{X}\mathbf{X}^{-1} = \mathbf{X}^{-1}\mathbf{X} = \mathbf{I}$, where $\mathbf{I}$ is the unity or identity matrix, in which the terms on the leading diagonal are all 1 and all other terms are 0; for a 3×3 identity matrix this is given by

$$\mathbf{I} = \begin{bmatrix} 1 & 0 & 0 \\ 0 & 1 & 0 \\ 0 & 0 & 1 \end{bmatrix}.$$

An inverse of $\mathbf{X}$ exists only if the determinant of $\mathbf{X}$ is non-zero. Such a matrix is said to be *invertible*. Therefore, it can be obtained only for some specific square matrices.

The final inverse matrix is given by transposing the cofactor matrix and dividing each element by the determinant calculated for the $\mathbf{X}$ matrix:

$$\mathbf{X}^{-1} = \frac{1}{|\mathbf{X}|}\mathbf{C}^{T}.$$

Example

Let us find the inverse of the matrix $\mathbf{X} = \begin{bmatrix} 1 & 4 & 2 \\ 1 & 4 & 1 \\ 1 & 2 & 2 \end{bmatrix}$.

Step 1. We begin by creating the matrix of minors (**M**) from matrix **X**.

The matrix of minors is formed by replacing each element of **X** by the determinant of the submatrix formed by removing the column and row the element lies in. For instance, M_{13} is the determinant of the matrix obtained by removing the first row and the third column of **X**, and its value is placed in the first row and third column of **M**:

$$\mathbf{X} = \begin{bmatrix} 1 & 4 & 2 \\ 1 & 4 & 1 \\ 1 & 2 & 2 \end{bmatrix} \longrightarrow \mathbf{M} = \begin{bmatrix} \begin{vmatrix} - & - \\ - & - \end{vmatrix} & \begin{vmatrix} - & - \\ - & - \end{vmatrix} & \begin{vmatrix} 1 & 4 \\ 1 & 2 \end{vmatrix} \\ \begin{vmatrix} - & - \\ - & - \end{vmatrix} & \begin{vmatrix} - & - \\ - & - \end{vmatrix} & \begin{vmatrix} - & - \\ - & - \end{vmatrix} \\ \begin{vmatrix} - & - \\ - & - \end{vmatrix} & \begin{vmatrix} - & - \\ - & - \end{vmatrix} & \begin{vmatrix} - & - \\ - & - \end{vmatrix} \end{bmatrix}$$

$$\longrightarrow \mathbf{M} = \begin{bmatrix} - & - & -2 \\ - & - & - \\ - & - & - \end{bmatrix}.$$

So the fully filled minor matrix for the **X** matrix is as follows:

$$\mathbf{X} = \begin{bmatrix} 1 & 4 & 2 \\ 1 & 4 & 1 \\ 1 & 2 & 2 \end{bmatrix} \longrightarrow \mathbf{M} = \begin{bmatrix} \begin{vmatrix} 4 & 1 \\ 2 & 2 \end{vmatrix} & \begin{vmatrix} 1 & 1 \\ 1 & 2 \end{vmatrix} & \begin{vmatrix} 1 & 4 \\ 1 & 2 \end{vmatrix} \\ \begin{vmatrix} 4 & 2 \\ 2 & 2 \end{vmatrix} & \begin{vmatrix} 1 & 2 \\ 1 & 2 \end{vmatrix} & \begin{vmatrix} 1 & 4 \\ 1 & 2 \end{vmatrix} \\ \begin{vmatrix} 4 & 2 \\ 4 & 1 \end{vmatrix} & \begin{vmatrix} 1 & 2 \\ 1 & 1 \end{vmatrix} & \begin{vmatrix} 1 & 4 \\ 1 & 4 \end{vmatrix} \end{bmatrix}$$

$$\longrightarrow \mathbf{M} = \begin{bmatrix} 6 & 1 & -2 \\ 4 & 0 & -2 \\ -4 & -1 & 0 \end{bmatrix}.$$

Step 2. We now create the cofactor matrix, **C**.

For even values of the sum $i + j$ (obtained when i and j are both odd or both even), $C_{ij} = M_{ij}$, and for odd values of the sum (obtained when either i or j is odd), $C_{ij} = -M_{ij}$. Thus, from the **M** matrix found in step 1 we obtain:

$$\mathbf{C} = \begin{bmatrix} (-1)^{1+1} \cdot 6 & (-1)^{1+2} \cdot 1 & (-1)^{1+3} \cdot (-2) \\ (-1)^{2+1} \cdot 4 & (-1)^{2+2} \cdot 0 & (-1)^{2+3} \cdot (-2) \\ (-1)^{3+1} \cdot (-4) & (-1)^{3+2} \cdot (-1) & (-1)^{3+3} \cdot 0 \end{bmatrix}$$

$$= \begin{bmatrix} (-1)^2 \cdot 6 & (-1)^3 \cdot 1 & (-1)^4 \cdot (-2) \\ (-1)^3 \cdot 4 & (-1)^4 \cdot 0 & (-1)^5 \cdot (-2) \\ (-1)^4 \cdot (-4) & (-1)^5 \cdot (-1) & (-1)^6 \cdot 0 \end{bmatrix} = \begin{bmatrix} 6 & -1 & -2 \\ -4 & 0 & 2 \\ -4 & 1 & 0 \end{bmatrix}.$$

Step 3. We transpose the cofactor matrix:

$$\mathbf{C}^T = \begin{bmatrix} 6 & -1 & -2 \\ -4 & 0 & 2 \\ -4 & 1 & 0 \end{bmatrix}^T = \begin{bmatrix} 6 & -4 & -4 \\ -1 & 0 & 1 \\ -2 & 2 & 0 \end{bmatrix}.$$

Step 4. We can now calculate the inverse matrix $\mathbf{X}^{-1}$.
Recall that $\mathbf{X}^{-1} = \frac{1}{|\mathbf{X}|}\mathbf{C}^T$. The determinant of $\mathbf{X}$ is:

$$|\mathbf{X}| = (-1)^{1+1} \cdot 1 \cdot \begin{vmatrix} 4 & 1 \\ 2 & 2 \end{vmatrix} + (-1)^{1+2} \cdot 1 \cdot \begin{vmatrix} 4 & 2 \\ 2 & 2 \end{vmatrix} + (-1)^{1+3} \cdot 1 \cdot \begin{vmatrix} 4 & 2 \\ 4 & 1 \end{vmatrix}$$

$$= 6 - 4 - 4 = -2.$$

Hence $\mathbf{X}^{-1}$ is given by

$$\mathbf{Z} = \mathbf{X}^{-1} = \frac{1}{-2}\mathbf{C}^T = -\frac{1}{2} \cdot \begin{bmatrix} 6 & -4 & -4 \\ -1 & 0 & 1 \\ -2 & 2 & 0 \end{bmatrix} = \begin{bmatrix} -3 & 2 & 2 \\ 0.5 & 0 & -0.5 \\ 1 & -1 & 0 \end{bmatrix}.$$

In **R** we write simply:

```
> x = matrix(c(1,4,2,1,4,1,1,2,2), nrow = 3, byrow = TRUE)
> z = solve(x)
```

The output is as follows:

```
> z
      [,1] [,2] [,3]
[1,] -3.0    2  2.0
[2,]  0.5    0 -0.5
[3,]  1.0   -1  0.0
```

If the result is correct it must be true that $\mathbf{X}\mathbf{X}^{-1} = \mathbf{I}$:

```
> x %*% z
     [,1] [,2] [,3]
[1,]    1    0    0
[2,]    0    1    0
[3,]    0    0    1
```

B.7   Matrix equations

Let us consider solving the following simultaneous equations for x_1 and x_2:

$$\begin{cases} a_{11}x_1 + a_{12}x_2 = c_1 \\ a_{21}x_1 + a_{22}x_2 = c_2. \end{cases}$$

Their solution may be found by solving the matrix equation

$$\begin{bmatrix} a_{11} & a_{12} \\ a_{21} & a_{22} \end{bmatrix} \cdot \begin{bmatrix} x_1 \\ x_2 \end{bmatrix} = \begin{bmatrix} c_1 \\ c_2 \end{bmatrix},$$

where

$$\begin{bmatrix} a_{11} & a_{12} \\ a_{21} & a_{22} \end{bmatrix}$$

is a matrix of coefficients (denoted by **A**) and

$$\begin{bmatrix} c_1 \\ c_2 \end{bmatrix}$$

is a matrix of intercepts (denoted by **C**). According to the previously described properties of the operations on matrices, it is true that:

$$\mathbf{AX} = \mathbf{C}$$

and

$$\mathbf{A}^{-1}\mathbf{AX} = \mathbf{A}^{-1}\mathbf{C},$$

where $\mathbf{A}^{-1}\mathbf{A}$ is an identity matrix (**I**) provided $\mathbf{A}^{-1}$ exists. Note that $\mathbf{IX} = \mathbf{XI} = \mathbf{X}$. Then the equation takes the form $\mathbf{X} = \mathbf{A}^{-1}\mathbf{C}$ and can be easily solved.

Example

Consider the following simultaneous equations:

$$\begin{cases} x_1 + x_2 = 2 \\ 2x_1 + 4x_2 = 1. \end{cases}$$

Then the matrices **A** and **C** are given by

$$\mathbf{A} = \begin{bmatrix} 1 & 1 \\ 2 & 4 \end{bmatrix}$$

and

$$\mathbf{C} = \begin{bmatrix} 2 \\ 1 \end{bmatrix}.$$

In order to obtain the unknown $\mathbf{X}$ matrix satisfying the equation $\mathbf{AX} = \mathbf{C}$, it was shown that the equation $\mathbf{X} = \mathbf{A}^{-1}\mathbf{C}$ has to be solved:

$$\mathbf{X} = \mathbf{A}^{-1}\mathbf{C} = \begin{bmatrix} 1 & 1 \\ 2 & 4 \end{bmatrix}^{-1} \cdot \begin{bmatrix} 2 \\ 1 \end{bmatrix}$$

$$\mathbf{A}^{-1} = \begin{bmatrix} 1 & 1 \\ 2 & 4 \end{bmatrix}^{-1} = \frac{1}{|\mathbf{A}|} \begin{bmatrix} 4 & -2 \\ -1 & 1 \end{bmatrix}^T$$

$$= \frac{1}{1 \cdot 4 - 1 \cdot 2} \cdot \begin{bmatrix} 4 & -1 \\ -2 & 1 \end{bmatrix} = \frac{1}{2} \cdot \begin{bmatrix} 4 & -1 \\ -2 & 1 \end{bmatrix} = \begin{bmatrix} 2 & -0.5 \\ -1 & 0.5 \end{bmatrix}$$

$$\mathbf{X} = \mathbf{A}^{-1}\mathbf{C} = \begin{bmatrix} 2 & -0.5 \\ -1 & 0.5 \end{bmatrix} \cdot \begin{bmatrix} 2 \\ 1 \end{bmatrix} = \begin{bmatrix} 3.5 \\ -1.5 \end{bmatrix}.$$

In **R**:

```
> a = matrix(c(1,1,2,4), nrow = 2, byrow = TRUE)
> c = matrix(c(2,1), nrow = 2, byrow = TRUE)
> x = solve(a,c)
```

The output is as follows:

```
> x
     [,1]
[1,]  3.5
[2,] -1.5
```

We can check that our results are correct:

$$\begin{cases} 3.5 + (-1.5) = 2 \\ 2 \cdot 3.5 + 4 \cdot (-1.5) = 1. \end{cases}$$

B.8   Eigenvectors and eigenvalues

A square matrix $\mathbf{X}$ of size $i \times i$ has i orthogonal eigenvectors ($\mathbf{w}$) and i eigenvalues (λ). Each eigenvalue is associated directly with one eigenvector of the matrix. These three quantities are combined in the equation $\mathbf{Xw} = \lambda\mathbf{w}$. The eigenvalues λ are the solutions of the equation:

$$\det(\mathbf{X} - \lambda\mathbf{I}) = \left| \begin{bmatrix} x_{11} & \cdots & x_{1i} \\ \vdots & \ddots & \vdots \\ x_{i1} & \cdots & x_{ii} \end{bmatrix} - \lambda\mathbf{I} \right| = \begin{vmatrix} x_{11} - \lambda & \cdots & x_{1i} \\ \vdots & \ddots & \vdots \\ x_{i1} & \cdots & x_{ii} - \lambda \end{vmatrix} = 0.$$

The solutions of this equation constitute the eigenvalues (λ). The eigenvectors are computed by solving the equation $(\mathbf{X} - \lambda\mathbf{I})\mathbf{w} = 0$.

Example

Suppose that $\mathbf{X} = \begin{bmatrix} 1 & 2 \\ -1 & 4 \end{bmatrix}$. Its eigenvalues λ_1 and λ_2 are obtained as solutions of the equation

$$\det(\mathbf{X} - \lambda\mathbf{I}) = \left\| \begin{bmatrix} 1 & 2 \\ -1 & 4 \end{bmatrix} - \lambda\mathbf{I} \right\| = \left\| \begin{bmatrix} 1 & 2 \\ -1 & 4 \end{bmatrix} - \begin{bmatrix} \lambda & 0 \\ 0 & \lambda \end{bmatrix} \right\|$$

$$= \begin{vmatrix} 1 - \lambda & 2 \\ -1 & 4 - \lambda \end{vmatrix} = 0.$$

Then the eigenvalues can be calculated from:

$$(1 - \lambda)(4 - \lambda) + 2 = 0,$$

that is,

$$\lambda^2 - 5\lambda + 6 = 0,$$

with solutions $\lambda_1 = 3$ and $\lambda_2 = 2$.

Eigenvectors corresponding to each of the eigenvalues can be obtained by solving $(\mathbf{X} - \lambda\mathbf{I})\mathbf{w} = 0$:

$$\left(\begin{bmatrix} 1 & 2 \\ -1 & 4 \end{bmatrix} - \lambda\mathbf{I} \right) \mathbf{w} = 0,$$

$$\begin{bmatrix} 1 - \lambda & 2 \\ -1 & 4 - \lambda \end{bmatrix} \begin{bmatrix} w_1 \\ w_2 \end{bmatrix} = 0.$$

When $\lambda = 3$,

$$\begin{bmatrix} -2 & 2 \\ -1 & 1 \end{bmatrix} \begin{bmatrix} w_1 \\ w_2 \end{bmatrix} = 0,$$

$$\begin{cases} -2w_1 + 2w_2 = 0, \\ -w_1 + w_2 = 0. \end{cases}$$

When $\lambda = 2$,

$$\begin{bmatrix} -1 & 2 \\ -1 & 2 \end{bmatrix} \begin{bmatrix} w_1 \\ w_2 \end{bmatrix} = 0,$$

$$\begin{cases} -w_1 + 2w_2 = 0, \\ -w_1 + 2w_2 = 0. \end{cases}$$

The vectors that satisfy the equations are multiples of

$$w_1 = \begin{bmatrix} w_{11} \\ w_{12} \end{bmatrix} = \begin{bmatrix} 1 \\ 1 \end{bmatrix} \quad \text{and} \quad w_2 = \begin{bmatrix} w_{21} \\ w_{22} \end{bmatrix} = \begin{bmatrix} 2 \\ 1 \end{bmatrix}.$$

They create an eigenspace basis. A common way of generating the final eigenvectors is by normalisation, which involves dividing each vector element by the vector length:

$$\mathbf{v}_1 = \frac{\begin{bmatrix} 1 \\ 1 \end{bmatrix}}{\|\mathbf{w}_1\|} = \frac{\begin{bmatrix} 1 \\ 1 \end{bmatrix}}{\sqrt{w_{11}^2 + w_{12}^2}} = \frac{\begin{bmatrix} 1 \\ 1 \end{bmatrix}}{\sqrt{1^2 + 1^2}} = \frac{\begin{bmatrix} 1 \\ 1 \end{bmatrix}}{\sqrt{2}} = \begin{bmatrix} \frac{\sqrt{2}}{2} \\ \frac{\sqrt{2}}{2} \end{bmatrix} \approx \begin{bmatrix} 0.707 \\ 0.707 \end{bmatrix}$$

and

$$\mathbf{v}_2 = \frac{\begin{bmatrix} 2 \\ 1 \end{bmatrix}}{\|\mathbf{w}_2\|} = \frac{\begin{bmatrix} 2 \\ 1 \end{bmatrix}}{\sqrt{w_{21}^2 + w_{22}^2}} = \frac{\begin{bmatrix} 2 \\ 1 \end{bmatrix}}{\sqrt{2^2 + 1^2}} = \frac{\begin{bmatrix} 2 \\ 1 \end{bmatrix}}{\sqrt{5}} = \begin{bmatrix} \frac{2\sqrt{5}}{5} \\ \frac{\sqrt{5}}{5} \end{bmatrix} \approx \begin{bmatrix} 0.894 \\ 0.447 \end{bmatrix}.$$

In **R**:

```
> x = matrix(c(1,2,-1,4), nrow = 2, byrow = TRUE)
> e = eigen(x)
```

The output is as follows:

```
> e
$values
[1] 3 2
$vectors
           [,1]        [,2]
[1,] -0.7071068 -0.8944272
[2,] -0.7071068 -0.4472136
```

ignoring the sign.

More information about matrix algebra can be found in Stephenson (1973).

Reference

Stephenson G 1973 *Mathematical Methods for Science Students*, 2nd edn. Longman, London.

Appendix C

Pool adjacent violators algorithm

In this appendix we provide a brief description of the PAV algorithm used in Chapter 6. Assume that there exist a validation set of posterior probabilities (Appendix A) that H_1 is true, namely P_i, and the respective ground-truth labels of the proposition that is actually true in each case, namely $y_i \in \{0, 1\}$, where i is an index representing the case in the validation set for which a posterior probability P_i is assigned. This index $i \in \{1, \ldots, N\}$, where $N = N_1 + N_2$, N_1 is the number of cases in the validation set for which H_1 is true, and N_2 is the number of cases in the validation set for which H_2 is true. Thus, N is the total number of cases in the validation set. With this notation, when $y_i = 1$ then H_1 is true for case i, and when $y_i = 0$ then H_2 is true for case i. Therefore, $P_i = P(y_i = 1)$. Moreover, without loss of generality, it is assumed that for two cases j and k in the validation set, $P_j \leq P_k$ if $j < k$, or in other words, that the probabilities P_i are sorted in ascending order by case index i. Each y_i has a value of $y_i = 1$ with probability P_i if H_1 is true and $y_i = 0$ with probability P_i if H_2 is true. Therefore, the y_i values are in fact the *oracle* probabilities that an oracle forecaster would elicit, as defined in Section 6.5.1. These variables are organised as (P_i, y_i) pairs.

For the PAV algorithm several new variables are defined:

- $m_{k,l}$ is a real number that represents the value of a real number between 0 and 1 assigned to P_i for i between k and l, where it is always assumed that $k < l$. It is implicitly assumed here that the value of $m_{k,l}$ is constant over all possible values of $i \in [k, l]$. Thus, $m_{i,i}$ is a real number between 0 and 1 assigned to P_i.

- $w_{k,l}$ is a real number that represents the number of cases between k and l, where it is always assumed that $k < l$. Thus, in fact, $w_{k,l} = l - k + 1$.

Under these conditions, the PAV gives a non-decreasing transformation $\mathbf{m} = (m_1, \ldots, m_N)$ for each P_i, such that P_i will be transformed into $m_{i,i} \equiv m_i$. It can be shown (Brümmer 2010) that m_i can be interpreted as probabilities, and are better calibrated than P_i

Statistical Analysis in Forensic Science: Evidential Value of Multivariate Physicochemical Data, First Edition.
Grzegorz Zadora, Agnieszka Martyna, Daniel Ramos and Colin Aitken.
© 2014 John Wiley & Sons, Ltd. Published 2014 by John Wiley & Sons, Ltd.
Companion website: www.wiley.com/go/physicochemical

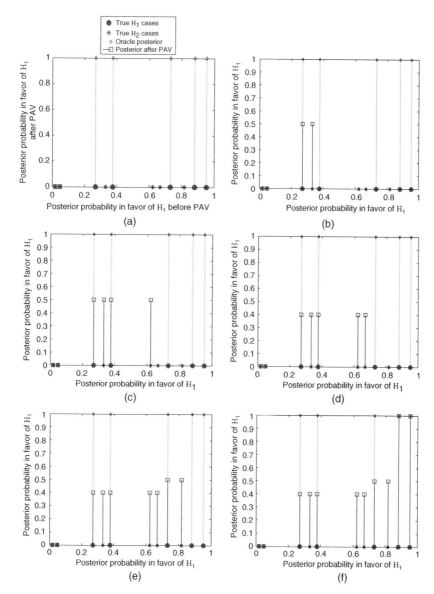

Figure C.1   PAV algorithm with a set of synthetic (P_i, y_i) pairs. In the diagrams, H_1 and H_2 refer to the propositions considered in evidence evaluation (e.g. in a comparison problem, they may respectively refer to the same-source and different-source propositions). P_i refers to the posterior probability of H_1 for case i in the validation set of posterior probabilities (x axis), that is, the probabilities before the PAV is used. On the other hand, y_i is 1 if H_1 is true in case i, and 0 if H_2 is true in case i, and therefore y_i is the *oracle* posterior probability for case i in the validation set. Initialisation is shown in (a). From (b) to (g) the different stages of the algorithm are detailed. The final stepwise transformation is highlighted in (h) with a dashed line.

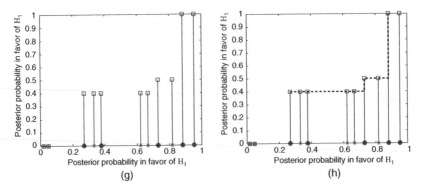

Figure C.1 (*Continued*)

with $i \in \{1, \ldots, N\}$. Therefore, the aim of the transformation **m** is to minimise the value of the empirical cross entropy (ECE) of the transformed set of posterior probabilities.

The PAV works as a *pooling* process, where the regions of i values for which the monotonicity condition of y_i is broken are pooled together in a new, wider region. Therefore, the value of the transformation in the new, pooled region is updated according to the values of the transformation in the original regions. Hence the name PAV.

The oracle set of probabilities $\{y_i\}$ is the aim of the transformation **m**, but the non-decreasing condition makes the final probability set **m** different from $\{y_i\}$ in general. However, the final probability set defined by the non-decreasing transformation **m** is better calibrated than P_i for all $i \in \{1, N\}$.

The PAV algorithm is as follows:

1. Input: (P_i, y_i) set, where, P_i is assumed sorted in increasing order according to i

2. Initialise: $m_{i,i} = y_i$
   while $\exists i$ *s.t.* $m_{k,i-1} \geq m_{i,l}$ **do**

   - Set $w_{k,l} = w_{k,i-1} + w_{i,l}$

   - Set $m_{k,l} = (w_{k,i-1}m_{k,i-1} + w_{i,l}m_{i,l})/w_{k,l}$

   - Replace $m_{k,i-1}$ and $m_{i,l}$ with $m_{k,l}$

   end while

3. Output the stepwise-constant function generated by **m**

An example of the transformation generated by the PAV algorithm using artificial data is given in Figure C.1. Every step of the algorithm is shown in the figure, where it can be seen that the final transformation tends to yield a posterior probability which is the proportion of cases in which H_1 is true in each of the regions (or *bins*) determined by the algorithm. This is in accordance with the definition of calibration (deGroot and Fienberg 1983; Ramos and Gonzalez-Rodriguez 2013).

In Brümmer (2010), a version of the PAV that works on *LR* values rather than posterior probabilities is proposed, and its analogy with the conventional PAV is shown. An implementation of that version of the PAV algorithm for Matlab^{TM} can be found in the FoCal toolkit (niko.brummer.googlepages.com/focal; last accessed in October 2013).

References

Brümmer N 2010 *Measuring, refining and calibrating speaker and language information extracted from speech* PhD thesis School of Electrical Engineering, University of Stellenbosch, South Africa. Available at http://sites.google.com/site/nikobrummer/ (last accessed 2 October 2013).

deGroot MH and Fienberg SE 1983 The comparison and evaluation of forecasters. *The Statistician* **32**, 12–22.

Ramos D and Gonzalez-Rodriguez J 2013 Reliable support: measuring calibration of likelihood ratios. *Forensic Science International* **230**(1–3), 156–169.

Appendix D

Introduction to R software

D.1 Becoming familiar with R

R is a free software environment, widely used for statistical calculations and graphics. It runs under various operating systems and can be downloaded from the website www.r-project.org. After installing the program in the Windows operating system and clicking on the **R** icon, the **R** Console appears as in Figure D.1. The routines provided in this book have been prepared in **R** version *2.15.2*.

The **R** Console is an editable window, in which users can write commands and look at the results of calculations. When graphics are generated, they appear in a separate window. Commands are displayed in red, while all results are in blue, though this colour scheme can be changed by the user if desired.

One of the most important things to remember when working with the **R** Console is the choice of workspace. This refers to a file folder in which the data files, **R** code fragments, and results are stored. This can be set by clicking on File in the main menu and selecting Change dir... as in Figure D.2.

R offers plenty of different packages designed for numerous applications, which may be loaded either from the Internet website www.r-project.org, or from the main menu as shown in Figure D.3.

Comments on the **R** code can be written by putting the # sign, a symbol which can be inserted anywhere in the line, not just at the beginning, so it can follow on the same line as a command. Everything that is written after this sign is neglected and treated as a comment. It is good practice to put a # sign at the beginning of a line and write a comment explaining what is to be done. Sections of code can be separated by a # sign for ease of reading.

In order to look at the specification of functions, for example the cos() function, ?cos should be entered; this will provide some helpful information.

Statistical Analysis in Forensic Science: Evidential Value of Multivariate Physicochemical Data, First Edition.
Grzegorz Zadora, Agnieszka Martyna, Daniel Ramos and Colin Aitken.
© 2014 John Wiley & Sons, Ltd. Published 2014 by John Wiley & Sons, Ltd.
Companion website: www.wiley.com/go/physicochemical

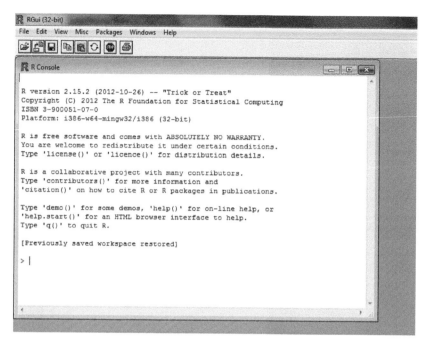

Figure D.1   **R** Console for the Windows operating system.

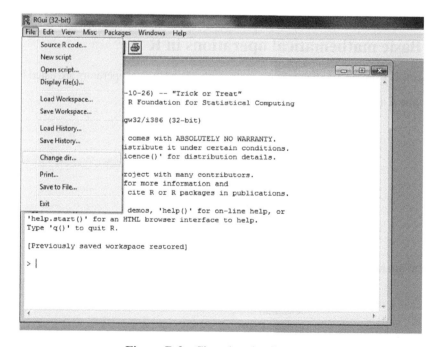

Figure D.2   Changing the directory.

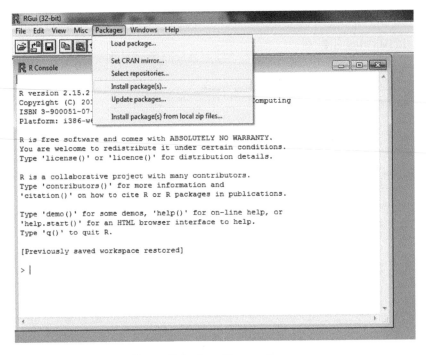

Figure D.3   Installing packages.

D.2   Basic mathematical operations in R

R may work as a simple calculator. Some mathematical and logical operators are listed below:

Mathematical operators:

addition	$+$
subtraction	$-$
multiplication	$*$
division	$/$
matrix multiplication	$\% * \%$

Relational operators:

less than	$<$
greater than	$>$
less than or equal to	$<=$
greater than or equal to	$>=$
equal to	$==$

Logical operators:

logical negation (NOT)   !
logical AND   & *or* &&
logical OR | *or* ||

Using these operators, calculations may be carried out either by entering the numbers via the keyboard (note that the prompt symbol > is already entered at the beginning of each line):

```
> 3 + 5
```

(and after pressing Enter)

```
[1] 8
```

or by creating new objects, named a and b for example, to which 3 and 5 would be assigned:

```
> a = 3
> b = 5
```

Typing a displays 3 and typing b displays 5.

```
> a
[1] 3
> b
[1] 5
```

The calculation $3 + 5$ can be performed by operating on the objects introduced:

```
> a + b
[1] 8

> a - b
[1] -2

> a * b
[1] 15

> a/b
[1] 0.6

> exp(a)
[1] 20.08554

> log10(a)
[1] 0.4771213

> sin(a)
[1] 0.14112
```

```
> cos(a)
[1] -0.9899925

> tan(a)
[1] -0.1425465
```

If rounded values are required, the `round` command is used. This allows the number of decimal places to be chosen. The command `signif()` carries out rounding to specified number of significant figures.

```
> round(1.2314, digits=2)
[1] 1.23

> signif(1.2314, digits=2)
[1] 1.2
```

When using logical operators, the program will only return logical responses if the statement is TRUE or FALSE.

```
> a >= b
[1] FALSE

> a <= b
[1] TRUE

> a == b
[1] FALSE

> a != b
[1] TRUE
```

To introducing conditions while performing calculations, the command `if()` is used.

```
> a=3
> if (a == 3) ##introducing the condition
+ {print("TRUE") ##prints "TRUE" when the condition is fulfilled
+ } else
+ {print("FALSE")} ##prints "FALSE" when the condition is not fulfilled
[1] "TRUE"

> a=100
> if (a == 3) {print("TRUE")} else {print("FALSE")}
[1] "FALSE"
```

D.2.1   Vector algebra

In **R**, vectors are denoted by `c()`. The elements of the vector are entered in the brackets, separated by commas.

```
#a vector created from elements 2, 3, and 6
> c(2,3,6)
[1] 2 3 6
```

```
#a vector combining a, b, and a*b elements
> a=3
> b=5
> c(a, b, a*b)
[1]  3  5 15
```

Here are some example of vector algebra on x and y:

```
> x = c(2,3,6)
> y = c(a, b, a*b)

> y
[1]  3  5 15

#adding vectors: corresponding elements of two vectors x and y are added
> x + y
[1]  5  8 21

#adding a constant to a vector: 10 is added to each element of vector x
> x + 10
[1]  12 13 16

#each element of vector x is multiplied by -1
> x * (-1)
[1]  -2 -3 -6
```

A vector may also be given as a sequence (seq()). Here is a sequence starting from 2, ending at 29, with a step equal to 2:

```
> seq(from = 2, to = 29, by = 2)
[1]  2  4  6  8 10 12 14 16 18 20 22 24 26 28
```

The default for by is 1. Note that the sequence stops at 28, since the highest even number is lower than 29. Sequences can also be generated as a series of repeated elements using rep(), which in the example creates a vector containing a series of values from 1 to 5 (by 1) repeated 3 times:

```
> rep(c(1:5), times = 3)
[1]  1  2  3  4  5  1  2  3  4  5  1  2  3  4  5
```

The next example shows the rep() function used to obtain a slightly different series of data. It creates a vector containing a series of values from 1 to 5 each repeated 3 times in a row:

```
> rep(c(1:5), each = 3)
[1]  1  1  1  2  2  2  3  3  3  4  4  4  5  5  5
```

Useful commands related to vectors are length(), unique(), and table(). The command length() returns the number of elements in the vector. The command unique()

gives the elements appearing in the vector at least once. The command table() shows how many times each element occurs. For the previous sequence:

```
> r = rep(c(1:5), each = 3)
> length(r)
[1] 15

> unique(r)
[1] 1 2 3 4 5

> table(r)
r
1 2 3 4 5
3 3 3 3 3
```

D.2.2   Matrix algebra

A matrix in **R** is created by using the matrix() function. This function needs a few parameters, such as the elements of the matrix, number of rows and columns of the matrix. Let us create a matrix with elements 1 to 12 arranged in 3 rows and 4 columns.

```
> matrix(c(1:12), nrow = 3, ncol = 4)
     [,1] [,2] [,3] [,4]
[1,]   1    4    7   10
[2,]   2    5    8   11
[3,]   3    6    9   12
```

If the user wants to fill the matrix with the same elements, but organised by rows, the command byrow = TRUE must be added to the function parameters:

```
> matrix(c(1:12), nrow = 3, ncol = 4, byrow = TRUE)
     [,1] [,2] [,3] [,4]
[1,]   1    2    3    4
[2,]   5    6    7    8
[3,]   9   10   11   12
```

To assign row and column names, the rownames() and colnames() commands are useful.

```
> R = matrix(c(1:12), nrow = 3, ncol = 4, byrow = TRUE)

> rownames(R) = c("(row 1)","(row 2)","(row 3)")
> colnames(R) = c("(col 1)","(col 2)","(col 3)","(col 4)")
> R
        (col 1) (col 2) (col 3) (col 4)
(row 1)     1       2       3       4
(row 2)     5       6       7       8
(row 3)     9      10      11      12
```

`dim()` is a useful command for checking the dimensions of a matrix. It will return two numbers: the number of rows followed by the number of columns:

```
> dim(R)
[1] 3 4
```

To obtain the number of rows or columns separately, use `nrow()` and `ncol()`:

```
> nrow(R)
[1] 3

> ncol(R)
[1] 4
```

R is a powerful tool for matrix algebra. This is illustrated with some examples. Let us create matrices **A** and **B**. Remember that adding (or subtracting) matrices requires that they are of the same size.

```
> A = matrix(c(1:12), nrow = 3, ncol = 4)
> A

     [,1] [,2] [,3] [,4]
[1,]    1    4    7   10
[2,]    2    5    8   11
[3,]    3    6    9   12

> B = matrix(c(1:12), nrow = 3, ncol = 4, byrow = TRUE)
> B
     [,1] [,2] [,3] [,4]
[1,]    1    2    3    4
[2,]    5    6    7    8
[3,]    9   10   11   12

> A + B
     [,1] [,2] [,3] [,4]
[1,]    2    6   10   14
[2,]    7   11   15   19
[3,]   12   16   20   24
```

When using simple `*` as a multiplication operator only elements described by the same set of indices are multiplied:

```
> A * B
     [,1] [,2] [,3] [,4]
[1,]    1    8   21   40
[2,]   10   30   56   88
[3,]   27   60   99  144
```

The classical operation of matrix multiplication is realised by the `%*%` operator. This requires that the number of columns in the first matrix is equal to the number of rows in the second. To carry out such an operation on the **A** and **B** matrices, one of them must be transposed. This means swapping the rows with columns, which is achieved by typing t().

```
#transposing a matrix
> t(B)
     [,1] [,2] [,3]
[1,]    1    5    9
[2,]    2    6   10
[3,]    3    7   11
[4,]    4    8   12

#matrix multiplication
> A %*% t(B)
     [,1] [,2] [,3]
[1,]   70  158  246
[2,]   80  184  288
[3,]   90  210  330
```

Given a square matrix **C**, its determinant and inverse are found using the functions det() and solve(), respectively:

```
> C = matrix(c(0,2,2,4,-2,0,4,1,-1), nrow = 3)
> C
     [,1] [,2] [,3]
[1,]    0    4    4
[2,]    2   -2    1
[3,]    2    0   -1

> det(C)
[1] 32

> solve(C)
         [,1]    [,2]    [,3]
[1,] 0.0625   0.125   0.375
[2,] 0.1250  -0.250   0.250
[3,] 0.1250   0.250  -0.250
```

If the result is correct, the equation $\mathbf{CC}^{-1} = \mathbf{C}^{-1}\mathbf{C} = \mathbf{I}$ is true, and this can be verified in **R** by typing:

```
> C%*%solve(C)
     [,1] [,2] [,3]
[1,]    1    0    0
[2,]    0    1    0
[3,]    0    0    1
```

D.3   Data input

As **R** is usually used for statistical computations, it can involve working with a huge amount of data. Therefore, it may be problematic to enter it from the keyboard. A solution to this may be to load tabular data from a text file using the read.table() command and assigning them to some object. In many cases working with lists of data is rather problematic and it is recommended to change the list type into a data frame type by using data.frame().

Prior to starting a new task in **R** it is recommended that all objects from previous **R** sessions are removed using `rm(list = ls())` or clicking on `Remove all objects` in the Console menu (Figure D.4). Checking what objects are stored in **R**'s memory is done by typing `ls()`.

Data are loaded from a file (`glass_data.txt` for example) as follows:

```
> data = read.table(file = "glass_data.txt", header = TRUE)
```

Remember that you need to be in the proper directory (choose `Change dir` from the `File` menu), which is a folder (say, `Appendix D`) containing the relevant data. The function parameter `header = TRUE` assures that the first row of the data is skipped, as it is usually not of a numerical type and contains, for example, letters or other data types. It is possible to look at the data by simply entering `data` or `head(data)`, which shows only the first six rows of the dataset loaded:

```
> head(data)
  item fragment  logNaO  logMgO  logAlO  logSiO  logKO   logCaO  logFeO
1   s1       f1 -0.6603 -1.4683 -1.4683 -0.1463 -1.7047 -1.1096 -5.6778
2   s1       f1 -0.6658 -1.4705 -1.4814 -0.1429 -1.7183 -1.1115 -5.3763
3   s1       f1 -0.6560 -1.4523 -1.4789 -0.1477 -1.6864 -1.1118 -5.6776
4   s1       f2 -0.6309 -1.4707 -1.5121 -0.1823 -1.7743 -1.1306 -2.6090
5   s1       f2 -0.6332 -1.4516 -1.4996 -0.1792 -1.7577 -1.1332 -5.6871
6   s1       f2 -0.6315 -1.4641 -1.4883 -0.1710 -1.7548 -1.1291 -5.6842
```

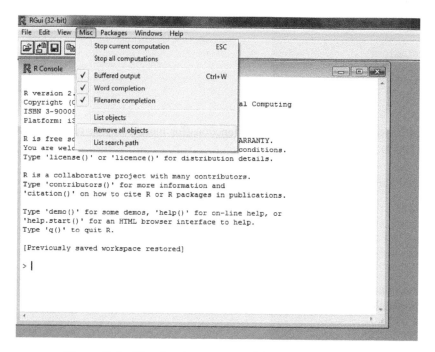

Figure D.4   Clearing the memory of the current **R** session.

The same data can downloaded from the *comparison* package (version 1.0–2) and imported into **R** by typing:

```
> install.packages("comparison")
> library("comparison")
> data(glass)
> data = data.frame(glass)
> attach(data)
```

The attach() command enables the **R** to keep the data in memory, so that, for example, matrix columns may be referenced only by their names such that, instead of data$logNaO, logNaO is sufficient.

D.4   Functions in R

A simpler way to perform calculations in **R** is to write all the required commands in a text file (e.g. name_of_the_file.txt) and then type source("name_of_the_file.txt") to run them. Writing the script in a text file allows the user to make any alterations needed without the necessity of running all the lines of the code written after the altered line separately. Instead, the whole program can be run using only the source() command. However, in this case if the user wants to see the value of a particular object, the command print() must be typed in the file containing the commands.

The source() command is useful for loading the functions, which are separate pieces of code. The scheme for creating a function is given below (this bit of code will not work when typed into **R**, as it is only a scheme):

```
function.name = function(PARAMETERS THAT THE FUNCTION IS DEPENDENT ON)
{
   CALCULATIONS
   result = list()
   return (result)
}
```

Let us consider the following simple function (for more complex functions, refer to Sections D.9–D.10):

```
concentration = function(n, V)
{
   C = n/V
   result = list(C)
   return (result)
}
```

After saving the function as function.R, it can be used to perform the calculations in a different code file:

```
> n=1.5
> V=0.5
```

```
> source("function.R") ##loading the relevant function from the same file
    folder
> results.c = concentration(n,V) ##performing the calculations
> concentration = results.c ##referring to the results of the calculations
> print(concentration) ##printing the results
[[1]]
[1] 3
```

The `for(){}` command is useful for iterative calculations (for explanation of `X[i]` see Section D.5):

```
> X=c(1,4,2,6)
> for (i in 1:length(X)) ##i is the iteration index
+ {
+    result = 2*X[i] + 5
+    print(result)
+ }
```

The output is as follows:

```
[1] 7
[1] 13
[1] 9
[1] 17
```

D.5 Dereferencing

In some cases it may be necessary to work with a subset of the data. The procedure of selecting a part of the dataset is called *dereferencing*. This may be done in many ways, and is shown for the glass data introduced in Section D.3.

```
#selecting the first row of the data
> data[1,]
item fragment  logNaO  logMgO  logAlO  logSiO   logKO  logCaO  logFeO
1    s1       f1 -0.6603 -1.4683 -1.4683 -0.1463 -1.7047 -1.1096 -5.6778

#selecting the fifth column of the data
> data[,5]
[1] -1.4683 -1.4814 -1.4789 -1.5121 -1.4996 -1.4883 -1.4708 -1.4664 -1.4612
[10] -1.4956 -1.4944 -1.4987 -1.6540 -1.6622 -1.6935 -1.7202 -1.6831
        -1.7269...

#selecting the third row and sixth column, which means only one cell
    of the dataset is extracted
> data[3,6]
[1] -0.1477

#selecting rows 3 to 8 but skipping columns 2 to 5
> data[3:8,-(2:5)]
  item logSiO   logKO  logCaO  logFeO
3    s1 -0.1477 -1.6864 -1.1118 -5.6776
```

```
4    s1 -0.1823 -1.7743 -1.1306 -2.6090
5    s1 -0.1792 -1.7577 -1.1332 -5.6871
6    s1 -0.1710 -1.7548 -1.1291 -5.6842
7    s1 -0.1165 -1.6789 -1.0358 -2.4307
8    s1 -0.1048 -1.6568 -1.0094 -5.6568
```

Another way of dereferencing uses the $ sign, so as to extract data by column names. Let us choose only those rows which refer to the s1 item (note that 1 is a number and not a letter l).

```
#choosing column named 'item'
> data$item
[1] s1 s1 s1 s1 s1 s1 s1 s1 s1 s1 s1 s1 s2 s2 s2 s2 s2 s2 s2 s2...
200 Levels: s1 s10 s100 s101 s102 s103 s104 s105 s106 s107 s108 s109 ...
   s99

#returns TRUE if the row refers to the 's1' or FALSE if not
> data$item == "s1"
 [1]   TRUE   TRUE   TRUE   TRUE   TRUE   TRUE   TRUE   TRUE   TRUE   TRUE   TRUE   TRUE
[13]  FALSE  FALSE  FALSE  FALSE  FALSE  FALSE  FALSE  FALSE...

#returns indices of rows fulfilling the condition
> which(data$item == "s1")
 [1]  1  2  3  4  5  6  7  8  9 10 11 12

#retrieving 's1' object measurements and assigning them to the 'data_s1'
   object
> data_s1 = data[which(data$item == "s1"),]
> data_s1
   item fragment  logNaO  logMgO  logAlO  logSiO   logKO  logCaO  logFeO
1    s1       f1 -0.6603 -1.4683 -1.4683 -0.1463 -1.7047 -1.1096 -5.6778
2    s1       f1 -0.6658 -1.4705 -1.4814 -0.1429 -1.7183 -1.1115 -5.3763
3    s1       f1 -0.6560 -1.4523 -1.4789 -0.1477 -1.6864 -1.1118 -5.6776
4    s1       f2 -0.6309 -1.4707 -1.5121 -0.1823 -1.7743 -1.1306 -2.6090
5    s1       f2 -0.6332 -1.4516 -1.4996 -0.1792 -1.7577 -1.1332 -5.6871
6    s1       f2 -0.6315 -1.4641 -1.4883 -0.1710 -1.7548 -1.1291 -5.6842
7    s1       f3 -0.6420 -1.4437 -1.4708 -0.1165 -1.6789 -1.0358 -2.4307
8    s1       f3 -0.6431 -1.4692 -1.4664 -0.1048 -1.6568 -1.0094 -5.6568
9    s1       f3 -0.6580 -1.4698 -1.4612 -0.0843 -1.5881 -0.9888 -5.6488
10   s1       f4 -0.6477 -1.4630 -1.4956 -0.1653 -1.7289 -1.1138 -5.6831
11   s1       f4 -0.6330 -1.4563 -1.4944 -0.1579 -1.7297 -1.1074 -5.6791
12   s1       f4 -0.6424 -1.4607 -1.4987 -0.1669 -1.7591 -1.1199 -2.6042
```

The data can also be divided up by factors (e.g. item name) using the split() function:

```
> split(data, data$item)
$s1
   item fragment  logNaO  logMgO  logAlO  logSiO   logKO  logCaO  logFeO
1    s1       f1 -0.6603 -1.4683 -1.4683 -0.1463 -1.7047 -1.1096 -5.6778
2    s1       f1 -0.6658 -1.4705 -1.4814 -0.1429 -1.7183 -1.1115 -5.3763
3    s1       f1 -0.6560 -1.4523 -1.4789 -0.1477 -1.6864 -1.1118 -5.6776
4    s1       f2 -0.6309 -1.4707 -1.5121 -0.1823 -1.7743 -1.1306 -2.6090
5    s1       f2 -0.6332 -1.4516 -1.4996 -0.1792 -1.7577 -1.1332 -5.6871
6    s1       f2 -0.6315 -1.4641 -1.4883 -0.1710 -1.7548 -1.1291 -5.6842
7    s1       f3 -0.6420 -1.4437 -1.4708 -0.1165 -1.6789 -1.0358 -2.4307
```

```
8     s1        f3 -0.6431 -1.4692 -1.4664 -0.1048 -1.6568 -1.0094 -5.6568
9     s1        f3 -0.6580 -1.4698 -1.4612 -0.0843 -1.5881 -0.9888 -5.6488
10    s1        f4 -0.6477 -1.4630 -1.4956 -0.1653 -1.7289 -1.1138 -5.6831
11    s1        f4 -0.6330 -1.4563 -1.4944 -0.1579 -1.7297 -1.1074 -5.6791
12    s1        f4 -0.6424 -1.4607 -1.4987 -0.1669 -1.7591 -1.1199 -2.6042



.
.
.


$s10
        item fragment  logNaO   logMgO   logAlO   logSiO    logKO   logCaO   logFeO
109   s10        f1 -0.8398 -5.3241 -1.7275 -0.1912 -1.0848 -5.3241 -5.6251
110   s10        f1 -0.8268 -5.6079 -1.7566 -0.1618 -1.0516 -5.6079 -5.3069
...
```

D.6   Basic statistical functions

Let us apply some statistics to the glass data from the *comparison* package (version 1.0–2).

```
#calculating the mean of data_s1 accounting for logSiO content
> mean(data_s1$logSiO)
[1] -0.1470917
```

The same result, but for all the variables in one calculation run, is obtained as follows:

```
> apply(data_s1[,3:9], 2, mean)
  logNaO       logMgO       logAlO       logSiO        logKO       logCaO       logFeO
-0.6453250 -1.4616833 -1.4846417 -0.1470917 -1.7114750 -1.0917417
    -4.8678917
##compare the fourth result above with the one when using command 'mean(
    data_s1$logSiO)'

#calculating the standard deviation for the data
> sd(data_s1$logSiO)
[1] 0.03070195

#calculating the variance for the data
> var(data_s1$logSiO)
[1] 0.0009426099

#calculating the median for the data
> median(data_s1$logSiO)
[1] -0.1528

#calculating the Q2 (which is median) for the data
> quantile(data_s1$logSiO, prob = 0.5)
    50%
-0.1528

#calculating the Q1 (lower quartile) for the data
> quantile(data_s1$logSiO, prob = 0.25)
     25%
-0.167925
```

```
#calculating the Q3 (upper quartile) for the data
> quantile(data_s1$logSiO, prob = 0.75)
    75%
-0.1363

#calculating the interquartile range for the data
> IQR(data_s1$logSiO)
[1] 0.031625

#finding the maximum value for the data
> max(data_s1$logSiO)
[1] -0.0843

#finding the minimum value for the data
> min(data_s1$logSiO)
[1] -0.1823
```

D.7   Graphics with R

R software is a useful tool for graphics. In this section only the basics are presented: many more complex charts, plots, and graphics may be generated.

D.7.1   Box-plots

The simplest way to present descriptive statistics for a dataset is to draw box-plots (Section 3.3.3, Figure D.5). For the glass data:

```
> boxplot(data[,3:9])
```

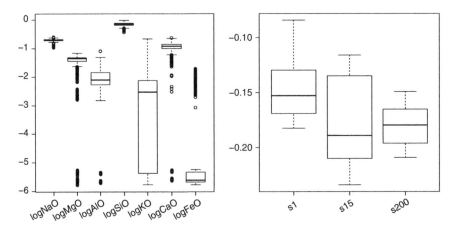

Figure D.5   Box-plots (left) for all glass objects with respect to the seven variables and (right) only for objects s1, s15, and s200 with respect to the logSiO variable.

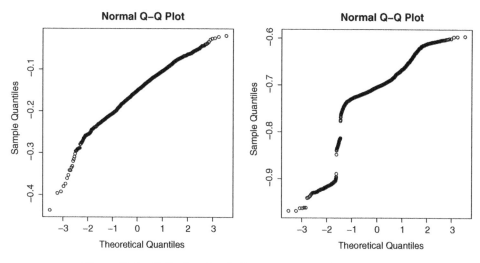

Figure D.6   Q-Q plots for (left) `logSiO` and (right) `logNaO`.

On the left in Figure D.5, the box-plots, one for each of the variables in columns 3 to 9, represent all the measurements on the particular variable and illustrate the parameters characterising these variables. Information about all the variables is displayed in one chart.

If the user wants to visualise the `logSiO` data structure for the objects s1, s15, and s200, then the following commands should be used (Figure D.5, right). The user needs to extract the data corresponding to the items mentioned and create a submatrix which is assigned to `subdata`.

```
> subdata = data[data$item %in% c("s1","s15","s200"),]
> boxplot(subdata$logSiO~as.character(subdata$item))
```

`subdata$logSiO` corresponds to the vertical axis, and `as.character(subdata$item)` divides the data with respect to the items s1, s15, and s200.

D.7.2   Q-Q plots

When analysing data it is important to know whether they show a distribution similar to normal. This can be investigated by plotting Q-Q plots (Section 3.3.5; Figure D.6).

In this method the quantiles characterising the data are plotted against theoretical standard normal distribution quantiles. If the points lie along the diagonal of the plot, the data may be assumed to be normally distributed. If not, then other ways of computing the probability density function must be employed. Based on the plots the distribution of `logSiO` (Figure D.6, left) can be assumed normal, but not that of `logNaO` (Figure D.6, right).

```
> qqnorm(data$logSiO)
```

```
> qqnorm(data$logNaO)
```

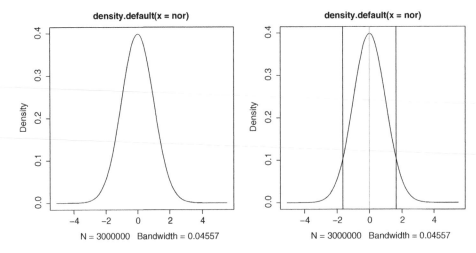

Figure D.7   Standard normal distribution curves (a detailed description is provided in the text).

D.7.3   Normal distribution

A set of random numbers from a normal distribution can be generated by the rnorm() function, which requires the mean and standard deviation as well as the number of observations to be given. Let us create $3 \cdot 10^6$ normally distributed observations with mean 0 and standard deviation 1, that is, a standard normal distribution (Section 3.3.5).

```
> nor = rnorm(3000000, mean = 0, sd = 1)
```

The probability density function for such data may be plotted as follows (Figure D.7, left):

```
> plot(density(nor))
```

Z values (equation (3.1)) corresponding to the probability of observing a value as high as Z (Figure D.7, right), may be calculated by qnorm(). By default the quantiles correspond to the standard normal distribution $N(0, 1)$.

For one-sided tests (Section 3.4) when α is set to 0.05 this can be done by typing:

```
> qnorm(0.95, mean = 0, sd = 1)
[1] 1.644854

> qnorm(0.05, mean = 0, sd = 1)
[1] -1.644854
```

and for normal distributions (Section 3.3.5) other than the standard normal distribution, for example $N(2, 1.4884)$:

```
> qnorm(0.95, mean = 2, sd = sqrt(1.4884))
[1] 4.006721
```

For two-sided tests when α is set on 0.05 this can be done by typing:

```
> qnorm(0.975, mean = 0, sd = 1)
[1] 1.959964
```

```
> qnorm(0.025, mean = 0, sd = 1)
[1] -1.959964
```

To find the probability of observing a value as high as Z, use `pnorm()`:

```
> pnorm(-1.959964, mean = 0, sd = 1)
[1] 0.025
```

For other distributions, such as the t distribution, F distribution, and χ^2 distribution, similar commands for finding the probabilities and the associated observations are available. For the t distribution (Section 3.3.5):

```
> qt(0.95, df = 5) ##where 5 is the number of degrees of freedom
[1] 2.015048
```

```
> pt(2.015048, df = 5) ##where 5 is the number of degrees of freedom
[1] 0.95
```

For the F distribution (Section 3.3.5):

```
> qf(0.95, df1 = 5, df2 = 10) ##where 5 and 10 are the numerator and
    denominator degrees of freedom
[1] 3.325835
```

```
> pf(3.325835, df1 = 5, df2 = 10) ##where 5 and 10 are the numerator and
    denominator degrees of freedom
[1] 0.95
```

For the χ^2 distribution (Section 3.3.5):

```
> qchisq(0.05, df = 5) ##where 5 is the number of degrees of freedom
[1] 1.145476
```

```
> pchisq(1.145476, df = 5) ##where 5 is the number of degrees of freedom
[1] 0.04999998
```

If α is set to 0.1 and the test is two-sided, vertical (v) lines cutting off the level of significance may be added to a density plot with the `abline()` command (Figure D.7, right). The line depicting the maximum of the Gaussian curve is drawn in grey in the figure).

```
> abline(v = 0, col = "gray")
> abline(v = qnorm(0.95))
> abline(v = qnorm(0.05))
```

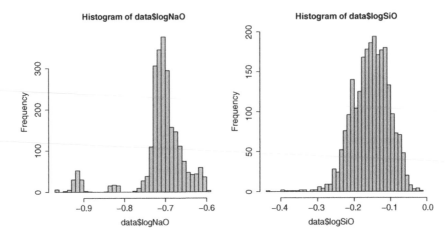

Figure D.8   Histograms for (left) `logNaO` and (right) `logSiO`.

D.7.4   Histograms

A simple way of investigating data structure is by plotting histograms (Section 3.3.3). However, their structure is strongly dependent on the location, number, and width of the bars. Nevertheless, they are very useful for examining the data distribution, especially if the data are described, for example, by multimodal distributions.

In **R** it is possible to decide on the number of bars (e.g. breaks = 30), so that the histogram becomes more informative for the user (Figure D.8).

```
> hist(data$logSiO, col = "gray", breaks = 30)

> hist(data$logNaO, col = "gray", breaks = 30)
```

Histograms can also be drawn with the probability density instead of frequency as in Figure D.9 (right).

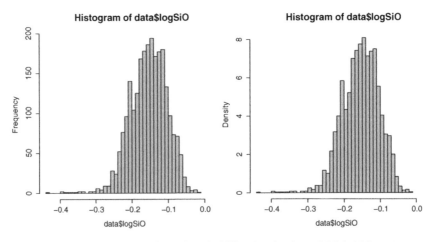

Figure D.9   Frequency (left) and probability density-based (right) histograms.

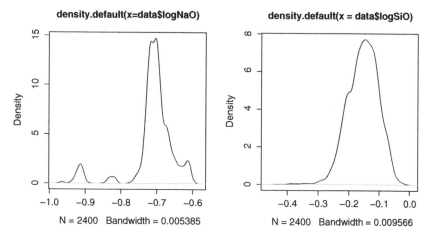

Figure D.10   Kernel density estimation for (left) `logNaO` and (right) `logSiO`.

D.7.5   Kernel density estimation

If the data distribution cannot be assumed normal, other methods for obtaining the probability density function should be applied, such as kernel density estimation (KDE; Section 3.3.5). This can be done with the `density()` function (Figure D.10).

```
> plot(density(data$logNaO))
> plot(density(data$logSiO))
```

It can be seen (Figure D.10) that `logNaO` cannot be estimated by normal distribution as the data structure is multimodal.

By calling the `density()` function, a summary of the data is given. There are two columns; x refers to the parameters for the data under consideration (in this case the `logNaO` variable) and y refers to the parameters of the corresponding probability density values.

```
> density(data$logNaO)
Call:
    density.default(x = data$logNaO)
Data: data$logNaO (2400 obs.); Bandwidth 'bw' = 0.005385
        x                 y
Min.    :-0.9854   Min.    :4.642e-05
1st Qu. :-0.8840   1st Qu. :7.014e-02
Median  :-0.7827   Median  :6.613e-01
Mean    :-0.7827   Mean    :2.465e+00
3rd Qu. :-0.6814   3rd Qu. :2.275e+00
Max.    :-0.5800   Max.    :1.472e+01
```

D.7.6   Correlation between variables

In order to look deeper into the data structure, the correlations between variables (Section 3.3.4) must be considered. The correlation matrix can be computed by the `cor()` command. It can also be visualised by plotting the data of each variable against all the others

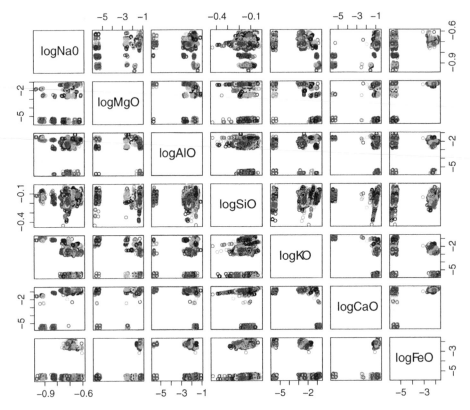

Figure D.11   Correlation between variables.

(Figure D.11) and arranging the plots in a matrix. The `round()` command rounds the data so that only a particular number of decimal places is given, which is specified in the brackets. The parameter `col = data$item` enables for assigning the colours to the points describing different items.

```
> round(cor(data[,3:9]), 2)
        logNaO logMgO logAlO logSiO logKO logCaO logFeO
logNaO    1.00   0.45  -0.08   0.08 -0.33   0.68   0.13
logMgO    0.45   1.00  -0.17  -0.08 -0.36   0.52   0.21
logAlO   -0.08  -0.17   1.00  -0.05  0.43  -0.16  -0.24
logSiO    0.08  -0.08  -0.05   1.00  0.07  -0.01   0.11
logKO    -0.33  -0.36   0.43   0.07  1.00  -0.33  -0.13
logCaO    0.68   0.52  -0.16  -0.01 -0.33   1.00   0.13
logFeO    0.13   0.21  -0.24   0.11 -0.13   0.13   1.00

> plot(data[,3:9], col = data$item)
```

It is also possible to extract one of the correlation plots, e.g. between `logMgO` and `logNaO` (Figure D.12).

```
> plot(data$logMgO, data$logNaO, col = data$item)
```

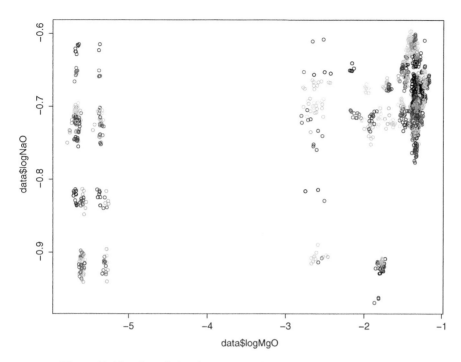

Figure D.12 Correlation between variables logMgO and logNaO.

If the value of correlation coefficient between, say, logMgO and logNaO is required, cor() is again used (Figure D.12).

```
> cor(data$logMgO, data$logNaO)
[1] 0.4524087
```

The *t*-test (Section 3.3.4) may be used to test for evidence of a non-zero correlation between a pair of normally distributed random variables. One way to test the significance of Pearson's correlation coefficient (Section 3.3.4) between the variables is to use the cor.test command.

```
> cor.test(data$logMgO, data$logNaO, alternative="two.sided", method="
    pearson", conf.level = 0.95)

        Pearson's product-moment correlation

data:  data$logMgO and data$logNaO
t = 24.8418,   df = 2398,   p-value < 2.2e-16
alternative hypothesis: true correlation is not equal to 0
95 percent confidence interval:
 0.4200000 0.4836649
sample estimates:
      cor
0.4524087
```

Section 3.4.8 discusses the interpretation of the results of using the cor.test command.

D.8   Saving data

To save text in a new file a simple command `write()` should be sufficient:

```
write("The text that has to be saved", file = "THE_FILE_NAME.txt")
```

This creates the file `THE_FILE_NAME.txt` containing the required text (`The text that has to be saved`).

To save matrices, the command `write.table()` should be used:

```
write.table(matrix, file = "THE_FILE_NAME.txt", append=TRUE)
```

The `append=TRUE` argument allows this bit of the results to be appended to the existing file content. The command `dev.copy()` enables graphics with a suitable file extension to be saved:

```
> plot(data$logMgO, data$logNaO, col = data$item)
> dev.copy(postscript, file="figure.eps")
> dev.off()
```

More detailed descriptions of all the commands and functions mentioned, as well as information on those not introduced in this Appendix, are available in Curran (2011), from the website `cran.r-project.org/doc/manuals/R-intro.pdf`, or from the **R** Console (using commands such as `?matrix`).

D.9   R codes used in Chapters 4 and 5

D.9.1   Comparison problems in casework studies

The **R** routines presented in this section[1] are suitable for solving comparison problems in casework studies for datasets such as are described in Chapter 4.

D.9.1.1   Function UC

The function UC can be found in the file `UC_comparison_calculations.R`.

```
##The function 'UC' provides the distributional parameters for likelihood
##ratio computation in comparison problems.
##For such parameters estimation it is necessary to use a database of similar
##objects described by the same variables gathered during analyses of the
##same type.
##The function calculates the within-object variability (denoted by 'U') and
##between-object variability (denoted by 'C') of the variables of interest.
```

---

[1] All the files are available from www.wiley.com/go/physicochemical

```
##The database consists of m objects for which n measurements were obtained
##as a result of the analysis.
##In the case of univariate data, the number of variables (p) is 1, whereas
##in the case of multivariate data, the number of variables is p>1.
##The files containing data MUST be organised in such a way that rows
##correspond to n measurements performed on m objects (which means nm rows)
##and columns correspond to p variables.

##Terms used in the function:
##population - a matrix (dimensions:(m*n) x p) modified from a database
##according to the jackknife procedure;
##variables - a vector indicating which columns are taken into account for
##calculations (columns depict the variables measured in the analysis).

UC = function(population, variables, p, n)
{
    items = unique(population$Item)
    m = length(items) ##'m' corresponds to the number of objects
##creating a population

    ##Defining 'S.star' and 'Sw' matrices initially filled with 0 at the
##beginning of the loops
    S.star = matrix(0, nrow = p, ncol = p)
    variable.names = colnames(population[,variables])
    rownames(S.star) = variable.names
    colnames(S.star) = variable.names

    Sw = matrix(0, nrow = p, ncol = p)
    rownames(Sw) = variable.names
    colnames(Sw) = variable.names

 ##Dealing with multivariate data (p>1)
 if (p>1)
 {
   ##'mean.all' corresponds to the vector of means of all variables
##calculated using all measurements for all objects
    mean.all = matrix(apply(population[,variables],2,mean), nrow = 1)
    colnames(mean.all) = variable.names

    ##'object.mean' corresponds to a matrix of means of all variables
##calculated using all measurements for each object
    object.mean= matrix(0, nrow = m, ncol = p)
    rownames(object.mean) = as.character(items)
    colnames(object.mean) = variable.names

    ##'i' runs through all objects from the population
      for (i in 1:m)
      {
        Sw2 = matrix(0, nrow = p, ncol = p)

        ##creating a matrix of measurements for the ith object
        ith.object = as.matrix(population[which(population$Item ==
items [i]),variables])

      ##calculating the 'object.mean'
        object.mean = matrix(apply(ith.object,2,mean),nrow = 1)
```

```
    ##'j' runs through n measurements for the chosen ith object
    for(j in 1:nrow(ith.object))
    {
        xij.minus.xi = ith.object[j,] - object.mean
        Sw2 = Sw2 + t(xij.minus.xi) %*% xij.minus.xi
    }

  ##creating 'S.star' matrix
    S.star = S.star + t(object.mean-mean.all) %*% (object.mean-
mean.all)

  ##creating 'Sw' matrix
    Sw = Sw + Sw2
  }

   ##creating 'U' and 'C' matrices
   U = Sw/(m*(n-1))
   C = S.star/(m-1)-U/n
}

##Dealing with univariate data (p=1)
if (p==1)
{
    mean.all = matrix(mean(population[,variables], nrow = 1))
    colnames(mean.all) = variable.names

    object.mean = matrix(0, nrow = m, ncol = p)
    rownames(object.mean) = as.character(items)
    colnames(object.mean) = variable.names

    for (i in 1:m)
    {
      Sw2 = matrix(0, nrow = p, ncol = p)

      ith.object = as.matrix(population[which(population$Item ==
items[i]),variables])
      object.mean = matrix(mean(ith.object), nrow = 1)
      for(j in 1:nrow(ith.object))
      {
          xij.minus.xi = ith.object[j,] - object.mean
          Sw2 = Sw2 + t(xij.minus.xi) %*% xij.minus.xi
      }

      S.star = S.star + t(object.mean-mean.all) %*% (object.mean-
mean.all)
      Sw = Sw + Sw2
    }

   U = Sw/(m*(n-1))
   C = S.star/(m-1)-U/n
 }

 result = list(U = U, C = C, mean.all = mean.all, object.mean =
object.mean, m = m)
   return (result)
}
```

D.9.1.2 Function `LR.Nor.function`

The function `LR.Nor.function` can be found in the file `LR_comparison_Nor.R`.

```
##The function 'LR.Nor.function' calculates likelihood ratio results in
##comparison problems assuming a normal between-object distribution.

LR.Nor.function = function(y.mean.1, y.mean.2, y.star, mean.all, U, C, n.1,
    n.2, p)
{
        ##Numerator calculation

        nom1 = (2*pi)^(-p)*exp(-1/2*(y.mean.1-y.mean.2) %*% solve(U/n.1+U/n
            .2) %*% t(y.mean.1-y.mean.2))*(det(U/n.1+U/n.2))^(-1/2)
        nom2 = exp(-1/2*(y.star-mean.all) %*% solve(U/(n.1+n.2)+C) %*% t(y.
            star-mean.all))*(det(U/(n.1+n.2)+C))^(-1/2)
        nom = nom1*nom2

        ##Denominator calculation

        denom1 = (2*pi)^(-p)*exp(-1/2*(y.mean.1-mean.all) %*% solve(U/n.1+C)
            %*% t(y.mean.1-mean.all))*(det(U/n.1+C))^(-1/2)
        denom2 = exp(-1/2*(y.mean.2-mean.all) %*% solve(U/n.2+C) %*% t(y.
            mean.2-mean.all))*(det(U/n.2+C))^(-1/2)

        denom = denom1*denom2

        LR.Nor = nom/denom

        result = list(LR.Nor = LR.Nor)
        return (result)
}
```

D.9.1.3 Function `LR.KDE.function`

The function `LR.KDE.function` can be found in the file `LR_comparison_KDE.R`.

```
##The function 'LR.KDE.function' calculates likelihood ratio results in
##comparison problems when KDE is used for probability density function
##estimation.

LR.KDE.function = function(y.mean.1, y.mean.2, y.star, U, C, h,
population, variables, p, m, n.1, n.2)
{
    ##Numerator calculation

  nom2 = (2*pi)^(-p/2)*exp(-1/2*(y.mean.1-y.mean.2) %*%
solve(U/n.1+U/n.2) %*% t(y.mean.1-y.mean.2))*(det(U/n.1+U/n.2))^(-1/2)

        nom1 = 0

        ##'i' runs through all objects from the population
        for(i in 1:m)
        {
          items = unique(population$Item)

          ##creating a matrix of measurements for the ith object
```

```r
        ith.object = as.matrix(population[which(population$Item ==
items[i]),variables])

        ##calculating the 'object.mean'
      object.mean = matrix(apply(ith.object,2,mean), nrow = 1)

        exp.1.1 = exp(-(y.star-object.mean) %*%
(solve(U/(n.1+n.2)+C*h^2)) %*% t(y.star-object.mean)/2)

        nom1 = nom1 + exp.1.1
    }

    nom2.1 = nom1/m

    exp.1.2 = (2*pi)^(-p/2)*det(U/(n.1+n.2)+C*h^2)^(-1/2)
    nom3 = nom2.1*exp.1.2

    nom = nom2*nom3

    ##Denominator calculation
    denom1 = 0

    for(i in 1:m)
    {
      items = unique(population$Item)
        ith.object = as.matrix(population[which(population$Item ==
items[i]),variables])
      object.mean = matrix(apply(ith.object,2,mean), nrow = 1)

        exp.2.1 = exp(-(y.mean.1-object.mean) %*%
(solve(U/n.1+C*h^2)) %*% t(y.mean.1-object.mean)/2)

        denom1 = denom1 + exp.2.1
    }
    denom2 = denom1/m

    exp.2.2 = (2*pi)^(-p/2)*det(U/n.1+C*h^2)^(-1/2)
    denom3 = denom2*exp.2.2

    denom4 = 0

    for(i in 1:m)
    {
      items = unique(population$Item)
        ith.object = as.matrix(population[which(population$Item ==
items[i]),variables])
      object.mean = matrix(apply(ith.object,2,mean),nrow = 1)

        exp.2.3 = exp(-(y.mean.2-object.mean) %*%
(solve(U/n.2+C*h^2)) %*% t(y.mean.2-object.mean)/2)

        denom4 = denom4 + exp.2.3
    }

    denom5 = denom4/m
```

```
        exp.2.4 = (2*pi)^(-p/2)*det(U/n.2+C*h^2)^(-1/2)
        denom6 = denom5*exp.2.4

        denom = denom3*denom6

        LR.KDE = nom/denom

        result = list(LR.KDE = LR.KDE)
        return (result)
}
```

D.9.1.4   Code used in casework applications

To run the code the user needs to type the command:

```
> source("comparison_casework_code.R")
```

Remember to have all the necessary functions in the same working directory. For example, see the organisation in the folders prepared for Chapters 4 and 5.

```
##Function 'source()' loads the functions needed for calculations
source("UC_comparison_calculations.R")
source("LR_comparison_Nor.R")
source("LR_comparison_KDE.R")

##Choosing the data sets containing measurements of control ('data.control')
##and recovered ('data.recovered') material as well as the relevant
##database ('population') for parameters estimation
population = read.table("glass_database.txt", header = TRUE) ##ENTER THE
##FILE NAME
data.recovered = read.table("data_recovered.txt", header = TRUE) ##ENTER THE
##FILE NAME
data.control = read.table("data_control.txt", header = TRUE) ##ENTER THE
##FILE NAME

m = length(unique(population$Item)) ##number of objects creating a database
m.control = length(unique(data.control$Item)) ##number of control objects
m.recovered = length(unique(data.recovered$Item)) ##number of recovered
##objects

##In the 'variables' it is necessary to enter the index of the column in the
##univariate problem (or columns in the multivariate problem) for the
##relevant variable(s)
variables = c(4) ##ENTER THE NUMBER OF THE COLUMN(S) THE CONSIDERED VARIABLE
## (S) IS (ARE) LOCATED IN, E.G. 4 FOR UNIVARIATE OR 4,5 FOR MULTIVARIATE
##MODELS
variables.names = colnames(population[variables])
variable.name = paste(variables.names, collapse = "_")
p = length(variables) ##number of variables considered; gives the idea of
##the problem dimensionality
n = length(unique(population$Piece)) ##number of measurements per object
##from the database

##'UC' function gives information about within- (U) and between- (C) object
##variability matrices
results.UC = UC(population, variables, p, n)
U = results.UC$U
C = results.UC$C
```

```
mean.all = results.UC$mean.all ##mean of all measurements performed on all
##objects from a database

##Calculating smoothing parameter ('h') as a bandwidth for KDE procedure
h = (4/(m*(2*p+1)))^(1/(p+4))

##Defining two matrices of LR results which are to be saved in a .txt file
##The first is related to results assuming a normal distribution, the second
##to those when KDE is used
##The matrices are organised in such a way that all the LR values for each
##comparison may be found at the intersection of the row and column marked
##by the relevant item names
output.matrix.Nor = matrix(0, ncol = m.control, nrow = m.recovered)
rownames(output.matrix.Nor) = as.character(unique(data.recovered$Name))
colnames(output.matrix.Nor) = as.character(unique(data.control$Name))

output.matrix.KDE = matrix(0, ncol = m.control, nrow = m.recovered)
rownames(output.matrix.KDE) = as.character(unique(data.recovered$Name))
colnames(output.matrix.KDE) = as.character(unique(data.control$Name))

##'i' and 'j' indices run through recovered and control objects denoted by
##'y.1' and 'y.2' with 'n.1' and 'n.2' measurements, respectively
for (i in 1:m.recovered)
{
   ##choosing 'y.1'
   y.1 = data.frame(data.recovered[which(data.recovered$Item == i),])

   for (j in 1:m.control)
   {
     ##choosing 'y.2'
     y.2 = data.frame(data.control[which(data.control$Item == j),])

     n.1 = length(y.1$Item) ##number of measurements on 'y.1'
     n.2 = length(y.2$Item) ##number of measurements on 'y.2'

     y.mean.1 = matrix(apply(as.matrix(y.1[,variables]), 2, mean), nrow = 1)
         ##mean for 'y.1'
     y.mean.2 = matrix(apply(as.matrix(y.2[,variables]), 2, mean), nrow = 1)
         ##mean for 'y.2'
     y.star = (n.1*y.mean.1+n.2*y.mean.2)/(n.1+n.2)

     ##Calculating LR when between-object variability is assumed normal (
##denoted by 'Nor')
     results.LR.Nor = LR.Nor.function(y.mean.1, y.mean.2, y.star, mean.all, U
         , C, n.1, n.2, p)
     LR.Nor = results.LR.Nor$LR.Nor

     ##Calculating the LR when between-object variability is estimated by KDE
     results.LR.KDE= LR.KDE.function(y.mean.1, y.mean.2, y.star, U, C, h,
         population, variables, p, m, n.1, n.2)
     LR.KDE = results.LR.KDE$LR.KDE

     ##Filling 'output matrix.Nor' and 'output matrix.KDE' with LR results
     output.matrix.Nor[i,j] = LR.Nor
     output.matrix.KDE[i,j] = LR.KDE
   }
}
```

```
##Saving calculations' results to a file
write(variable.name, file = paste(variable.name, "_comparison_casework_Nor.
    txt", sep=""), sep = "\t", append = TRUE)
write.table(signif(output.matrix.Nor, digits = 4), file = paste(variable.
    name, "_comparison_casework_Nor.txt", sep=""), quote = FALSE, sep = "\t"
    , col.names = TRUE, row.names = TRUE, dec = ".", append = TRUE)
write(variable.name, file = paste(variable.name, "_comparison_casework_KDE.
    txt", sep=""), sep = "\t", append = TRUE)
write.table(signif(output.matrix.KDE, digits = 4), file = paste(variable.
    name, "_comparison_casework_KDE.txt", sep=""), quote = FALSE, sep = "\t"
    , col.names = TRUE, row.names = TRUE, dec = ".", append = TRUE)

##Graphical presentation of descriptive statistics (box-plots, Q-Q plots,
##KDE density functions)
layout(matrix(c(rep(1, times = p), seq(from = 2, to = 1+2*p), 1), p, 3),
    widths = c(1.8,1,1))

##Box-plots
names = c()
for(k in 1:p) names = c(names, c(variables.names[k]))
if (p == 1) {boxplot(population[,variables], data.recovered[,variables],
    data.control[,variables], names = c(paste("pop_", names, sep=""), paste(
    "rec_", names, sep=""), paste("contr_", names, sep="")), cex.xlab = 0.5)
    } else
{boxplot(c(population[,variables], data.recovered[,variables], data.control
    [,variables]), names = c(paste("pop_", names, sep=""), paste("rec_",
    names, sep=""), paste("contr_", names, sep="")), cex.xlab = 0.5)}

##Q-Q plots
for (k in 1:p) qqnorm(sapply(split(population[,variables[k]],population$Name),
    mean), main = variables.names[k])

##KDE probability density
install.packages("KernSmooth")
library(KernSmooth)
p = 1
h = (4/(m*(2*p+1)))^(1/(p+4))
for (k in 1:length(variables))
plot(bkde(sapply(split(population[,variables[k]],population$Name),mean),
    kernel = "normal", bandwidth = h), type = "l", ylab = "probability
    density", xlab = "data", main = variables.names[k])

##Saving graphics to a .eps file
dev.copy(postscript, paste(variable.name,"_casework_descriptive_statistics.
    eps", sep=""))
dev.off()
```

D.9.2   Comparison problems in research studies

The **R** routines presented in this section are suitable for solving comparison problems in research studies for datasets such as are described in Chapter 4.

The code in this section is used to determine false answer rates. To run the code the user needs to type the command:

```
> source("comparison_research_code.R")
```

Remember to have all the necessary functions (along with UC, LR.Nor.function, and LR.KDE.function given in Section D.9.1) in the same working directory. For example, see the organisation in the folders prepared for Chapters 4 and 5.

```
##Loading the functions needed for calculations
source("UC_comparison_calculations.R")
source("LR_comparison_Nor.R")
source("LR_comparison_KDE.R")

##Choosing the dataset ('data.analysed') on which calculations are carried
##out as well as the relevant database for parameters estimation ('
##population')
population = read.table("glass_database.txt", header = TRUE) ##ENTER THE
##FILE NAME
data.analysed = read.table("glass_database.txt", header = TRUE) ##ENTER THE
##FILE NAME

##'m.all.analysed' corresponds to the number of objects on which
##calculations are carried out
m.all.analysed = length(unique(data.analysed$Item))

##In the 'variables' it is necessary to enter the index of the column in the
##univariate problem (or columns in the multivariate problem) for the
##relevant variable(s)
variables = c(4) ##ENTER THE NUMBER OF THE COLUMN(S) THE VARIABLE(S) IS (ARE
##) LOCATED IN, E.G. 4 FOR UNIVARIATE OR 4,5 FOR MULTIVARIATE MODELS
variables.names = colnames(data.analysed[variables])
variable.name = paste(variables.names, collapse = "_")
p = length(variables) ##number of variables considered; gives the idea of
##the problem dimensionality
n = length(unique(data.analysed$Piece)) ##number of measurements per object
fp.Nor = 0
fp.KDE = 0
fn.Nor = 0
fn.KDE = 0

##Defining two matrices of LR results which are to be saved in a .txt file
##The first is related to results assuming a normal distribution, the second
##to those when KDE is used
##The matrices are organised in such a way that all the LR values for each
##comparison may be found at the intersection of the row and column marked
##by the relevant item names
output.matrix.Nor = matrix(0,ncol = m.all.analysed, nrow = m.all.analysed)
rownames(output.matrix.Nor) = as.character(unique(data.analysed$Name))
colnames(output.matrix.Nor) = as.character(unique(data.analysed$Name))

output.matrix.KDE = matrix(0,ncol = m.all.analysed, nrow = m.all.analysed)
rownames(output.matrix.KDE) = as.character(unique(data.analysed$Name))
colnames(output.matrix.KDE) = as.character(unique(data.analysed$Name))

##'i' and 'j' indices run through all the compared items denoted by 'y.1' and
##'y.2' with 'n.1' and 'n.2' measurements, respectively
##'i' is always smaller than or equal to 'j' so as to avoid repeating
##calculations
for (i in 1:m.all.analysed)
```

```
{
  for (j in i:m.all.analysed)
  {
    if (i == j) ##When i = j, then the samples from the same object are
##compared (delivering the rates of false negative answers)
    {
      y.1.2 = data.analysed[which(data.analysed$Item == i),] ##choosing an
##object described by n measurements which will be divided into two
##samples 'y.1' and 'y.2'
      y.1 = data.frame(y.1.2[1:6,]) ##choosing 'y.1' ##ENTER THE NUMBER OF
##THE ROW(S) CREATING THE CONTROL SAMPLE, E.G. 1:6
      y.2 = data.frame(y.1.2[7:12,]) ##choosing 'y.2' ##ENTER THE NUMBER OF
##THE ROW(S) CREATING THE RECOVERED SAMPLE, E.G. 7:12

      population = data.analysed[which(data.analysed$Item != i),] ##choosing
##the relevant population by employing the jackknife procedure
      m = length(unique(population$Item)) ##number of objects creating a
##database
    }
    else ##When i is not equal to j, samples from different objects are
##compared (giving the false positive answer rates)
    {
      ##choosing 'y.1'
      y.1 = data.frame(data.analysed[which(data.analysed$Item == i),])
      ##choosing 'y.2'
      y.2 = data.frame(data.analysed[which(data.analysed$Item == j),])

      population = data.analysed[which(data.analysed$Item != i),]
      population = population[which(population$Item != j),] ##choosing the
##relevant population by employing the jackknife procedure
      m = length(unique(population$Item)) ##number of objects creating a
##database
    }

    ##'UC' function gives information about within- (U) and between- (C)
##object variability matrices
    results.UC = UC(population, variables, p, n)
    U = results.UC$U
    C = results.UC$C
    mean.all = results.UC$mean.all ##mean of all measurements performed on
##all objects from a database

    n.1 = length(y.1$Item) ##number of measurements performed on 'y.1'
    n.2 = length(y.2$Item) ##number of measurements performed on 'y.2'

    y.mean.1 = matrix(apply(as.matrix(y.1[,variables]), 2, mean), nrow = 1)
        ##mean for 'y.1'
    y.mean.2 = matrix(apply(as.matrix(y.2[,variables]), 2, mean), nrow = 1)
        ##mean for 'y.2'
    y.star = (n.1*y.mean.1+n.2*y.mean.2)/(n.1+n.2)

    ##Calculating smoothing parameter ('h') as a bandwidth for KDE procedure
    h = (4/(m*(2*p+1)))^(1/(p+4))

    ##Calculating LR when between-object variability is assumed normal (
##denoted by 'Nor')
```

```
    results.LR.Nor = LR.Nor.function(y.mean.1, y.mean.2, y.star, mean.all, U
        , C, n.1, n.2, p)
    LR.Nor = results.LR.Nor$LR.Nor

    ##Calculating the LR when between-object variability is estimated by KDE
    results.LR.KDE= LR.KDE.function(y.mean.1, y.mean.2, y.star, U, C, h,
        population, variables, p, m, n.1, n.2)
    LR.KDE = results.LR.KDE$LR.KDE

    ##Filling 'output matrix.Nor' and 'output.matrix.KDE' with LR results
    output.matrix.Nor[j,i] = LR.Nor
    output.matrix.KDE[j,i] = LR.KDE

    if(i != j & LR.Nor > 1) {fp.Nor = fp.Nor + 1} ##false positive answers (
##fp) when Nor is assumed
    if(i != j & LR.KDE > 1) {fp.KDE = fp.KDE + 1} ##false positive answers (
##fp) when KDE is used
    if(i == j & LR.Nor < 1) {fn.Nor = fn.Nor + 1} ##false negative answers (
##fn) when Nor is assumed
    if(i == j & LR.KDE < 1) {fn.KDE = fn.KDE + 1} ##false negative answers (
##fn) when KDE is used
    }
}

##Saving calculations' results to a file
write.table(signif(output.matrix.Nor, digits = 4), file = paste(variable.
    name,"comparison_research_Nor.txt", sep=""), quote = FALSE, sep = "\t",
    col.names = TRUE, row.names = TRUE, dec = ".", append = TRUE)
write.table(signif(output.matrix.KDE, digits = 4), file = paste(variable.
    name,"comparison_research_KDE.txt", sep=""), quote = FALSE, sep = "\t",
    col.names = TRUE, row.names = TRUE, dec = ".", append = TRUE)

##This part of code transforms the way the results are displayed (from an m
##x m matrix to one column of results)

fp.all = m.all.analysed*(m.all.analysed-1)/2 ##number of all possible
##results for different objects comparisons
LR.different.Nor = matrix(0, nrow=fp.all, ncol=1)
LR.different.KDE = matrix(0, nrow=fp.all, ncol=1)
q=0

for(s in 1:(m.all.analysed-1))
{
  for(r in (s+1):m.all.analysed)
  {
    q = 1 + q
    LR.different.Nor[q,1] = output.matrix.Nor[r,s]
    LR.different.KDE[q,1] = output.matrix.KDE[r,s]
  }
}
write.table(signif(LR.different.Nor, digits = 4), file = paste(variable.name
    ,"_comparison_research_Nor_different.txt", sep=""), quote = FALSE, sep =
    " ", col.names = TRUE, row.names = TRUE, dec = ".")
write.table(signif(LR.different.KDE, digits = 4), file = paste(variable.name
    ,"_comparison_research_KDE_different.txt", sep=""), quote = FALSE, sep =
    " ", col.names = TRUE, row.names = TRUE, dec = ".")
```

```
LR.same.Nor = matrix(0, nrow=m.all.analysed, ncol=1)
LR.same.KDE = matrix(0, nrow=m.all.analysed, ncol=1)

for(t in 1:m.all.analysed)
{
  LR.same.Nor[t,1] = output.matrix.Nor[t,t]
  LR.same.KDE[t,1] = output.matrix.KDE[t,t]
}
write.table(signif(LR.same.Nor, digits = 4), file = paste(variable.name,"_
    comparison_research_Nor_same.txt", sep=""), quote = FALSE, sep = "\t",
    col.names = TRUE, row.names = TRUE, dec = ".")
write.table(signif(LR.same.KDE, digits = 4), file = paste(variable.name,"_
    comparison_research_KDE_same.txt", sep=""), quote = FALSE, sep = "\t",
    col.names = TRUE, row.names = TRUE, dec = ".")

##Providing summary results expressed by levels of false positive and false
##negative model responses

error_rates = matrix(0, nrow = 4, ncol = 1)
rownames(error_rates) = c("fp_Nor", "fp_KDE", "fn_Nor", "fn_KDE")
colnames(error_rates) = variable.name
error_rates[1,1] = fp.Nor/fp.all*100
error_rates[2,1] = fp.KDE/fp.all*100
error_rates[3,1] = fn.Nor/m.all.analysed*100
error_rates[4,1] = fn.KDE/m.all.analysed*100
write.table(signif(error_rates, digits = 3), file = "comparison_research_
    error_rate.txt", append = TRUE, quote = FALSE, sep = "\t", col.names =
    TRUE, row.names = TRUE, dec = ".")

##Graphical presentation of descriptive statistics (box-plots, Q-Q plots,
##KDE density functions) and LR value distributions
layout(matrix(c(rep(1,times=p),seq(2,(2*p+1))), 3, p, byrow=TRUE))

##Box-plots
boxplot(data.analysed[,variables], ylab = "data")

##Q-Q plots
for (k in 1:p) qqnorm(sapply(split(data.analysed[,variables[k]],data.
    analysed$Name),mean), main = variables.names[k])
##KDE probability density
install.packages("KernSmooth")
require(KernSmooth)
p = 1
h = (4/(m.all.analysed*(2*p+1)))^(1/(p+4))
for (k in 1:length(variables))
plot(bkde(sapply(split(data.analysed[,variables[k]],data.analysed$Name),mean),
    kernel = "normal", bandwidth=h), type = "l", ylab = "probability
    density", xlab = "data", main = variables.names[k])

##Saving graphics to a .eps file
dev.copy(postscript, paste(variable.name,"_research_descriptive_statistics.
    eps", sep=""))
dev.off()

par(mfrow = c(2,2))

##Histograms illustrating LR distributions
```

```
hist(log10(LR.different.Nor), main = expression(paste("true-",H[d]," LR
    distribution assuming Nor", sep="")), col = "gray", xlab = "logLR",
    breaks = 50, cex.main = 0.7)
hist(log10(LR.different.KDE), main = expression(paste("true-",H[d]," LR
    distribution assuming KDE", sep="")), col = "gray", xlab = "logLR",
    breaks = 50, cex.main = 0.7)
hist(log10(LR.same.Nor), main = expression(paste("true-",H[p]," LR
    distribution assuming Nor", sep="")), col = "gray", xlab = "logLR",
    breaks = 50, cex.main = 0.7)
hist(log10(LR.same.KDE), main = expression(paste("true-",H[p]," LR
    distribution assuming KDE", sep="")), col = "gray", xlab = "logLR",
    breaks = 50, cex.main = 0.7)

##Saving graphics to a .eps file
dev.copy(postscript, paste(variable.name,"_research_LR_distribution.eps",
    sep=""))
dev.off()
```

D.9.3   Classification problems in casework studies

The **R** routines presented in this section are suitable for solving classification problems in casework studies for datasets such as are described in Chapter 5.

D.9.3.1   Function UC.1

The function UC.1 for two-level models is available in the file UC1_two_level_calculations.R.

```
##The functions 'UC.1' and 'UC.2' provide the distributional parameters for
##likelihood ratio computation in classification problems.
##For such parameters estimation it is necessary to use a database of similar
##objects described by the same variables gathered during analyses of the
##same type.
##The function calculates between-object variability (denoted by 'C') and
##within-object variability (denoted by 'U') of the relevant variables, which
##is a two-level approach.
##The population.1 consists of m.1 objects for which n measurements were
##obtained as a result of the analysis.
##In the case of univariate data, the number of variables (p) is 1, whereas
##in the case of multivariate data, the number of variables is p>1.
##The files containing data MUST be organised in such a way that rows
##correspond to n measurements performed on m objects (which means nm rows
##) and columns correspond to p variables.
##Terms used in the function:
##population - a matrix (dimensions:(m*n) x p), a modification of a database
##according to jackknife procedure;
##variables - a vector indicating which columns are taken into account for
##calculations (columns depict the variables measured in the analysis);
##'.1' indicates that the values calculated are based on the results of the
##analysis of objects from category 1.

UC.1 = function(population.1, variables, p)
{
```

```
      items.1 = unique(population.1$Item)
      m.1 = length(items.1) ##'m.1' corresponds to the number of objects
##creating population.1


      ##Defining 'S.star' and 'Sw' matrices initially filled with 0 at the
##beginning of the loops
      S.star.1 = matrix(0, nrow = p, ncol = p)
      variable.names = colnames(population.1[,variables])
      rownames(S.star.1) = variable.names
      colnames(S.star.1) = variable.names

  Sw.1 = matrix(0, nrow = p, ncol = p)
      rownames(Sw.1) = variable.names
      colnames(Sw.1) = variable.names


      ##Dealing with multivariate data (p>1)
      if (p>1)
      {
         ##'mean.all.1' corresponds to the vector of means of all
##variables calculated using all measurements for all objects belonging to
##category 1
                mean.all.1 = matrix(apply(population.1[,variables],2,mean),
nrow = 1)
                colnames(mean.all.1) = variable.names


                ##'object.mean.1' corresponds to a matrix of means of all
##variables calculated using all measurements for each object from category 1
                object.mean.1 = matrix(0, nrow = m.1, ncol = p)
                rownames(object.mean.1) = as.character(items.1)
                colnames(object.mean.1) = variable.names


                ##'i' runs through all objects from the population.1
                for (i in 1:m.1)
                {
                      ##creating a matrix of measurements for the ith object
                      ith.object.1 =
as.matrix(population.1[which(population.1$Item == items.1[i]),variables])


                      ##calculating the mean for each object from category 1
                      object.mean.1 = matrix(apply(ith.object.1,2,mean), nrow
= 1)

      ##'j' runs through all measurements for a particular ith.object.1
                      for(j in 1:nrow(ith.object.1))
                      {
                            xij.minus.xi.1 = ith.object.1[j,] - object.mean.1
                            Sw.1 = Sw.1 + t(xij.minus.xi.1) %*% xij.minus.xi.1
                      }


                      S.star.1 = S.star.1 + t(object.mean.1-mean.all.1) %*%
(object.mean.1-mean.all.1)
                }
      ##creating 'U' and 'C' matrices
      U.1 = Sw.1/(m.1*(n-1))
      C.1 = S.star.1/(m.1-1)-U.1/n
      }
```

```
    ##Dealing with univariate data (p=1)
    if (p==1)
    {
        mean.all.1 = matrix(mean(population.1[,variables], nrow = 1))
        colnames(mean.all.1) = variable.names

        object.mean.1 = matrix(0, nrow = m.1, ncol = p)
        rownames(object.mean.1) = as.character(items.1)
        colnames(object.mean.1) = variable.names

        for (i in 1:m.1)
        {
            ith.object.1 =
as.matrix(population.1[which(population.1$Item == items.1[i]),variables])
            object.mean.1 = matrix(mean(ith.object.1), nrow = 1)

            for(j in 1:nrow(ith.object.1))
            {
                xij.minus.xi.1 = ith.object.1[j,] - object.mean.1
                Sw.1 = Sw.1 + t(xij.minus.xi.1) %*% xij.minus.xi.1
            }

            S.star.1 = S.star.1 + t(object.mean.1-mean.all.1) %*%
(object.mean.1-mean.all.1)
        }

      U.1 = Sw.1/(m.1*(n-1))
      C.1 = S.star.1/(m.1-1)-U.1/n
    }

    result = list(U.1 = U.1, C.1 = C.1, mean.all.1 = mean.all.1,
object.mean.1 = object.mean.1, m.1 = m.1)
    return (result)
}
```

D.9.3.2 Function LR.Nor.function

The function LR.Nor.function is available in the file LR_classification_two_level_Nor.R.

##The function 'LR.Nor.function' calculates LR for two-level classification
##problems assuming a normal between-object data distrubution

```
LR.Nor.function = function(y.mean, mean.all.1, mean.all.2, U.1, U.2, C.1,
C.2, p)
{
    ##Numerator calculation
  nom = (2*pi)^(-p/2)*det(U.1/n+C.1)^(-1/2)*exp(-1/2*(y.mean-mean.all.1)
%*% solve(U.1/n+C.1) %*% t(y.mean-mean.all.1))

    ##Denominator calculation
  denom = (2*pi)^(-p/2)*det(U.2/n+C.2)^(-1/2)*exp(-1/2*(y.mean-
mean.all.2) %*% solve(U.2/n+C.2) %*% t(y.mean-mean.all.2))

    LR.Nor = nom/denom
```

```
    result = list(LR.Nor = LR.Nor)
    return (result)
}
```

D.9.3.3 Function `LR.KDE.function`

The function `LR.KDE.function` is available in the file `LR_classification_two_level_KDE.R`.

```
##The function 'LR.KDE.function' calculates LR for two-level
##classification problems when KDE is used for probability density function
##estimation.

LR.KDE.function = function(population.1, population.2, variables, y.mean,
U.1, U.2, C.1, C.2, h.1, h.2, p, m.1, m.2)
{
    ##Numerator calculation
    nom1 = 0

    ##'i' runs through all objects from the population.1
    for(i in 1:m.1)
  {
      items.1 = unique(population.1$Item)

        ##creating a matrix of measurements for the ith object
        ith.object.1 = as.matrix(population.1[which(population.1$Item
== items.1[i]),variables])

        ##calculating the 'object.mean.1'
      object.mean.1 = matrix(apply(ith.object.1,2,mean), nrow = 1)

        exp.1.1 = exp(-(y.mean-object.mean.1) %*%
solve(U.1/n+C.1*h.1^2) %*% t(y.mean-object.mean.1)/2)

        nom1 = nom1 + exp.1.1
  }

    nom2 = nom1/m.1

    exp.1.2 = (2*pi)^(-p/2) * (det(U.1/n+C.1*h.1^2))^(-1/2)
    nom = nom2*exp.1.2

    ##Denominator calculation
    denom1 = 0

    for(i in 1:m.2)
    {
      items.2 = unique(population.2$Item)
        ith.object.2 = as.matrix(population.2[which(population.2$Item
== items.2[i]),variables])
      object.mean.2 = matrix(apply(ith.object.2,2,mean), nrow = 1)

        exp.2.1 = exp(-(y.mean-object.mean.2) %*%
solve(U.2/n+C.2*h.2^2) %*% t(y.mean-object.mean.2)/2)
```

```
            denom1 = denom1 + exp.2.1
    }

  denom2 = denom1/m.2

    exp.2.2 = (2*pi)^(-p/2) * (det(U.2/n+C.2*h.2^2))^(-1/2)
    denom = denom2*exp.2.2

    LR.KDE = nom/denom

    result = list(LR.KDE = LR.KDE)
    return (result)
}
```

D.9.3.4   Code used in casework applications

To run the code the user needs to type the command:

```
> source("classification_casework_code.R")
```

Remember to have all the necessary functions in the same working directory. For example, see the organisation in the folders prepared for Chapters 4 and 5.

```
##Choosing the relevant dataset and database for parameters estimation
data = read.table("evidence.txt", header = TRUE) ##ENTER THE FILE NAME
population = read.table("cwp_database.txt", header = TRUE) ##ENTER THE FILE
##NAME

## Establishing categories
category.1.name = unique(population$Factor)[1]
category.2.name = unique(population$Factor)[2]
categories = paste(unique(population$Factor), collapse = "_")

##In the 'variables' it is necessary to enter the index of the column in the
##univariate problem (or columns in the multivariate problem) for the
##relevant variable(s)
variables = c(6) ##ENTER THE NUMBER OF THE COLUMN(S) THE RELEVANT VARIABLE(S
##) IS (ARE) LOCATED IN, E.G. 4 FOR UNIVARIATE OR 4,5 FOR MULTIVARIATE
##MODELS
variables.names = colnames(population[variables])
variable.name = paste(variables.names, collapse = "_")
m.all.analysed = length(unique(data$Item)) ##'m.all.analysed' corresponds to
##the number of all objects subjected to calculations
p = length(variables) ##number of variables considered; gives the idea of
##the problem dimensionality
n = length(unique(population$Piece)) ##number of measurements per object

##'population.1' and 'population.2' refer to the populations of objects
##belonging to category 1 and 2 respectively
population.1 = population[which(population$Factor == category.1.name),]
population.2 = population[which(population$Factor == category.2.name),]

##Function 'source()' loads the functions needed for calculations
if (n == 1){
  source("UC1_one_level_calculations.r")
  source("UC2_one_level_calculations.r")
```

```
  source("LR_classification_one_level_Nor.r")
  source("LR_classification_one_level_KDE.r")
} else {
  source("UC1_two_level_calculations.r")
  source("UC2_two_level_calculations.r")
  source("LR_classification_two_level_Nor.r")
  source("LR_classification_two_level_KDE.r")
}


##'UC' function provides information on within- (U) and between- (C) object
##variability matrices
results.UC.1 = UC.1(population.1, variables, p)
U.1 = results.UC.1$U.1
C.1 = results.UC.1$C.1
mean.all.1 = results.UC.1$mean.all.1 ##mean of all measurements performed on
##all objects from population.1
m.1 = results.UC.1$m.1 ##number of objects in population.1

results.UC.2 = UC.2(population.2, variables, p)
U.2 = results.UC.2$U.2
C.2 = results.UC.2$C.2
mean.all.2 = results.UC.2$mean.all.2
m.2 = results.UC.2$m.2


##Calculating smoothing parameters ('h.1' and 'h.2') as bandwidths for KDE
##procedure
h.1 = (4/(m.1*(2*p+1)))^(1/(p+4))
h.2 = (4/(m.2*(2*p+1)))^(1/(p+4))


##Defining matrices of LR results which are to be saved in a .txt file
output.matrix.Nor = matrix(0, ncol = 2, nrow = m.all.analysed)
rownames(output.matrix.Nor) = unique(data$Name)
colnames(output.matrix.Nor) = c(paste(variable.name,"_Nor", sep=""), "
    category_Nor")

output.matrix.KDE = matrix(0, ncol = 2, nrow = m.all.analysed)
rownames(output.matrix.KDE) = unique(data$Name)
colnames(output.matrix.KDE) = c(paste(variable.name,"_KDE", sep=""), "
    category_KDE")


##'i' index runs through all the objects denoted by 'analysed.object'
for (i in 1:m.all.analysed)
{
      analysed.object = data[which(data$Item == i),]

      ##Calculating parameters for the 'analysed.object', which is to be
##classified
      y = data.frame(analysed.object[,variables])
      y.mean = matrix(apply(y,2,mean), nrow = 1) ##mean for the analysed
##object

      ##Calculating LR when between-object distribution is assumed normal
## (denoted by 'Nor')
      results.LR.Nor = LR.Nor.function(y.mean, mean.all.1, mean.all.2,
U.1, U.2, C.1, C.2, p)
      LR.Nor = results.LR.Nor$LR.Nor
```

```
    ##Calculating the LR when between-object distribution is estimated by KDE
    results.LR.KDE = LR.KDE.function(population.1, population.2,
variables, y.mean, U.1, U.2, C.1, C.2, h.1, h.2, p, m.1, m.2)
    LR.KDE = results.LR.KDE$LR.KDE

    ##Filling 'output.matrix.Nor' and 'output.matrix.KDE' with LR results
    output.matrix.Nor[i,1] = signif(LR.Nor, digits = 4)
    output.matrix.KDE[i,1] = signif(LR.KDE, digits = 4)

  if (LR.Nor > 1) {output.matrix.Nor[i,2] = as.character(category.1.name)}
  else {output.matrix.Nor[i,2] = as.character(category.2.name)}
  if (LR.KDE > 1) {output.matrix.KDE[i,2] = as.character(category.1.name)}
  else {output.matrix.KDE[i,2] = as.character(category.2.name)}
}

##Saving results to a file
write.table(output.matrix.Nor, file = paste(variable.name, "_", categories,"
    _classification_casework_Nor.txt", sep=""), quote = FALSE, sep = "\t",
    col.names = TRUE, row.names = TRUE, dec = ".", append = TRUE)
write.table(output.matrix.KDE, file = paste(variable.name, "_", categories,"
    _classification_casework_KDE.txt", sep=""), quote = FALSE, sep = "\t",
    col.names = TRUE, row.names = TRUE, dec = ".", append = TRUE)

##Graphical presentation of descriptive statistics (box-plots, Q-Q plots,
##KDE density functions)
layout(matrix(c(1:6), 2, 3))

for (k in 1:length(variables))
{
  ##Box-plots
  boxplot(split(data[,variables[k]], data$Name), names=c(as.character(unique
      (data$Name))),las = 2, cex.xlab = 0.5)
  boxplot(population.1[,variables[k]],population.2[,variables[k]], names=c("
      pop.1","pop.2"),las = 2, cex.xlab = 0.5)

  ##Q-Q plots
  qqnorm(sapply(split(population.1[,variables[k]],population.1$Item),mean),
    main = paste(variables.names[k],"_",category.1.name,sep=""),
    cex.main = 0.9)
  qqnorm(sapply(split(population.2[,variables[k]],population.2$Item),mean),
    main = paste(variables.names[k],"_",category.2.name,sep=""),
    cex.main = 0.9)

  ##KDE probability density
install.packages("KernSmooth")
  require(KernSmooth)
  p = 1
  h.1 = (4/(m.1*(2*p+1)))^(1/(p+4))
  h.2 = (4/(m.2*(2*p+1)))^(1/(p+4))
  plot(bkde(sapply(split(population.1[,variables[k]],population.1$Item),mean),
    kernel = "normal", bandwidth=h.1), type = "l", ylab = "probability
    density", xlab = "data-category 1", main = paste(variables.names[k],
    "_",category.1.name,sep=""), cex.main = 0.9)
  plot(bkde(sapply(split(population.2[,variables[k]],population.2$Item),mean),
    kernel = "normal", bandwidth=h.2), type = "l", ylab = "probability
    density", xlab = "data-category 2", main = paste(variables.names[k],
    "_",category.2.name,sep=""), cex.main = 0.9)
```

```
##Saving graphics to a .eps file
dev.copy(postscript, paste(variables.names[k], "_",categories,"_casework_
    descriptive_statistics.eps", sep=""))
dev.off()
}
```

D.9.4   Classification problems in research studies

The **R** routines presented in this section are suitable for solving classification problems in research studies for datasets such as are described in Chapter 5.

To run the code the user needs to type the command:

```
> source("classification_research_code.R")
```

Remember to have all the necessary functions (along with UC.1, UC.2, LR.Nor. function, and LR.KDE.function given in Section D.9.3) in the same working directory. For example, see the organisation in the folders prepared for Chapters 4 and 5.

```
##Choosing the dataset on which calculations are carried out
data = read.table("cwp_database.txt", header = TRUE) ##ENTER THE FILE NAME

##Establishing categories
category.1.name = unique(data$Factor)[1]
category.2.name = unique(data$Factor)[2]
categories = paste(unique(data$Factor), collapse = "_")

##In the 'variables' it is necessary to enter the index of the column in the
##univariate problem (or columns in the multivariate problem) for the
##relevant variable(s)
variables = c(6) ##ENTER THE NUMBER OF THE COLUMN(S) THE CONSIDERED VARIABLE
##(S) IS (ARE) LOCATED IN, E.G. 4 FOR UNIVARIATE OR 4,5 FOR MULTIVARIATE
##MODELS
variables.names = colnames(data[variables])
variable.name = paste(variables.names, collapse = "_")

m.all.analysed = length(unique(data$Item)) ##'m.all.analysed' corresponds to
##the number of all objects subjected to calculations
p = length(variables) ##number of variables considered; gives the idea of
##the problem dimensionality
n = length(unique(data$Piece)) ##number of measurements per object

category.column = data$Factor[seq(from = 1, to = length(data$Item), by = n)]

##Loading the functions needed for calculations
if (n == 1){
  source("UC1_one_level_calculations.r")
  source("UC2_one_level_calculations.r")
  source("LR_classification_one_level_Nor.r")
  source("LR_classification_one_level_KDE.r")
} else {
  source("UC1_two_level_calculations.r")
  source("UC2_two_level_calculations.r")
```

```
  source("LR_classification_two_level_Nor.r")
  source("LR_classification_two_level_KDE.r")
}
##Defining matrices of LR results which will be saved in a .txt file
output.matrix.Nor = matrix(0, ncol = 1, nrow = m.all.analysed)
rownames(output.matrix.Nor) = c(unique(data$Item))
colnames(output.matrix.Nor) = c(paste(variable.name,"_Nor", sep=""))

output.matrix.KDE = matrix(0, ncol = 1, nrow = m.all.analysed)
rownames(output.matrix.KDE) = c(unique(data$Item))
colnames(output.matrix.KDE) = c(paste(variable.name,"_KDE", sep=""))

##'i' index runs through all the objects denoted by 'analysed.object'
for (i in 1:m.all.analysed)
{
    analysed.object = data[which(data$Item == i),]
    population = data[which(data$Item != i),] ##choosing relevant
##population by employing jackknife procedure

    ##'population.1' and 'population.2' refer to the populations of
##objects belonging to category 1 and 2 respectively
    population.1 = population[which(population$Factor ==
category.1.name),]
    population.2 = population[which(population$Factor ==
category.2.name),]

    ##'UC' function delivers information about within- (U) and between-
##(C) object variability matrices
    results.UC.1 = UC.1(population.1, variables, p)
    U.1 = results.UC.1$U.1
    C.1 = results.UC.1$C.1
    mean.all.1 = results.UC.1$mean.all.1 ##mean of all measurements
##performed on all objects from population.1
    m.1 = results.UC.1$m.1 ##number of objects in population.1

    results.UC.2 = UC.2(population.2, variables, p)
    U.2 = results.UC.2$U.2
    C.2 = results.UC.2$C.2
    mean.all.2 = results.UC.2$mean.all.2
    m.2 = results.UC.2$m.2

    ##Calculating smoothing parameters ('h.1' and 'h.2') as bandwidths
##for KDE procedure
    h.1 = (4/(m.1*(2*p+1)))^(1/(p+4))
    h.2 = (4/(m.2*(2*p+1)))^(1/(p+4))

    ##Calculating parameters for the 'analysed.object', which is to be
##classified
    y = data.frame(analysed.object[,variables])
    y.mean = matrix(apply(y,2,mean), nrow = 1) ##mean for the analysed
##object

    ##Calculating LR when between-object distribution is assumed normal
##(denoted by 'Nor')
    results.LR.Nor = LR.Nor.function(y.mean, mean.all.1, mean.all.2,
U.1, U.2, C.1, C.2, p)
```

```
     LR.Nor = results.LR.Nor$LR.Nor

     ##Calculating the LR when between-object distribution is estimated by KDE
       results.LR.KDE = LR.KDE.function(population.1, population.2,
variables, y.mean, U.1, U.2, C.1, C.2, h.1, h.2, p, m.1, m.2)
   LR.KDE = results.LR.KDE$LR.KDE

       ##Filling 'output.matrix.Nor' and 'output.matrix.KDE' with LR
##results
       output.matrix.Nor[i,1] = LR.Nor
       output.matrix.KDE[i,1] = LR.KDE

}

##Saving results to a file in the form of a 'summary.matrix'
summary.matrix = matrix(ncol = 2, nrow = 3)
rownames(summary.matrix) =c(paste("correct in ",category.1.name, sep=""),
     paste("correct in ", category.2.name, sep=""),"all correct")
colnames(summary.matrix) = c(paste(variable.name,"_Nor", sep=""), paste(
     variable.name,"_KDE", sep=""))

d.1 = length(which(data$Factor == category.1.name))/unique(table(population
     .1$Item))
d.2 = length(which(data$Factor == category.2.name))/unique(table(population
     .1$Item))
##Calculating the percentage of correct classifications of objects coming
##from category 1 assuming Nor
correct.Nor.1.percent = length(which(output.matrix.Nor[1:d.1,1] > 1))/d.1*
     100
summary.matrix[1,1] = correct.Nor.1.percent

##Calculating the percentage of correct classifications of objects coming
##from category 2 assuming Nor
correct.Nor.2.percent = length(which(output.matrix.Nor[(d.1+1):(d.1+d.2), 1]
     < 1))/d.2*100
summary.matrix[2,1] = correct.Nor.2.percent

##Calculating the percentage of correct classifications of objects coming
##from category 1 using KDE
correct.KDE.1.percent = length(which(output.matrix.KDE[1:d.1,1] > 1))/d.1*
     100
summary.matrix[1,2] = correct.KDE.1.percent

##Calculating the percentage of correct classifications of objects coming
##from category 2 using KDE
correct.KDE.2.percent = length(which(output.matrix.KDE[(d.1+1):(d.1+d.2),1]
     < 1))/d.2*100
summary.matrix[2,2] = correct.KDE.2.percent

##Calculating the percentage of correct classifications of all objects
##assuming Nor
correct.Nor.percent = (length(which(output.matrix.Nor[1:d.1,1] > 1)) +
     length(which(output.matrix.Nor[(d.1+1):(d.1+d.2), 1] < 1)))/(d.1+d.2)*
     100
summary.matrix[3,1] = correct.Nor.percent
```

```
##Calculating the percentage of correct classifications of all objects using
##KDE
correct.KDE.percent = (length(which(output.matrix.KDE[1:d.1,1] > 1)) +
    length(which(output.matrix.KDE[(d.1+1):(d.1+d.2), 1] < 1)))/(d.1+d.2)*
    100
summary.matrix[3,2] = correct.KDE.percent

write.table(cbind(signif(output.matrix.Nor, digits = 4), matrix(category.
    column, ncol=1)), file = paste(variable.name, "_", categories,"_
    classification_research_Nor.txt", sep=""), quote = FALSE, sep = "\t",
    col.names = TRUE, row.names = TRUE, dec = ".", append = TRUE)
write.table(cbind(signif(output.matrix.KDE, digits = 4), matrix(category.
    column, ncol=1)), file = paste(variable.name, "_", categories,"_
    classification_research_KDE.txt", sep=""), quote = FALSE, sep = "\t",
    col.names = TRUE, row.names = TRUE, dec = ".", append = TRUE)
write.table(signif(summary.matrix, digits = 3), file = paste(categories,"_
    classification_research_correct_rate.txt", sep=""), quote = FALSE, sep =
    "\t", col.names = TRUE, row.names = TRUE, dec = ".", append = TRUE)

##Graphical presentation of descriptive statistics (box-plots, Q-Q plots,
##KDE density functions) and LR value distributions
layout(matrix(c(rep(c(1,2), each = p),seq(from = 3, to = 4*p+2)), 2*p, 3))

data.1 = data[which(data$Factor == category.1.name),]
data.2 = data[which(data$Factor == category.2.name),]

##Box-plots
boxplot(data.1[,variables], ylab = category.1.name)
boxplot(data.2[,variables], ylab = category.2.name)

##Q-Q plots
for (k in 1:p) qqnorm(sapply(split(data.1[,variables[k]],data.1$Item),mean),
    main = paste(variables.names[k]," for ",category.1.name, sep=""),
    cex.main = 0.9)
for (k in 1:p) qqnorm(sapply(split(data.2[,variables[k]],data.2$Item),mean),
    main = paste(variables.names[k]," for ",category.2.name, sep=""),
    cex.main = 0.9)

##KDE probability density
install.packages("KernSmooth")
library(KernSmooth)
p = 1
m.1 = length(unique(data.1$Item))
m.2 = length(unique(data.2$Item))
h.1 = (4/(m.1*(2*p+1)))^(1/(p+4))
h.2 = (4/(m.2*(2*p+1)))^(1/(p+4))

for (k in 1:length(variables))
plot(bkde(sapply(split(data.1[,variables[k]],data.1$Item),mean), kernel =
    "normal", bandwidth=h.1), type = "l", ylab = "probability density",
    main = category.1.name, xlab = variables.names[k], cex.main = 0.9)

for (k in 1:length(variables))
plot(bkde(sapply(split(data.2[,variables[k]],data.2$Item),mean), kernel =
    "normal", bandwidth=h.2), type = "l", ylab = "probability density",
    main = category.2.name, xlab = variables.names[k], cex.main = 0.9)
##Saving graphics to a .eps file
```

```
dev.copy(postscript, paste(variable.name, "_", categories,"_research_
    descriptive_statistics.eps", sep=""))
dev.off()

par(mfrow = c(2,1))

##Histograms illustrating LR distributions
output.matrix.Nor.log.1 = log10((output.matrix.Nor[1:length(unique(data.1$
    Item)),1]))
output.matrix.Nor.log.2 = log10(1/(output.matrix.Nor[(length(unique(data.1$
    Item))+1):length(unique(data$Item)),1]))
output.matrix.KDE.log.1 = log10((output.matrix.KDE[1:length(unique(data.1$
    Item)),1]))
output.matrix.KDE.log.2 = log10(1/(output.matrix.KDE[(length(unique(data.1$
    Item))+1):length(unique(data$Item)),1]))
hist(output.matrix.Nor.log.1, main = paste("LR distribution assuming Nor for
    ",category.1.name, sep=""), col = "gray", xlab = "logLR", breaks = 20,
    cex.main = 0.8)
hist(output.matrix.Nor.log.2, main = paste("LR distribution assuming Nor for
    ",category.2.name, sep=""), col = "gray", xlab = "logLR", breaks = 20,
    cex.main = 0.8)
dev.copy(postscript, paste(variable.name, "_", categories,"_research_LR_
    distribution_Nor.eps", sep=""))
dev.off()

hist(output.matrix.KDE.log.1, main = paste("LR distribution using KDE for ",
    category.1.name, sep=""), col = "gray", xlab = "logLR", breaks = 20, cex
    .main = 0.8)
hist(output.matrix.KDE.log.2, main = paste("LR distribution using KDE for ",
    category.2.name, sep=""), col = "gray", xlab = "logLR", breaks = 20, cex
    .main = 0.8)
dev.copy(postscript, paste(variable.name, "_", categories,"_research_LR_
    distribution_KDE.eps", sep=""))
dev.off()
```

D.10   Evaluating the performance of LR models

The **R** code fragments (Appendix D) for generating histograms, Tippett plots, DET plots, and ECE plots are given in this section. Plotting graphics requires the different and same data, which are the LR results for true-H_2 and true-H_1 hypotheses. The **R** routines presented in this section are suitable for evaluating the performance (Chapter 6) of the models described in Chapters 4 and 5.

D.10.1   Histograms

```
histograms = function(LR.H1.exp, LR.H2.exp)
{
  log.LR.H1.exp = log10(LR.H1.exp)
  log.LR.H2.exp = log10(LR.H2.exp)

  log.LR.H1.exp[which(log.LR.H1.exp < -20),] = -20
  log.LR.H2.exp[which(log.LR.H2.exp < -20),] = -20
```

```
    log.LR.H1.exp[which(log.LR.H1.exp > 20),] = 20
    log.LR.H2.exp[which(log.LR.H2.exp > 20),] = 20

    min = min(log.LR.H1.exp,log.LR.H2.exp)
    max = max(log.LR.H1.exp,log.LR.H2.exp)
    breaks = 50
    stepbins = (max-min)/breaks
    xbars = seq(min, max, by=stepbins)

    log.LR.H1.exp = log.LR.H1.exp[(log.LR.H1.exp < (xbars[length(xbars)]-
        stepbins/2)) & log.LR.H1.exp > (xbars[1]-stepbins/2)]
    log.LR.H2.exp = log.LR.H2.exp[(log.LR.H2.exp < (xbars[length(xbars)]-
        stepbins/2)) & log.LR.H2.exp > (xbars[1]-stepbins/2)]

    set = par(mfrow=c(2,1), mar=c(4,4,1,2))
    hist(log.LR.H1.exp, breaks=xbars-stepbins/2, col="blue", main="", xlab="
        logLR")
    legend("topleft", expression(paste(H[1]," true")), fill="blue")

    hist(log.LR.H2.exp, breaks=xbars-stepbins/2, col="darkorange", main="",
        xlab="logLR")
    legend("topright", expression(paste(H[2]," true")), fill="darkorange")

    par(set)
}
```

D.10.2   Tippett plots

```
Tippett_plot = function(LR.H1.exp, LR.H2.exp)
{
    log.LR.H1.exp = log10(LR.H1.exp)
    log.LR.H2.exp = log10(LR.H2.exp)

    log.LR.H1.exp[which(log.LR.H1.exp < -20),] = -20
    log.LR.H2.exp[which(log.LR.H2.exp < -20),] = -20
    log.LR.H1.exp[which(log.LR.H1.exp > 20),] = 20
    log.LR.H2.exp[which(log.LR.H2.exp > 20),] = 20

    min = min(log.LR.H1.exp,log.LR.H2.exp)
    max = max(log.LR.H1.exp,log.LR.H2.exp)

    x.range.data = rbind(min-1,log.LR.H1.exp,log.LR.H2.exp, max+1)
    x.range = x.range.data[order(x.range.data),1]

    Tippett.2 = matrix(0, nrow = length(x.range), ncol = 1)
    Tippett.1 = matrix(0, nrow = length(x.range), ncol = 1)

    for (i in 1:length(x.range))
    {
        Tippett.2[i] = length(which(log.LR.H2.exp > x.range[i]))/nrow(log.LR.H2.
            exp)*100
        Tippett.1[i] = length(which(log.LR.H1.exp > x.range[i]))/nrow(log.LR.H1.
            exp)*100
    }
```

```
false.positives = round(length(which(log.LR.H2.exp > 0))/nrow(log.LR.H2.
    exp)*100,2)
false.negatives = round(length(which(log.LR.H1.exp < 0))/nrow(log.LR.H1.
    exp)*100,2)

plot(x.range, Tippett.2, type="s", xlab=expression(paste(log[10],"LR
    greater than")), ylab="Proportion of cases [%]", xlim=c(-3,3), ylim=c
    (0,100), lty=3)
par(new=TRUE)
plot(x.range, Tippett.1, type="s", xlab=expression(paste(log[10],"LR
    greater than")), ylab="Proportion of cases [%]", xlim=c(-3,3), ylim=c
    (0,100))
legend("bottomleft", c(expression(paste("true-",H[2]," LR values")),
    expression(paste("true-",H[1]," LR values"))), lty=c(3,1), bty="n")
abline(v=0, col="gray", lty=4)

}
```

D.10.3   DET plots

```
DET_plot = function(LR.H1.exp, LR.H2.exp)
{
  log.LR.H1.exp = log10(LR.H1.exp)
  log.LR.H2.exp = log10(LR.H2.exp)

  log.LR.H1.exp[which(log.LR.H1.exp < -20),] = -20
  log.LR.H2.exp[which(log.LR.H2.exp < -20),] = -20
  log.LR.H1.exp[which(log.LR.H1.exp > 20),] = 20
  log.LR.H2.exp[which(log.LR.H2.exp > 20),] = 20

  min = min(log.LR.H1.exp,log.LR.H2.exp)
  max = max(log.LR.H1.exp,log.LR.H2.exp)

  treshold.range = rbind(min-1,log.LR.H1.exp,log.LR.H2.exp, max+1)
  treshold = treshold.range[order(treshold.range),1]

  false.positives = matrix(0, nrow = length(treshold), ncol = 1)
  false.negatives = matrix(0, nrow = length(treshold), ncol = 1)

  for (i in 1:length(treshold))
  {
    tmp.treshold = treshold[i]
    false.positives[i] = length(which(log.LR.H2.exp > tmp.treshold))/nrow(
        log.LR.H2.exp)*100
    false.negatives[i] = length(which(log.LR.H1.exp <= tmp.treshold))/nrow(
        log.LR.H1.exp)*100
  }

  x = qnorm(false.positives/100)
  y = qnorm(false.negatives/100)

  x[which(x == -Inf)] = qnorm(0.000001)
  y[which(y == -Inf)] = qnorm(0.000001)
  x[which(x == Inf)] = qnorm(0.999999)
  y[which(y == Inf)] = qnorm(0.999999)
```

```
plot(x, y, type="S", xlab="false positives [%]", ylab="false negatives [%]
    ", xaxt="n", yaxt="n", xlim=c(qnorm(0.0001),qnorm(0.5)), ylim=c(qnorm
    (0.0001),qnorm(0.5)))
axis.range = c(0.0001, 0.001, 0.01, 0.02, 0.05, 0.1, 0.2, 0.5, 1, 2, 5,
    10, 20, 40, 50)
axis.gauss = qnorm(axis.range/100)
abline(h=axis.gauss, lty=3, col="gray")
abline(v=axis.gauss, lty=3, col="gray")
axis(side=1,at=axis.gauss,labels=axis.range)
axis(side=2,at=axis.gauss,labels=axis.range)
}
```

D.10.4   ECE plots

```
ECE_plot = function(LR.H1.exp, LR.H2.exp)
{
  P.H1 = seq(from=0.01, to=0.99, by=0.01)
  P.H2 = 1 - P.H1
  a.priori.odds = P.H1/P.H2

  N.H1 = nrow(LR.H1.exp)
  N.H2 = nrow(LR.H2.exp)

  set = c("null","exp","cal")

  for (x in 1:length(set))
  {
    if(x == 1) ## LR for null method
    {LR.H1 = rep(1, times=N.H1)
    LR.H2 = rep(1, times=N.H2)}

    if(x == 2) ## experimental LR
    {LR.H1 = as.matrix(LR.H1.exp)
    LR.H2 = as.matrix(LR.H2.exp)
    LR.H1[which(LR.H1 < 10^-20)] = 10^-20
    LR.H2[which(LR.H2 < 10^-20)] = 10^-20
    LR.H1[which(LR.H1 > 10^20)] = 10^20
    LR.H2[which(LR.H2 > 10^20)] = 10^20}

    if(x == 3) ## calibrated LR according to PAV
    {require("isotone")

    LR.H1.H2.exp = c(LR.H1,LR.H2)
    indices = order(LR.H1.H2.exp)

    LR.H1.H2.exp.sorted = sort(LR.H1.H2.exp)
    posterior.prob = c(rep(1, times=N.H1), rep(0, times=N.H2))
    posterior.prob.sorted = posterior.prob[indices]

    posterior.H1.H2.cal = data.frame(gpava(LR.H1.H2.exp.sorted, posterior.
        prob.sorted)[1])
    LR.H1.H2.cal = posterior.H1.H2.cal/(1-posterior.H1.H2.cal)/(N.H1/N.H2)
    LR.H1 = LR.H1.H2.cal[which(indices %in% c(1:N.H1)),]
    LR.H2 = LR.H1.H2.cal[which(indices %in% (N.H1+1):(N.H1+N.H2)),]
```

```
LR.H1[which(LR.H1 < 10^-20)] = 10^-20
LR.H2[which(LR.H2 < 10^-20)] = 10^-20
LR.H1[which(LR.H1 > 10^20)] = 10^20
LR.H2[which(LR.H2 > 10^20)] = 10^20}

penalty.H1 = 0
penalty.H2 = 0

for (i in 1:N.H1)
{
  a = -log2(LR.H1[i]*a.priori.odds/(1+LR.H1[i]*a.priori.odds))
  penalty.H1 = penalty.H1 + a
}

for (j in 1:N.H2)
{
   b = log2(1 + LR.H2[j]*a.priori.odds)
   penalty.H2 = penalty.H2 + b
}

ECE= P.H1/N.H1*penalty.H1 + P.H2/N.H2*penalty.H2

col = c("black","red","blue")
lty = c(3,1,2)
if (x %in% c(2,3)) {par(new=TRUE)}
plot(log10(a.priori.odds), ECE, xlim=c(-2,2), ylim=c(0,1), type="l", col
    =col[x], lty=lty[x], xlab = expression(paste("prior log"[10],"(Odds)
    ")))
abline(v=0, lty=2, col="gray")
legend("topleft", c("null method with LR=1", "experimental LR", "
    calibrated LR after PAV"), col=col, lty=lty, bty="n", cex=0.8)
  }
}
```

Reference

Curran JM 2011 *Introduction to Data Analysis with R for Forensic Scientists*. CRC Press, Boca Raton, FL.

Appendix E

Bayesian network models

E.1 Introduction to Bayesian networks

There are plenty of LR models that could be used for measuring the evidential value of samples, for example in comparison and classification problems (Chapters 4 and 5). These models allow us to evaluate univariate and multivariate data and they also take into account various sources of errors, for example the variation between- and within-objects (Section 3.3.2). Moreover, these LR models allow us to apply a kernel density estimation procedure when the data are not normally distributed (Section 3.3.5). Such a situation is especially common for the between-object distribution in forensic databases. The disadvantage of applying these models is that they require a relatively large database in order to evaluate all the model parameters. This point is especially crucial for the evaluation of multivariate data. Moreover, there is little software available which can be used relatively easily by people without experience in programming to determine likelihood ratios (but see **calcuLatoR**; Appendix F). Therefore, if scientists want to use these models they have to write their own case-specific routines using the **R** software introduced in Appendix D, for example. Also, there is limited understanding of these evaluation procedures, such as those related to LR models, among the representatives of judicial systems, including forensic experts. Therefore, the possibility of applying models based on Bayesian networks (BNs) for the evaluation of physicochemical data has been considered (Zadora 2009, 2010; Zadora and Wilk 2009). Bayesian networks graphically represent factors such as those considered in LR models, and the uncertainties related to them (Taroni *et al.* 2006), in the process of evaluating evidence from physicochemical data. This makes LR models more intelligible to non-statisticians.

A Bayesian network is a probabilistic graphical model that represents a set of variables along with their probabilistic dependencies. The network comprises the following elements:

- nodes, which represent the variables (continuous and discrete; Chapter 3) and hypotheses;

Statistical Analysis in Forensic Science: Evidential Value of Multivariate Physicochemical Data, First Edition.
Grzegorz Zadora, Agnieszka Martyna, Daniel Ramos and Colin Aitken.
© 2014 John Wiley & Sons, Ltd. Published 2014 by John Wiley & Sons, Ltd.
Companion website: www.wiley.com/go/physicochemical

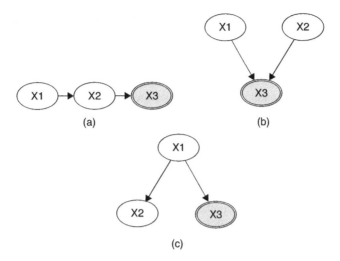

Figure E.1 Basic connections used in Bayesian networks and the joint probability distributions ascribed to them: (a) serial connection $(P(X1, X2, X3) = P(X1)P(X2|X1)$ $P(X3|X2))$; (b) converging connection $(P(X1, X2, X3) = P(X1)P(X2)P(X3|X1, X2))$; (c) diverging connection $(P(X1, X2, X3) = P(X1)P(X2|X1)P(X3|X1))$. (Reproduced by permission of Wiley.)

- arrows between nodes, which represent dependence links between different variables and hypotheses (e.g. Chapter 2);

- marginal and conditional probabilities (Appendix A) for nodes, depending on the status of the nodes.

Moreover, fundamental to the idea of a network is the combination of simpler parts and basic connections used in a BN (Figure E.1).

The key feature of a BN is the fact that it provides a method for decomposing a joint probability distribution of many variables $(X_1, \ldots, X_p)$ into a set of local distributions of a few variables. The multiplication law allows the decomposition of a joint probability distribution with p variables as a product of $p - 1$ conditional distributions and a marginal distribution:

$$P(X_1, \ldots, X_p) = \left[\prod_{i=2}^{p} P(X_i|X_1, \ldots, X_{i-1}) \right] P(X_1).$$

From the Markov property, a BN can be factorised as the product of the probabilities of all variables in the network, conditional only on their parents (**PA**):

$$P(X_1, \ldots, X_p) = \left[\prod_{i=1}^{p} P(X_i|\mathbf{PA}(X_i)) \right].$$

Therefore, suitable joint probability distributions can be ascribed to each of the basic connections (Figure E.1).

Denote a node associated with X_i by Xi. Information obtained from suitable databases is put into probability tables designed for each node. If there is an arrow pointing from node X1 to node X2 it is said that X1 is a *parent* of X2 and X2 is a *child* of X1 – in other words, X2 is *dependent* on X1. A node with no parents is called a *root* node. A table of probabilities associated with a root node contains marginal prior probabilities. Tables relating to child node(s) contain conditional probabilities (Appendix A), probabilities conditional on the values taken by the parents. When the current state of a particular node is known, for example the value of the variable determined for the recovered sample, it is entered into suitable nodes. This value is called *hard evidence* and the procedure is known as *instantiation*. After entering the hard evidence, information obtained in the discrete type node is in the form of posterior probabilities. For instance, if H_1 and H_2 are the hypotheses in a case, and E is the hard evidence observed in a case, then $P(H_1|E)$ and $P(H_2|E)$ are the posterior probabilities for the (discrete) value of the true hypothesis in the case. As already mentioned, the forensic expert should assign $P(E|H_1)$ and $P(E|H_2)$, expressed in the form of a likelihood ratio. These probabilities can also be obtained by BN, by using equation (2.1) (also Appendix A).

E.2   Introduction to Hugin Researcher™ software

Hugin Researcher™ is a graphical user interface which allows users to draw and analyse various BN models. A demonstration version can be downloaded free of charge from www.hugin.com. This version is available for personal use only and enables the analysis of the BN models presented in this book. A full version is required if one wishes to work with more complex BN models used in forensic analysis, for example as described in Taroni *et al.* (2006), or for more than personal use.

The analysis of BNs using **Hugin Researcher**™ in this section refers to two models. The first (Figure E.2(a)) is constructed of two discrete type nodes, H and E. Suppose H has two states: H1 (e.g. the suspect broke a glass object) and H2 (another person broke the glass object). Suppose E has three states: 0 (e.g. no glass fragments recovered from the suspect's clothes reveal similarity to a control material), 1 (one glass fragment recovered from the suspect's clothes reveals similarity to a control material), and more1 (more than one glass fragment recovered from suspect clothes reveals similarity to a control material).

The second model (Figure E.2(b)) is constructed of two continuous nodes, V1 and V2, under an assumption that the distribution of the relevant variables can be estimated by normal

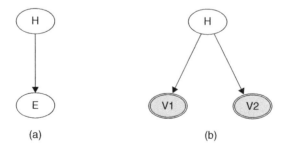

(a)   (b)

Figure E.2   Bayesian network models discussed in this Appendix: (a) constructed of two discrete type nodes; (b) constructed of one discrete and two continuous type nodes.

Figure E.3 Hugin Researcher[TM] icon.

distributions, and a discrete type node, H. Node H is taken to have two states: H1 (e.g. a glass fragment originates from category 1) and H2 (a glass fragment originates from category 2).

E.2.1 Basic functions

E.2.1.1 Opening and closing

Hugin Researcher[TM] can be opened by double-clicking on the relevant icon (Figure E.3). This opens a window, Class:unnamed1 (Figure E.4) in which a new BN model can be created. The software can be closed by clicking on the cross in the top righ-hand corner of the screen or by choosing the Exit option from the File menu.

E.2.1.2 Saving a Bayesian network

A newly created BN can be saved by selecting File (Figure E.5) from the main menu and clicking on Save as. Alternatively, the Save icon can be selected. To name a file, simply type the desired text in the appropriate field within the Save network window. To select a location to save the file, click the Save in field and choose the required location. Finally, click on the Save button or press Enter. The file is saved with the file extension *.oobn.

E.2.1.3 Opening a Bayesian network

To open a BN saved as *.oobn select File from the main menu and click on Open. Highlight the file you wish to open and press Enter.

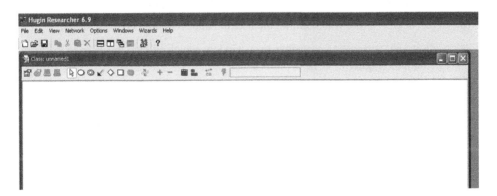

Figure E.4 The Class:unnamed1 window.

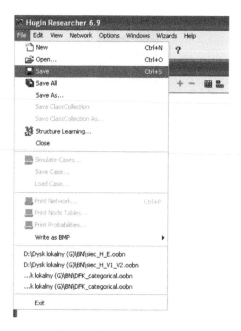

Figure E.5 `File` menu.

E.2.2 Creating a new Bayesian network

E.2.2.1 Basic functions

The tools shown in Figure E.6 are used to create new networks in the `Class:unnamed1` window (Figure E.4):

(a) `Select tool` – highlighting and moving objects, i.e. the entire BN model or parts of it (node(s) or arrow(s); Figure E.6(a));

(b) `Discrete chance tool` – creation of discrete type nodes (Figure E.6(b));

(c) `Continuous chance tool` – creation of continuous type nodes (Figure E.6(c));

(d) `Link tool` – a tool which allows the user to connect nodes (Figure E.6(d));

(e) `Switch to Run Mode` – switch to a window which allows the user to perform calculations (Figure E.6(e)).

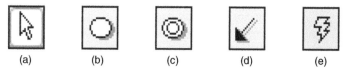

Figure E.6 Icons used in creating Bayesian networks. Details are provided in text.

Figure E.7 A discrete type node.

To construct the BN shown in Figure E.2(b), proceed as follows:

1. Select the `Discrete chance tool` icon (Figure E.6(b)) and click anywhere within the `Class:unnamed1` window. This creates the discrete type node (Figure E.7).

2. To add a continuous type node, select the `Continuous chance tool` (Figure E.6(c)). Click within the `Class:unnamed1` window to add the node (Figure E.8). Repeat this process to add additional continuous type nodes.

3. To join, or connect, these nodes, select the `Link tool` (Figure E.6(d)). Click on the `Link tool` icon. Move the cursor to the node from which the link starts (the parent node; Figure E.9(a)). Click on the left-hand mouse button and keep the button depressed. Move (a dashed line appears) to the node at which the link ends (the child node). Release and the link appears (Figure E.9(b)).

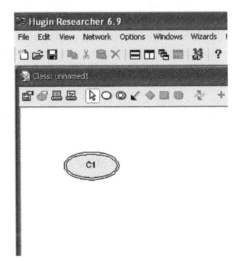

Figure E.8 A continuous type node.

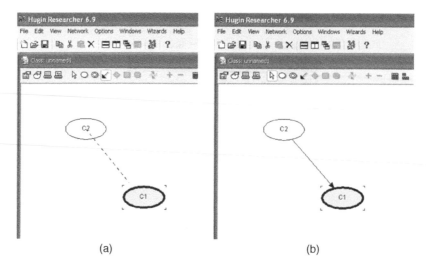

(a)   (b)

Figure E.9   Creating a connection between two nodes.

In order to remove a connection between nodes do the following:

1. Click on the desired arrow (it becomes bold).

2. Either press the Delete button on the keyboard or click the right-hand mouse button and select Delete from the menu that appears (Figure E.10).

E.2.2.2   Changing node properties

In order to change the properties of a node, click and highlight it. Then press the right-hand mouse button to open the menu shown in Figure E.11. Select the Node Properties option which opens an additional window (Figure E.12(a)). Here you can change the name of a node by typing the new name in the Name field and clicking on Apply. If the node is continuous then press OK. If the node is discrete then select the States tab (each new discrete type node is described by State 1 and State 2). Select and highlight the State name you would like to change (Figure E.12(b)). Type the new name in the middle part of the window and click on Rename. Follow this procedure for all of the states you wish to rename. Finally, click on Apply and then OK. If more than two states are required then additional states can be added by clicking on the Add Before or Add After buttons on the right-hand side of the window. Any unwanted states can be removed by clicking on the Remove button. The Up and Down buttons can be used to change the order of the states within the list.

E.2.2.3   Typing of numerical data into nodes

Select the entire BN (e.g. select by keeping the left-hand mouse button depressed and moving across the BN). This highlights each node and arrow. Then right-click and select the Open Tables option from the menu that appears (Figure E.11). Select a node for which values of probabilities should be added. Type your values in a suitable format, for examples tables

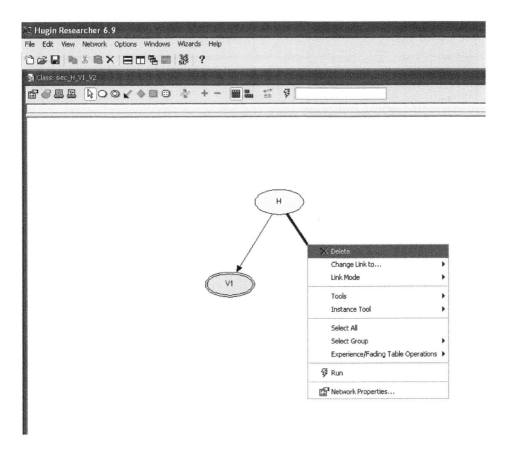

Figure E.10   Removing a connection between nodes.

Figure E.11   Menu for changing node properties.

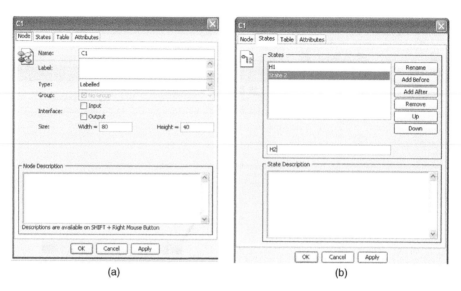

Figure E.12 The `Node Properties` window for discrete type node `C1`: (a) `Node` tab;
(b) `States` tab.

in Figures E.13 and E.14. In the case of a continuous type node characterised by a normal
distribution it is necessary to type the values of normal distribution parameters – mean and
variance (Section 3.3.5). For example:

(a) node H (Figure E.13(a)): $P(H_1) = P(H_2) = 0.5$ for hypotheses H_1 and H_2;

(b) node V1 (Figure E.13(b)): $N(1.5194, 5.5 \cdot 10^{-6})$ for H_1 and $N(1.5203, 6.8 \cdot 10^{-6})$
for H_2;

(c) node V2 (Figure E.13(c)): $N(2.8342, 3.7 \cdot 10^{-2})$ for H_1 and $N(3.6674, 7.5 \cdot 10^{-2})$ for
H_2.

In the case of a BN constructed only of discrete type nodes the analysis would look like
Figure E.14, after entering the following probabilities:

(a) node H (Figure E.14(a)): $P(H_1) = P(H_2) = 0.5$ for H_1 and H_2 hypotheses,

(b) node E (Figure E.14(b)): $P(E = 0|H_1) = 0.1$, $P(E = 1|H_1) = 0.2$, $P(E \geq 1|H_1) =$
0.7 for H_1, $P(E = 0|H_2) = 0.8$, $P(E = 1|H_2) = 0.15$, $P(E \geq 1|H_2) = 0.05$ for H_2.

E.2.3   Calculations

E.2.3.1   Basic functions

Select the `Switch to Run Mode` icon (Figure E.6(e)). A new window appears (Figure E.15),
in which the basic functions are shown (Figure E.16):

(a) `Expand node list` (Figure E.16(a)): expand a list of nodes from the state marked
in Figure E.15 to the state marked in Figure E.17.

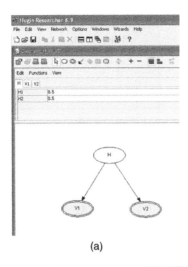

(a)

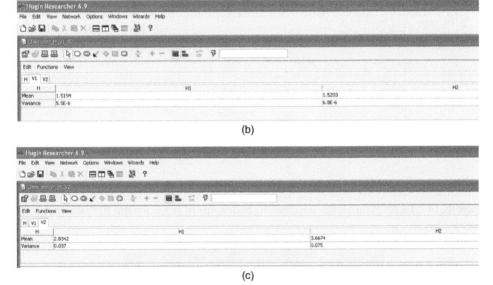

(b)

(c)

Figure E.13 Tables with values assigned to nodes: (a) H; (b) V1; (c) V2 for the Bayesian network presented in (a).

(b) Collapse node list (Figure E.16(b)): collapse a list of nodes from the state marked in Figure E.17 to the state marked in Figure E.15.

(c) Update monitor graphs (Figure E.16(c)): actualisation of calculations.

(d) Initialize network (Figure E.16(d)): initialise the starting values of the network.

(e) Switch to Edit Mode (Figure E.16(e)): switch to a window where the network can be edited.

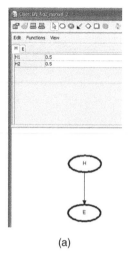

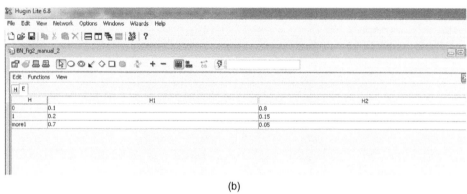

(b)

Figure E.14   Tables with values assigned to nodes: (a) H; (b) E for the Bayesian network presented in (a).

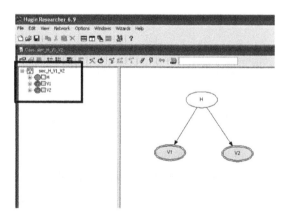

Figure E.15   Window obtained by clicking on the icon Switch to Run Mode. An area in the square is extended in Figure E.17 after clicking on the Expand node list icon.

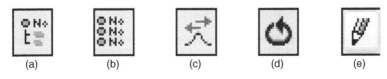

(a)   (b)   (c)   (d)   (e)

Figure E.16   Icons used when calculations are carried out in `Run Mode`.

E.2.3.2   Entering of the hard evidence

Expand a list to the form illustrated in Figure E.17 by clicking on the `Expand node list` icon (Figure E.16(a)) and then the `Update monitor graph` icon (this should be done whenever entering new data):

(a) To enter continuous type hard evidence, select a node from the list (not from the diagram) into which hard evidence is to be entered (Figure E.18(a)) and right-click. Select the `Insert Continuous Finding` option and type the value of the hard evidence in the `Evidence/Value` field, for example 1.5169 for node `V1` (Figure E.18(b)) or 3.5010 for node `V2`. Then click on `Enter Evidence`. The result of entering both pieces of hard evidence is shown in Figure E.19.

(b) To enter discrete type hard evidence, double-click on the state of the node to enter the hard evidence value, for example `more1` for node `E` in Figure E.20.

Successful entry of the hard evidence is confirmed by highlighting the state of the particular node (Figures E.19 and E.20) and by the appearance of the letter e on the node into which the hard evidence was entered. New values appear in all nodes as in Figures E.19 and E.20.

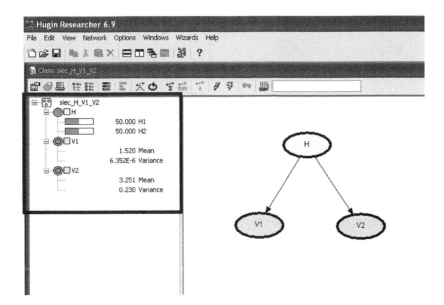

Figure E.17   Window obtained by clicking on the `Expand node list` icon (Figure E.16a). See also the area marked by a square in Figure E.15.

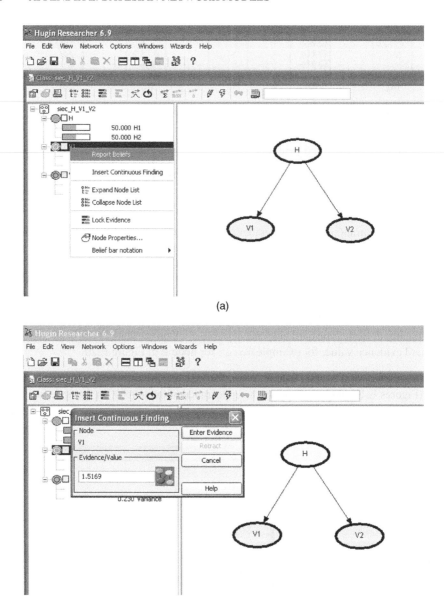

(a)

(b)

Figure E.18 Entering continuous type hard evidence: (a) selection the `Insert Continuous Finding` option; (b) value 1.5169 typed as hard evidence into node `V1`.

E.2.3.3 LR calculation

Values in the parent type node `H` are posterior probabilities (Figures E.19 and E.20). They can be used in order to evaluate the evidential value of the hard evidence of interest, for example for calculation of the likelihood ratio.

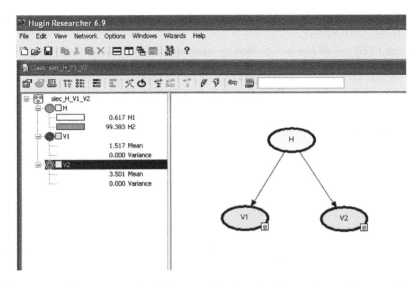

Figure E.19 Results obtained after entering hard evidence into nodes V1 and V2.

From Figure E.19 it is concluded that:

(a) $P(H_1|E) = 0.00617$,
    $P(H_2|E) = 0.99383$.
    When $P(H_1) = P(H_2) = 0.5$, then by equation (2.1),

$$LR = \frac{P(H_1|E)}{P(H_2|E)} = \frac{0.00617}{0.99383} = 0.006$$

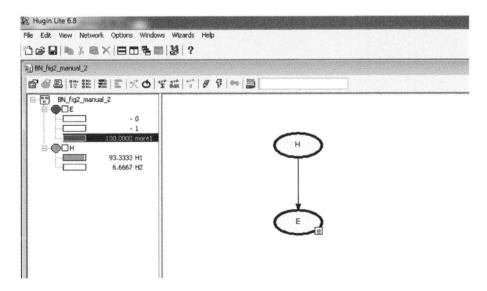

Figure E.20 Results obtained after entering hard evidence into the discrete node E.

(b) $P(H_1|E) = 0.933333,$
$P(H_2|E) = 0.066667.$
When $P(H_1) = P(H_2) = 0.5$, then by equation (2.1),

$$LR = \frac{P(H_1|E)}{P(H_2|E)} = \frac{0.933333}{0.066667} = 14.$$

E.2.3.4   Performing new calculations

Click on the `Initialize network` icon to return to the original values of the network in order to perform new calculations (for new hard evidence) as desired.

References

Taroni F, Aitken CGG, Garbolino P, Biedermann A 2006 *Bayesian Networks and Probability Inference in Forensic Science*. John Wiley & Sons, Ltd, Chichester.

Zadora G 2009 Evaluation of evidence value of glass fragments by likelihood ratio and Bayesian network approaches. *Analytica Chimica Acta* **642**, 279–290.

Zadora G 2010 Evaluation of evidential value of physicochemical data by a Bayesian network approach. *Journal of Chemometrics* **24**, 346–366.

Zadora G, Wilk D 2009 Evaluation of evidence value of refractive index measured before and after annealing for container and float glass fragments. *Problems of Forensic Science* **80**, 365–377.

Appendix F

Introduction to calcuLatoR software

F.1 Introduction

A graphical user interface called **calcuLatoR** for the calculation of likelihood ratios for univariate data (such as refractive index data; Section 1.2.2) in comparison problems has been developed and described in Zadora and Surma (2010). The LR model assumes two sources of variation: between replicates within the same object (within-object variability) and between replicates between various objects (between-object variability). It was assumed that the within-object distribution is normal with constant variance (equation (3.3)). The between-object distribution was modelled by a univariate kernel density estimation (equations (4.7) and (4.8)).

The **calcuLatoR** can also be used for evaluating the evidential value of multivariate data, but it must be assumed that the variables concerned are independent. In this situation, *LR* should be calculated for each variable separately and the final *LR* value is equal to their product.

F.2 Manual

Using the data files included in the starting package should return the same result for anyone using this software (Section 4.4.4). Therefore, these files can be used to check the accuracy of the software's performance before an *LR* calculation is performed in real casework.

The starting package for Windows and Linux users is available in the relevant file folder named `Appendix F` on the website[1]. During the first execution, for example using

---

[1] All the files are available from www.wiley.com/go/physicochemical

Statistical Analysis in Forensic Science: Evidential Value of Multivariate Physicochemical Data, First Edition.
Grzegorz Zadora, Agnieszka Martyna, Daniel Ramos and Colin Aitken.
© 2014 John Wiley & Sons, Ltd. Published 2014 by John Wiley & Sons, Ltd.
Companion website: www.wiley.com/go/physicochemical

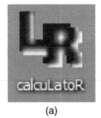

(a)

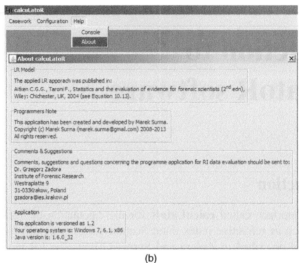

(b)

Figure F.1   Installation of **calcuLatoR**. (a) An LR icon is created during first execution of **calcuLatoR**. (b) Running **calcuLatoR** opens the window shown. Selecting Help->About opens an About calcuLatoR window which provides general information about the software.

calcuLatoR.exe for Windows users, an LR icon (Figure F.1(a)) and an LR Data folder are created. Opening **calcuLatoR**, for example by double-clicking on the icon shown in Figure F.1(a), opens a window as shown in Figure F.1(b). General information about **calcuLatoR** can be found by selecting Help->About (Figure F.1(b)).

First-time users need to add an operator name (Figure F.2), which is included in reports produced by the software. Other names can be added at any time.

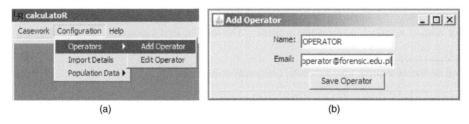

(a)   (b)

Figure F.2   Adding a new operator: (a) location of Add Operator and Edit Operator options; (b) Add Operator dialogue box.

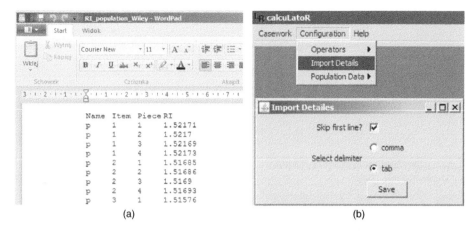

(a)   (b)

Figure F.3   Entering details about the structure and type of population, recovered, and control data files: (a) the required structure of the population, recovered, and control data (the file either has or does not have a row with names of columns); (b) settings when `*.txt` files with rows separated by tabs are used and files have a row giving column names.

Figure F.3(a) shows the required format for the population data (e.g. the `RI_population_Wiley.txt` file in the starting package), as well as the control (`RI_control.txt`), and recovered (`RI_recovered.txt`) object data files. Details on the format of the population, recovered, and control data files should be input before importing the data (Figure F.3(b)). In Figure F.3(a), a text file with rows separated by tabs, and with a header row showing the names of columns, is used. The settings for this file are shown in Figure F.3(b). If no header row is present, the `Skip first line?` option should not be ticked. The software also works with comma separated value files (`*.csv`). In this case, select `comma` in the `Select delimiter` option.

Options available in the `Configuration` menu (Figure F.4(a)) should be used in order to load a file containing background population data (e.g. `RI_population_Wiley.txt`). An `Edit Data Set` option (Figure F.4(a)) allows the user to open and edit the data in the population database selected (Figure F.4(b)). Various files with different population data can be loaded in order to carry out calculations. The file to be used in a particular case can be selected in various ways, such as `Population Data` in Figure F.4(a) and `Change Population Data Set` in Figures F.6(a),(b).

The `Recovered data` and `Control Data` sections of the window shown in Figure F.5 should be used to enter data (e.g. refractive index; Section 1.2.2) obtained during the analysis of recovered and control samples. This data can be:

(a) typed in by an operator;

(b) inserted using the `Import Recovered Data` and `Import Control Data` buttons (Figure F.5). The `RI_recovered.txt` and `RI_control.txt` files (included in the starting package) contain example RI data from five recovered samples, from `e1` to `e5`, and one control sample, `c1` (Figure F.5).

Note that each recovered or control sample must have a unique identifier (in the `Name` column in Figure F.5); this should not include dots, commas, or spaces. The `Evidences`

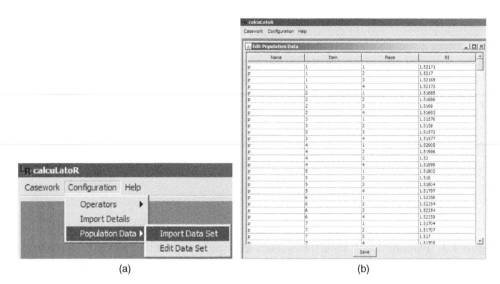

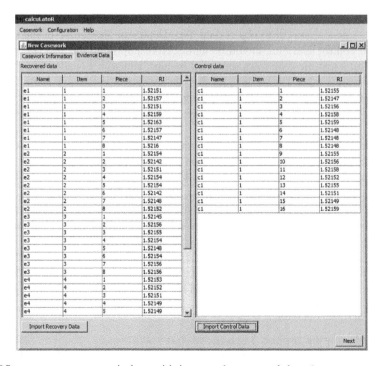

Figure F.4 (a) `Configuration` menu; (b) loading a file with population data, `RI_population_Wiley.txt`.

Figure F.5 `New Casework` window with imported recovered data (`RI_recovered.txt`) and control data (`RI_control.txt`).

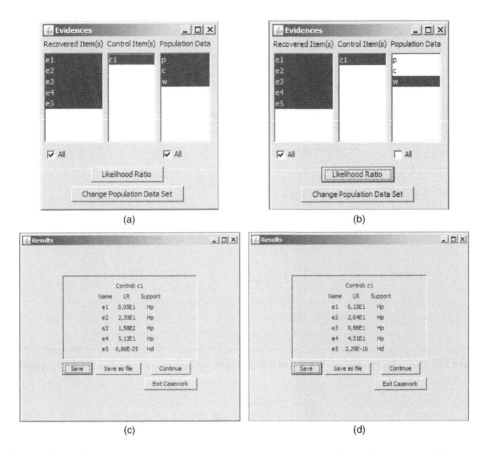

Figure F.6 The Evidences window allows the user to specify which recovered item(s) (Recovered Items) should be compared with each control item (Control Items). Recovered samples may be compared with only one control sample in one calculation run; however, more than one control sample may be available for analysis in the casework. The Population Data window provides an opportunity to select which categories in the database should be used in calculations, for example (a) all categories (categories c, w, p in RI_population_Wiley.txt), or (b) only category w. Windows (c) and (d) illustrate the results of calculations when the recovered samples and the control sample are compared with a database containing (c) samples from all categories in the database and (d) samples from category w in the database. Note: in this software H_p is used instead of the notation H_1 favoured in this book and H_d is used instead of H_2.

window (Figures F.6(a),(b)) opens when the Next button is clicked. More than one control sample can be imported for analysis in each case; however, a separate calculation run must be carried out for each control sample, and recovered samples can only be compared with one control sample per calculation run (Figures F.6(c),(d)). This is to prevent any possible mistakes when all possible comparisons between control and recovered samples were carried out in one calculation run.

The Population Data (Figure F.6) column allows the user to select which background population category or categories to use in calculations, for example all categories (Figure F.6(a)) or only one category w (Figure F.6(b)).

Figure F.7   File showing results saved as a *.txt file, calculator_results.txt, opened in WordPad.

The Change Population Data Set button (Figures F.6(a),(b)) is used to change the population data used in the calculations. The likelihood ratio can be calculated by clicking on the Likelihood Ratio button (Figures F.6(a),(b)). Figure F.6(c) shows that, when samples from all categories (c, w, p) were used, the *LR* calculated when, for example, recovered sample e1 was compared with the control c1 was 80.5 (8.05E1). Figure F.6(d) shows that an *LR* value of 61.0 was obtained for the same comparison when samples from the w category only were used.

Results can be saved by clicking the following buttons shown in Figures F.6(c),(d):

(a) Save: results saved by this option can be opened within **calcuLatoR** by selecting Open Result in the Casework menu.

(b) Save as file: results saved by this option are in the form of a *.txt file. These can be opened using any external text editor, such as WordPad, and can also be printed (Figure F.7).

Reference

Zadora G, Surma M 2010 The calcuLatoR – software for evaluation of evidence value of univariate continuous type data. *Chemometrics and Intelligent Laboratory Systems* **103**, 43–52.

Index

μ-X-ray fluorescence, μ-XRF, *see* glass,
    μ-X-ray fluorescence

accuracy, 192, 195, 204
analysis of variance, *see* ANOVA
annealing, *see* glass, annealing
ANOVA, 78–85
arithmetic mean, *see* data, central tendency,
    arithmetic mean
artificial neural network, ANN, *see*
    classification problem, artificial
    neural network

Bayes' theorem, 31–32
between-object variability, 26, 42–44,
    107–180
box-plots, 44, 114, 117, 121, 126, 131, 138,
    159, 162, 165, 170, 258

calibration, 192, 195
car paints, 10–13
   likelihood ratio, LR, 119–125, 205–209
   Py-GC/MS, 119–125
central tendency, *see* data, central
    tendency
classification problem, 2, 8, 20, 27–31, 32,
    82–84
   artificial neural network, ANN, 28–30
   discriminant analysis, DA, 27, 29
   likelihood ratio, LR, 31, 151–180

naïve Bayesian classifier, NBC, 28, 31
partial least-squares discriminant
    analysis, PLS-DA, 28
SIMCA, 28, 29, 88
support vector machines, SVM, 28, 31
cluster analysis, CA, 28, 85–92
   averaged linkage method, 89
   clustering, 89–91
   dendrogram, 89–91
   furthest neighbour, 89
   hierarchical cluster analysis, 89–92
   nearest neighbour, 89
   Ward's algorithm, 89–91
comparison problem, 2, 12, 13, 20, 21–27,
    32
   likelihood ratio, LR, 23–26, 107–150
   two-stage approach, 21–23, 26–27,
    69–71
confidence interval, 67–69

data
   central tendency, 39
    arithmetic mean, 39–41, 257
    median, 39–41, 257
    mode, 39–41
   continuous, 35–105
   covariance, 45–49
   descriptive statistics, 39–59
   discrete, 23, 35, 107, 119, 151, 225–227,
    296–300, 305

Statistical Analysis in Forensic Science: Evidential Value of Multivariate Physicochemical Data, First Edition.
Grzegorz Zadora, Agnieszka Martyna, Daniel Ramos and Colin Aitken.
© 2014 John Wiley & Sons, Ltd. Published 2014 by John Wiley & Sons, Ltd.
Companion website: www.wiley.com/go/physicochemical

Printed and bound by CPI Group (UK) Ltd, Croydon, CR0 4YY

27/10/2024

14580216-0003